Zero Carbon Emission with Nanomaterials for Environment Matrices

Editors

Tanuj Kumar
Central University of Jammu, Jammu
Jammu and Kashmir, India

Richa Kothari
Department of Environmental Sciences
Central University of Jammu, Jammu
Jammu and Kashmir, India

Anita Singh
Department of Environmental Sciences
Central University of Jammu, Jammu
Jammu and Kashmir, India
and
Department of Environmental Studies,
Central University of Haryana, Jant-Pali, Mahendergarh
Haryana, India

Deepak Pathania
Department of Environmental Sciences
Central University of Jammu, Jammu
Jammu and Kashmir, India

CRC Press
Taylor & Francis Group
Boca Raton London New York

CRC Press is an imprint of the
Taylor & Francis Group, an **informa** business
A SCIENCE PUBLISHERS BOOK

Cover credit: Cover image provided by the editors was designed by Dr. Tanuj Deswal and reproduced by his kind permission.

First edition published 2027
by CRC Press
2385 NW Executive Center Drive, Suite 320, Boca Raton FL 33431

and by CRC Press
4 Park Square, Milton Park, Abingdon, Oxon, OX14 4RN

CRC Press is an imprint of Taylor & Francis Group, LLC

Library of Congress Cataloging-in-Publication Data (applied for)

ISBN: 978-1-032-74599-2 (hbk)
ISBN: 978-1-032-79467-9 (pbk)
ISBN: 978-1-003-49218-4 (ebk)

DOI: 10.1201/9781003492184

Typeset in Times New Roman
by Prime Publishing Services

Preface

With the world facing the multidimensional challenges brought about by climate change, degradation of the environment, and increasing energy and natural resource demands, the science community continues to seek sustainable, innovative, and scalable solutions. Among the most exciting and active fields to appear as solutions to these challenges is nanotechnology, which has opened up incredible potential for designing and applying materials with superior properties. Particularly, nanomaterials, designed at the nanoscale, have revealed outstanding physical, chemical, and biological properties making them promising candidates for an impressive array of applications, from carbon capture and cleanup to energy conversion and beyond. This edited book is a compilation of work that collectively foreground the revolutionary contribution of nanotechnology to environmental sustainability, specifically carbon-based nanomaterials, and applications in climate change mitigation. The chapters provide concepts, new developments, applications, and directions for future research. They provide a complete picture of how nanomaterials can help work towards zero-emission targets, enhance environmental quality, and promote sustainable development.

Chapter 1 in this edited book sets the foundation for the remaining chapters in the book. The chapter presents the basic properties of nanomaterials such as zero-dimensional carbon nanostructures like graphene quantum dots and nanodiamonds and delves into their relevance in environmental applications. The chapter emphasizes the significance of the nanostructures for carbon emissions trapping and storage, a critical aspect in combating global warming. Readers are led through the distinctive physicochemical characteristics of nanomaterials which render them extraordinarily effective at carbon capture, and also their potential for incorporation into many industrial and environmental systems.

Expanding on this premise, Chapter 2 examines in further detail how nanomaterials can be used as capture agents for carbon. Then, it provides an all-inclusive background on many of their structural properties, such as high surface area and porosity, and touches on some of the major challenges concerning scale-up of synthesis and application. It also emphasizes the need for both interdisciplinary collaboration and long-term commitment to research and development effort in order to circumvent barriers of cost, efficiency, and long-term stability. Through a discussion of many types of nanomaterials and their roles in adsorption processes, it provides an all-inclusive framework for the effective quantification of carbon capture systems.

Chapter 3 broadly expands the context by addressing the function of nanomaterials for environmental monitoring and remediation. As soil, air, and water pollution becomes an increasing issue as a result of growing levels of industrialization and agricultural intensification, there is a pressing demand for reliable and more environmentally friendly alternatives to conventional pollution management strategies. Nanomaterials offer unique opportunities because of their reactivity and modifiability. This chapter addresses how nanomaterial-based sensor technologies can detect pollutants in the environment with high sensitivity and selectivity. It also discusses the prominence of electrochemical and optical sensors for detection and measurement of carbon-containing pollutants, thus opening a new area of environmental monitoring.

Chapter 4 discusses zero-dimensional carbon nanomaterials, presenting a detailed analysis of their surface modification methods for augmenting CO_2 capture efficiency. Different methods like acid treatment, amine and imine grafting, and heteroatom doping are discussed at length. The chapter presents both theoretical models and experimental findings showing how such modifications enhance adsorption traits. It also discusses the ecological implications of such methods and how they are to be incorporated in sustainable materials science. With more people turning to green technologies, functionalization of nanomaterials is important in order to modulate their functioning for particular use.

Chapter 5 provides a comprehensive view on the application of nanomaterials in carbon sequestration. It highlights the functions of advanced nanostructures, namely carbon nanotubes, graphene, and metal organic frameworks, in the capture and immobilization of carbon dioxide. These nanomaterials are known for their high surface chemistry and adsorption capacity. The chapter also accounts for the possible synergistic effects of mixing diverse nanomaterials to form composite structures with enhanced sequestration performance. Aside from the technological aspect, the authors delve into the wider economic and environmental consequences of such technologies, considering their feasibility for large scale industrial implementation. A holistic approach is essential in relocating laboratory-scale research into commercially viable solutions.

Chapter 6 shifts to the field of catalysis, discussing how nanomaterials can be used to drive the conversion of CO_2 into valuable products like fuels and chemicals. The chapter contends that carbon emissions being used to create resources is not only a way of curbing atmospheric CO_2 concentrations but also towards the development of a circular carbon economy. By taking advantage of the catalytic activity of nanomaterials, scientists can engineer systems that catalyze these conversions with high selectivity and efficiency. The chapter also points out recent advances in catalytic routes and underscores the significance of incorporating green chemistry principles into carbon abatement strategies.

Electrochemical conversion of CO_2 is the focus of Chapter 7. In this chapter, the authors explore the function of nanomaterials in improving the performance of electrocatalysts, which play a crucial role in facilitating electrochemical reactions that convert carbon dioxide into energy rich compounds. Due to their distinctive electrical properties and surface functionalities, nanomaterials have the ability to greatly enhance the selectivity, stability, and catalytic activity of these processes. The chapter emphasizes the importance of further investigation of nanomaterial design principles to bridge current limitations and tap new horizons in sustainable energy generation.

In Chapter 8, the authors highlighted on photocatalytical conversion of CO_2 using nanomaterials. Single atom nanoparticles, metal oxide-based photocatalysts, mesoporous oxide nanoparticles, nanostructured inorganic-organic hybrid systems for CO_2 conversion were discussed in this chapter. Different fuels production from CO_2 like formic acid, methanol, ethanol, methane, formaldehyde, hydrogen, olefins, and gasoline were presented in this chapter.

Chapters 9 and 10 focuses on broad range of environmental applications of nanomaterials, very specific to air and water contamination reduction measures. Chapter 9 discusses the role of utilization of nanomaterials in purifying strategies for air, including nanoadsorbents, nano catalysts, and sensor devices. These are mainly utilized for detection and neutralizing the airborne contaminants such as NO_2, H_2S, and particulate matter. The chapter critically analyzes the possible risks related to use of nanomaterials, such as ecotoxicity and related impacts on human health. Chapter 10, in contrast, explores nanomaterials in water treatment applications. It explores how nanocomposites and membranes can play a role in removal of contaminants such as heavy metals and organic contaminants present in wastewater. The chapter has reviewed conventional methods and discusses how nanotechnology provides cost effective, and environmentally friendly option.

Chapter 11 focuses on utilization of nanomaterials for sustainable agriculture and food production. Nanomaterials for soil improvement like nonfertilizer and nanopesticides were presented in this chapter. Technologies like nanoencapsulation, nanobiosenor were also discussed in this chapter. The chapter also covers application of nanotechnology in food processing and packaging aspect. The field studies and success stories related to nanotechnology in agriculture and food industry is also presented.

Chapter 12 illustrates an important perspective on the sustainability of nanomaterials in respect to Life Cycle Assessment (LCA). It discusses that while nanomaterials are capable to address several environmental benefits, their generation and disposal also give rise to crucial issues regarding their long-term ecological impacts. The chapter also outlines the challenges faced during conduct of LCA studies, such as the selection of functional units, inventory data limitations, and emission quantification. It puts forward the pathways for improving these assessments, such as the development of protocols and integrated approach for environmental modeling.

Finally, the last chapter of this edited book, Chapter 13 highlights the environmental and health safety aspects related to nanomaterials. The natural, incidental, engineered and anthropogenic sources of nanomaterials were discussed.

Collectively, these chapters constitute a multidisciplinary analysis of nanotechnology and nanomaterials to solve some of the most critical environmental and energy issues of our era. From fundamental principles to catalytic utilization and life cycle aspects, this edited book offers an integrated vision for the contribution of nanoscience to creating a sustainable future. This book is meant to be both a reference for academic researchers, industry experts, graduate students, and policymakers. By bringing into focus the gigantic potential of nanotechnology for environmental protection and carbon management, we seek to promote interdisciplinary cooperation and speed up the shift to a low-carbon, resource-efficient world economy.

We would like to extend our thanks to all contributing authors for their dedicated approach. Their combined wisdom and creativity not only offered a deeper understanding of nanomaterials but also a hopeful road map for how science can work for society at a moment of pressing need.

Editors
Tanuj Kumar
Richa Kothari
Anita Singh
Deepak Pathania

Contents

Chapter **1**

Introduction to Nanomaterials for Zero Carbon: Concepts and Applications

Arunima Verma
Department of Nanoscience and Materials,
Central University of Jammu, Jammu-181143, Jammu and Kashmir India

1.1 INTRODUCTION

The capacity of Carbon-Based Nanomaterials (CBN) to transport therapeutic compounds and enable the observation of cells and tissues has made a significant influence on the biomedical profession in recent decades. This is crucial for the diagnosis and treatment of damaged or diseased tissues. The 0D carbon nanomaterials category includes fullerenes, nanodiamonds, Carbon and Graphene Quantum Dots (CQDs and GQDs), and carbon nanotubes, whereas the 1D and 2D categories are carbon nanowalls and carbon nanotubes [1–3]. Bioimaging, cell fluorescent labelling, stem cell engineering [4, 5], biosensing [6, 7], drug/gene delivery [8, 9], photothermal [10], and photodynamic therapy are some of the potential biological uses of CBN shown in Figure 1.1 the years [11, 12]. Carbon Black Nanoparticles (CBNs) have great photostability, a limited emission spectrum that can be adjusted, and inherent fluorescence, all of which make them ideal for use in imaging and diagnostics of cells and tissues. It is also possible to improve their qualities by altering their surfaces with functional groups, such as carboxylic acid, hydroxyl, or epoxy. In addition to their exceptional

*For Correspondence: arunimaverma144@gmail.com

DOI: 10.1201/9781003492184-1

optical features, CBN are highly sought after and well suited for theragnostic applications due to their high surface areas, mechanical characteristics, and electrical characteristics [13]. One of the most important concerns for CBN's practical use in biomedicine is their biological safety, which is connected to their aqueous stability and interactions with cells and tissues. They have recently been the subject of research that suggests they could be useful as anti-inflammatory and anti-cancer medications. For instance, when cancer cells absorb CBN, it triggers Reactive Oxygen Species (ROS), which in turn cause cell death, damage to lipids and DNA [14]. In addition, graphene materials have an effect on macrophage metabolic activity, leading to an increase in reactive oxygen species (ROS) and mitochondrial membrane damage, both of which trigger cell death through apoptosis [15]. But biofunctionalisation of CBN improves delivery efficacy by decreasing clearance and increasing retention in the body. Furthermore, carbon is naturally occurring in its free form and has the greatest melting point of all the elements. The chemical offspring of the carbon tree are carbon nanomaterials. Researchers have been captivated by carbon nanostructures due to its unique electrical, structural, mechanical, optical, biological, and chemical capabilities. These materials have been extensively investigated and found applications in many different fields. In the ultraviolet (UV) area, CQDs and GQDs show their optical adsorption range at 230–320 nm and 270–390 nm, respectively. According the [16], the π-π^* and n-π^* transitions are responsible for this phenomenon in the C=C and C=O bonds, respectively. On the other hand, compared to other 0D carbon nanomaterials, CQDs appear to have more wavelength adsorption. The tunable photoluminescence (PL) properties of CQDs and GQDs make them exceptionally versatile among other 0D carbon nanomaterials, making them useful in a wide range of applications. In contrast to GQDs, which display a high PL Quantum Yield (QY) because of their layered structure and high crystallinity, CQDs have a low PLQY because of the emissive traps on their surfaces. In spite of claims that aromatic structures, emissive traps, oxygen-containing groups, edge defects, and Quantum Confinement Effects (QCEs) are to blame for PL, the precise process by which this phenomenon is produced remains unknown [17]. The PL of Combined Quantum Dots (CQDs) and Graphene Quantum Dots (GQDs) is affected by a wide range of parameters, including temperature, solubility, surface oxidation degree, polarity, size, surface functionalisation, hybrid doping, and pH. Based on the results of an experiment, it was found that the absorption peak energy of GQDs decreased from 6.2 to 4.6 eV as their size increased from 5 to 35 nm. Additionally, the shape of the GQDs changed from circular and elliptical to zigzag and armchair edges, then to hexagonal and mostly armchair-edged rectangular. It can be inferred that the QCEs have a significant impact on the PL characteristics of quantum dots [18] since the presence of Oxygen-Functional Groups (OFGs) like C–O, C=O, and COOH was discovered to enhance the size of GQDs. Furthermore, it was shown that GQDs displayed high PL under alkaline conditions as opposed to acidic settings, suggesting that PL is pH-dependent [19, 20]. It was observed, however, that the functional groups and surface states resulted in maximum PL intensity at neutral pH [21, 22]. The size, concentration, and pH of the solution are the other two variables that impact the

PL intensity of GQDs. Not only that, but the surface and molecular structure, PL, and functional groups of GQDs are all affected by the reaction temperature. Doping with nitrogen and the carbonisation process both cause the PL QY to rise with increasing temperature. The optical properties of QDs, such as the Aggregation-Induced Emission Enhancement (AIEE) effect, Aggregation-Induced Red-Shift emission (AIRS) effect, and Aggregation-Induced Emission Quenching (AIEQ) effect, are affected by aggregation, and these properties can be used in sensing applications. QDs also have a high surface free energy, which helps to reduce particle aggregation [23]. In addition, when exposed to electrical processes, the 0D carbon nanomaterials display electrochemiluminescence (ECL), a form of luminescence [24]. In excited-state luminescence switching (ECL), a voltage is used to excite an electron, resulting in a steady emission of luminescence [25]. Similar ECL behaviour to QDs has been noted in GQDs as well. Increasing the surface defects and functional groups can improve the ECL's solubility, which in turn improves the ECL's imaging and diagnostic capabilities, among others [26]. This chapter will cover the difficulties of reducing carbon emissions, the role of nanomaterials in achieving this goal, and their theoretical and practical implementations for zero-carbon emission capture and storage.

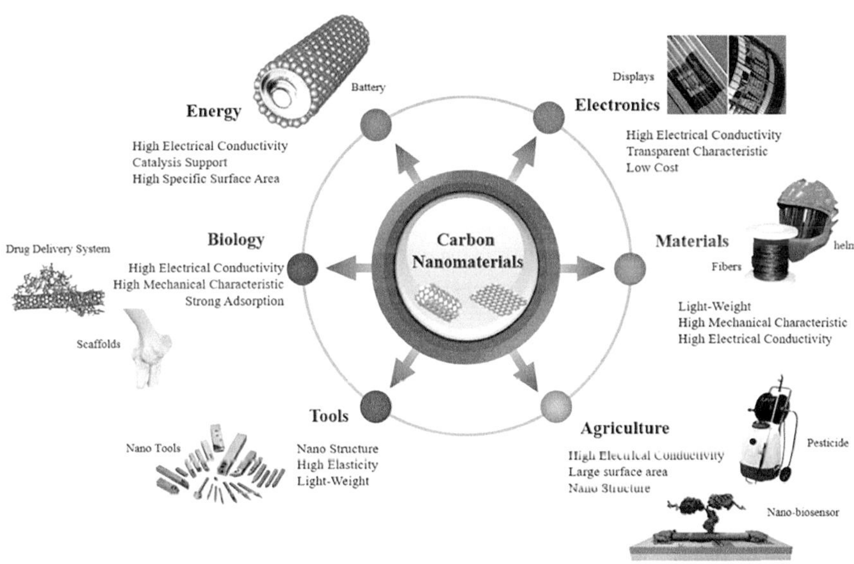

Figure 1.1 Carbon-based nanomaterials (CBNs) and their diverse applications [27].

1.2 APPLICATIONS OF NANOMATERIALS FOR ZERO CARBON EMISSION

1.2.1 Environment

Pollution occurs when undesirable chemical entities are present in the environment and either eliminate or significantly reduce the efficiency of natural processes.

Air pollution is rising greatly with other environmental crises caused by human activities such as population growth, urbanisation, and industry [28]. Consequently, it is very difficult to improve and enhance the quality of land, air, and water. Locating and studying environmental contaminants is a critical step in environmental protection. Treatment of pollution in the environment has been greatly enhanced by the use of nanotechnology in conjunction with material sciences. Among the many engineered nanomaterials (ENMs) utilised to filter water polluted with heavy metal ions, zero-dimensional carbon nanomaterials stand out for their extraordinary capabilities [29]. The benefits of 0D carbon nanomaterials in pollution detection over conventional approaches include their efficiency, speed, precision, and simplicity. And because of their clearly defined sub-nanometre structure, they are able to emit fluorescence, making them a very effective fluorescent nanoprobe for detecting a wide range of environmental contaminants [30]. For detecting pollutants, carbon dot-based biosensors either functionalise carbon dots with specific recognition molecules or analyse the change in fluorescence signals caused by direct contact between carbon dots and analytes. Even at low concentrations, heavy metal contamination is extremely harmful to humans and other living beings [31]. A new class of materials with great promise for the elimination of heavy metal contamination is zero-dimensional carbon nanomaterials. Heavy metals like Cu^{2+}, Co^{2+}, and Ni^{2+} were discovered in one experiment using hydrothermally produced CQDs. The ‘turn-off’ method was applied for sensing with the synthetic CQDs employed as nanoprobes. By utilising Photo-induced Electron Transfer (PET), it was noted that the fluorescence of the CQDs diminished with increasing metal concentration. As an added incentive, the examination of drinking water samples was also a successful use of these signals [32]. Because of their diverse surface charges, excellent water stability, good biocompatibility [33], and minimal cytotoxicity [34], CQDs are also used to detect a range of harmful bacteria [35]. As demonstrated in Figure 1.2, a turn-off fluorescence sensor array was developed for the purpose of detecting various bacteria. The array was able to identify six distinct bacteria with a precision rate of 91.6% [36].

1.2.2 Biomedical

0D carbon nanomaterials, as illustrated in Figure 1.3 have undoubtedly attracted a great deal of interest from biomedical researchers for applications such as cancer therapy, medication administration, gene therapy, biosensors, and so on. The biomedical applications and properties of carbon nanoparticles have been found to be improved through their conjugation with other nanomaterials [37]. Because cancer is fast becoming one of the deadliest diseases, finding better ways to detect and cure the disease is of the utmost importance. Specifically, carbon nanoparticles showed great promise in the detection and treatment of cancer. To cure cancer, CQDs and GQDs are a good option because of their great photoluminescence capabilities, minimal toxicity, and good biocompatibility [38]. In addition, GQDs and CQDs both have high photoluminescence, which aids in tracing the

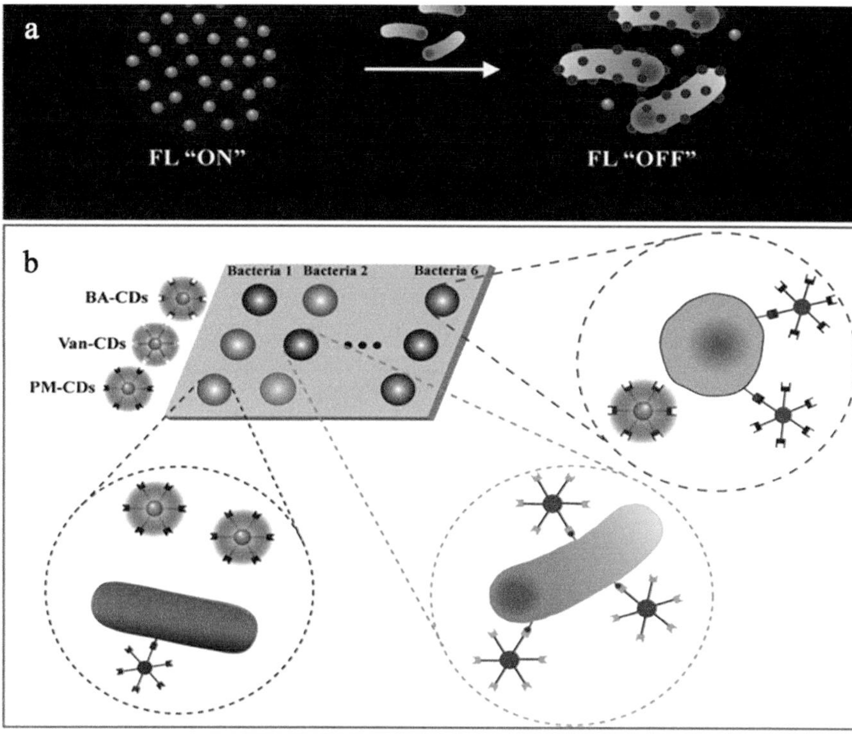

Figure 1.2 Schematic of pattern-based bacteria recognition using three receptors on carbon dots: (a) decreased fluorescence intensity from bacteria binding, and (b) fluorescence pattern from carbon dots responses to six bacteria [36].

nanoparticles inside the body and accurately monitoring the drug release to the intended organ. In addition, zero-dimensional carbon nanomaterials-based cancer photothermal therapy (PTT) is being used since it has been found that raising the temperature causes cancer cell proteins to denature, which ultimately kills the cancer cells. Because it is noninvasive, has few side effects, is biocompatible, and has minimal biotoxicity, PTT is among the best cancer treatments. While SWCNTs were the initial carbon-based nanomaterials used in PTT, their size was soon surpassed by that of CQDs and GQDs, which led to their replacement [39]. One striking method that has emerged from the proliferation of diagnostic tools made possible by modern science is the use of fluorescent probes. Fluorescent probes have been developed using a variety of nanomaterials. Despite drawbacks including low photostability and a short lifespan, carbon-based nanomaterials are nevertheless thought of being helpful fluorescent probes for cancer detection. For labelling and bio-imaging, photo-bleached CQDs and GQDs are used because of their unique cellular permeability, outstanding optical characteristics, low cytotoxicity, etc. [40]. The first step in treating any illness is, of course, making a diagnosis. The use of CQDs as a PL contrast agent was initially reported by Yang et al. [41]. After injecting PEG-glycated CQDs into mice, researchers were able to capture a range of FL pictures at different wavelengths. They found that PEG glytated CQDs had wide circulation durations, active addition at the tumour location, and great internalisation in cancer cells. The medicine was efficiently supplied to the targeted site. The high effectiveness of internalisation and the ease with which drugs might enter cancer cells were both attributed to the negative charge of the cell membrane. Targeted gene and medication delivery was another area where CQDs and GQDs proved effective. Due to its improved drug loading capacity and stability under various pH conditions, DNA-CQDs were an efficient delivery system for doxorubicin, an anticancer medication [42]. The photoluminescence of the CQDs was shown to be improved when N and P were present in DNA, suggesting that they can also be used for imaging.

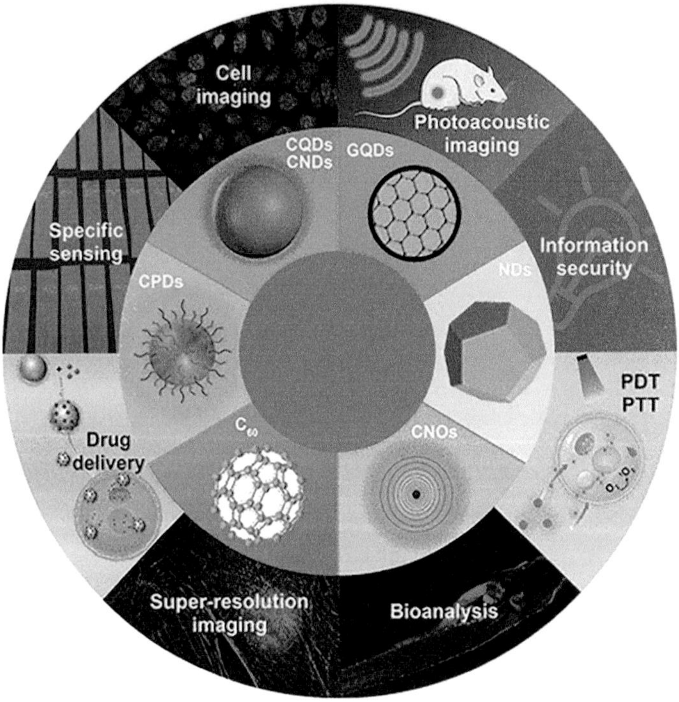

Figure 1.3 Schematic illustration of various biomedical utilities of zero-dimensional carbon nanomaterials.

1.2.3 Agriculture

Additionally, carbon dots have shown tremendous promise in the realm of agriculture. The cost of using carbon dots in agriculture is unsuitable due to several restrictions, such as the need for specialised equipment, limited yield, labourious synthesis, high temperatures, etc. To guarantee the safety of crops, people, and the environment, carbon dots are the most promising nanomaterials because they have excellent chemical inertness, low toxicity, high biocompatibility, water-solubility, environmentally friendly synthesis routes, and widely available precursors. Adding carbon dots to a nutritional solution makes them more stable and storable due to their high chemical inertness. Carbon dots can be readily absorbed and distributed via plant cells due to their incredibly small structures. Also, after absorption, carbon dots can be tracked and their action monitored by their fluorescence properties [43]. While research into carbon dots' potential agricultural uses is still in its early stages, a 2015 study by Tripathi and Sarkar et al. [44], found that 150 μg of CQDs applied to wheat roots had a beneficial effect. Nevertheless, the absence of emission spectra in the publication and the short culture duration of only ten days cast doubt on their findings. In a separate work from 2017, Zheng et al. [45] employed rapeseed pollen in a hydrothermal manner to synthesis carbon dots. These dots were then put directly to the Hoagland nutrient solution used for hydroponic growing, resulting in an environmentally friendly strategy. Figure 1.4 shows the results of a study that used this hydroponically grown Rome lettuce as a model to examine the effects of various doses of CQDs on the plant. By reducing the nitrate concentration, this experiment demonstrated that 20–30 mg L^{-1} CQDs can effectively boost production output. Additionally, it has been demonstrated that carbon dots are transported from nutrient solutions to vessels via vascular bundles and cellular pathways through transpiration stream. As a result, carbon dots synthesised using rapeseed pollen can be used to label cells *in vivo*, impacting plant physiology and yield [46].

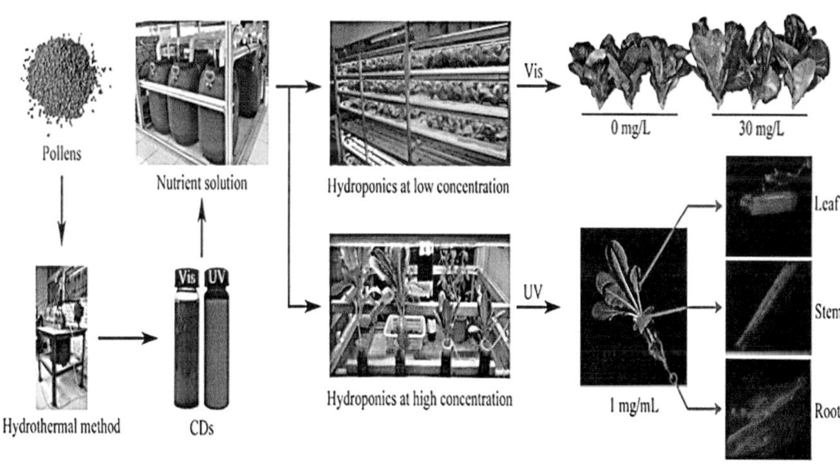

Figure 1.4 Schematic illustration of the experiment [45].

1.3 ACTIVATED AND PYROGENIC CARBON WITH THEIR DERIVATIVES FOR CARBON CAPTURE

One carbonaceous material with impressive properties is Activated Carbon (AC). It is very porous, very mechanically stable, and has a higher adsorptive capacity than other materials. Figure 1.5 shows that although AC can be produced from biomass by pyrolysis, it must first undergo a number of physical or chemical activation procedures. Animal feathers, coal, lignin, coffee grinds, coconut shells, nitrogen-rich chitosan, synthetic organic polymers, and other materials can be used to make AC [47, 48]. To make chemically activated carbon, the biomass feedstock must be treated with a salt, base, or acid. Various chemicals are utilised to chemically activate carbon in order to enhance the surface area of AC, remove or introduce specific functional groups, and more. By removing acidic oxygen groups and adding nitrogen groups to carbon at extremely high temperatures, a process known as "activation" occurs, resulting in an increase in basicity [49]. The nitrogen-containing functional groups that ACs react with include amines, nitric acids, and ammonia. Additionally, ACs have the ability to generate other functional groups such as lactam, pyridinic, pyrrolic, imide, and amide groups [50]. One way to oxidise a substance is to add an aqueous solution or atmospheric gas. Phenol and other weak acid groups can be integrated when the temperature is high enough. Through the application of acid digestion, which included HF, HCl, and HNO_3, the samples' unburned carbon and ash could be eliminated [51]. One of the two parts of physical activation is getting biomass to turn into pyrogenic carbon at a low enough temperature. Two, the carbon compounds were activated using gas steams at extremely high temperatures, including O_2, CO_2, and others. Some heteroatoms, such as sulphur, nitrogen, hydrogen, and oxygen, are also present. There are specific heteroatoms that define the AC chemical properties. Delocalised electrons in the carbon structure and surface functional groups dictated AC acidic and basic characteristics. This acidic quality of AC surfaces is caused by the oxygen-containing functional groups. Predominantly, these clusters can be seen on the exterior, or basel plane, of carbon structures. Inorganic functional groups that contain oxygen include ether, quinone, chromone, pyrone, carbonyl, phenol, lactone, and carboxylic. Common techniques for introducing functional groups containing oxygen include the use of aqueous and gas oxidants. In the gas-phase treatment, steam, carbon dioxide, and oxygen air can be used, while in the liquid-phase treatment, nitric and sulphuric acids can be employed [52]. Protons can bind to carbon aromatic rings with π electrons and basic surface functions that contain nitrogen, which can accompany the AC basicity [53]. For the purpose of CO_2 capture, Plaza et al. [54] heated almond and olive stones to 800°C using CO_2. Using a thermogravimetric analyser (Mettler Toledo), they varied the temperature of the produced BC from 25 to 100°C under ambient pressure to evaluate its CO_2 adsorption characteristics. The adsorption isotherm of the prepared samples was estimated using the Toth model. At 100°C, 25°C, and 0°C, respectively, under 101 kPa pressure, carbon derived from olive stones could absorb 0.7, 3.0, and 4.8 mmol/g of CO_2. Biochar was produced by Gao et al. [55] using sugar cane bagasse and hickory wood as feedstock at temperatures of 300, 450, and 600°C. At ambient temperature and 101 kPa pressure, the produced biochar exhibited the maximum CO_2 adsorption capacity of 73.55 mg g^{-1}. Wang et al. [56] prepared AC from celtuce leaves by activating it with KOH at 600°C. Figure 1.5 shows the activation procedure, which involves carbonising the dried lettuce leaves and then activating them with KOH. The CO_2 adsorption isotherms at 0 and 25°C were studied using a Micromeritics ASAP 2020 static volumetric analyser for the experiment. At 0 and 25°C under ambient pressure, the produced material demonstrates a CO_2 adsorption capacity of 6.04 and 4.36 mmol/g.

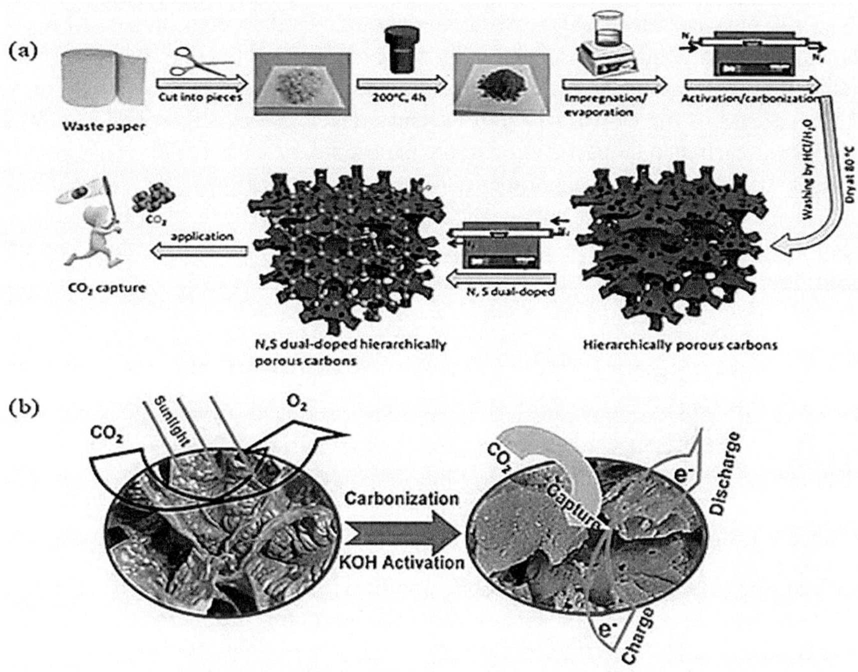

Figure 1.5 Synthesis process of activated carbon from biomass [56].

1.4 CHALLENGES IN REDUCING CARBON EMISSIONS

A great deal of research in the last several decades has focused on how greenhouse gases (GHGs) contribute to global warming. Carbon dioxide (CO_2) and other greenhouse gases have been shown by climate models to be essential in keeping the Earth's surface temperatures within liveable ranges for human habitation [57]. However, during the Industrial Revolution in the twentieth century, the same species disrupted the delicate equilibrium between atmospheric concentrations of greenhouse gases and global temperature. Now, we face a situation that we have brought upon ourselves, where an increase in CO_2 levels could threaten our very survival on Earth [58]. We have already experienced warm winters, prolonged droughts, unprecedented magnitudes of wildfires, rising sea levels, and an increase in the global temperature, as shown by 2020 being recorded as the hottest year. This is all because of the accelerated emission of CO_2 by humans, which can set off a global disaster cascade. We have already passed the threshold level (i.e., 350 ppm) of CO_2 in the atmosphere; the present level is 415 ppm, according to the Mauna Loa Observatory. The predicted socioeconomic impact of rising CO_2 levels in the near future is negative, and necessary worldwide action is required to mitigate the threat [59]. Some research into CCUS technologies has showed promise in terms of turning CO_2 and other big contributors to climate change into marketable goods. Because nanoparticles may possess better chemical, physical, electrical, and/or material qualities than their bulk equivalents, the value-added goods may receive even more attention if they are nanomaterials. Many areas can benefit from nanomaterials; recently, nanotechnology was crucial in the success of the COVID-19 [60] vaccines developed by Pfizer and Moderna. Graphene, carbon nanoparticles, carbon nano-onions, carbon nano-honeycomb/platelet structures, carbon nanotubes (CNTs), carbon nanofibres (CNFs), graphene, carbon nanosheets, carbon nanoparticles, and inorganic nanostructures have all been synthesised using CO_2 gas in recent decades. One specific method that has been developed for this purpose is phase separation.

1.4.1 Altering CO_2 into Carbon Nanomaterials

Varieties of carbon nanomaterials have been produced from CO_2 conversion, including CNTs, CNFs, carbon nanosheets, nanoparticles, nano-onions, porous nanomaterials, graphite honeycombs, and graphene-grafted CNFs:

1.4.1.1 Chemical Vapour Deposition-Based CNTs and CNFs

Carbon dioxide gas (CO_2) chemical synthesis initially relied on Chemical Vapour Deposition (CVD), a standard technique for producing high-quality semiconductors. By directly converting CO_2 with a Mg/Ni template, Wang et al. [61] were able to create three-dimensional (3D) carbon nanotube foams (CNTFs) using the CVD technique. In short, the cylindrical CNTs were produced by reacting CO_2 gas, which was a 1:3 mixture of CO_2 and argon gases, with Mg at 800°C for 1–2 hours. Ni nanoparticles served as the catalyst for the CNT production process. As shown in Figure 1.6(a) and (b) a 3D network of CNTs was discovered using electron microscopy investigation of the as-prepared CNTFs. The CNTFs also exhibit mechanical and electrical qualities that are well-suited to semiconductors. An economically viable CO_2 utilisation technique would not benefit from the energy-intensive processes of CVD fabrication, which include prolonged exposure to high temperatures and pressures and poor scalability. Therefore, using moderate reaction conditions and highly efficient catalysts to convert CO_2 into useful chemicals has been the focus of recent improvements in the CVD process. In a modified CVD process developed by Kim et al. [62] CO_2-derived CNTs (CCNTs) were produced in a controlled environment with mild parameters (500–700°C and 1 atm) by reacting 100% CO_2 gas with $NaBH_4$ and metallic Ni, which operated as a reducing agent and catalyst, respectively. In addition, unlike earlier CVD fabrications, the suggested synthetic technique only requires one phase of continuous CVD, solving the problem of low scalability. In comparison to commercially available activated carbon (YP-50F) employed in supercapacitors, the mesoporous structure of as-prepared CCNTs has contributed to their superior electrical characteristics (Figure 1.6(c) and (d)).

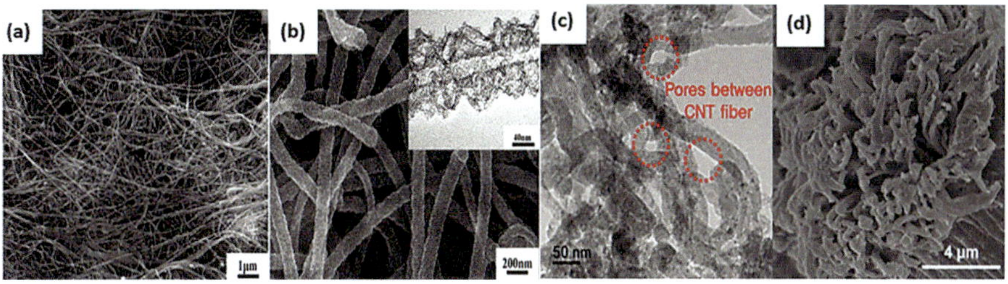

Figure 1.6 (a) Microstructures of CNTF; (b) High-resolution morphology of CNTF, Inset: TEM image of the single CNT [61] (c) TEM and (d) SEM images to show the pores between the CNT fibres in CC10 [62].

1.4.1.2 Various Nanostructures Based on Carbon

In addition to structures based on carbon nanotubes, other groups have reported the creation of additional nanostructures using carbon dioxide gas as the carbon source. Chen and colleagues [63, 64] used the same molten salt electrochemical technique to create a novel nanosheet porous structure in the form of a honeycomb (Figure 1.7(a) and (b)). They were able to employ a flue gas combination containing SO_2 without pretreatments because they used Li_2CO_3–Na_2CO_3–K_2CO_3–Li_2SO_4 as the electrolyte, which is a significant distinction from the other popular electrochemical methods. Because sulphur oxides (SO_x) gases can have corrosive reactions with the electrochemical components, making the process more expensive, desulphurisation of the flue gas mixture is typically required before it is introduced to the electrochemical system. The new electrolyte, in conjunction with moderate reaction conditions (475°C and Ni/SnO_2 electrodes), makes the cost effective approach of flue gas conversion into valuable nanomaterials proposed by Chen et al. [65] even more attractive. Figure 1.7(c) and (d) shows that the same group was able to produce crystalline graphite at 775°C in an electrochemical reaction mediated by a Li_2CO_3–Na_2CO_3–K_2CO_3–Li_2SO_4 electrolyte using a mixture of CO_2 and SO_2 flue gas [64]. Powder X-Ray Diffraction (PXRD) scans indicated the presence of crystalline features in the nanostructure as the reaction temperature increased, and the as-produced graphite exhibited a honeycomb structure identical to their earlier investigation (d–h of Figure 1.7). According to Liu et al. [66], the nano-onion structure in particular is 20,000 times more valuable than coal, demonstrating the substantial economic significance of these carbon nanostructures in comparison to coal. In addition, they calculated that preparing carbon nano onions using molten carbonate electrolysis would cost $1,000/ton, which is far less than other comparable approaches. When comparing the CO_2-based graphene manufacturing process to the industrial methods, comparable calculations for graphene platelets reveal a $1,000 profit within 30 seconds.

1.5 FUTURE PERSPECTIVES

High CO_2 adsorption capability is provided by a combination of a microporous network and a large surface area, as was shown in earlier research. Activated

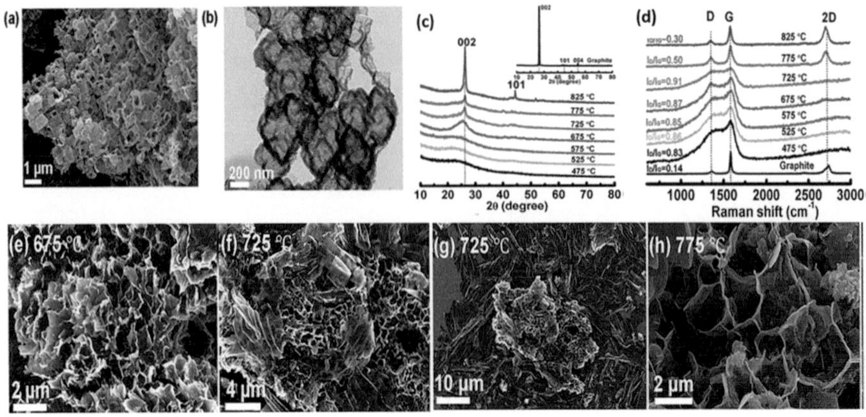

Figure 1.7 (a) SEM, (b) TEM [63]; (c) XRD patterns of the electrolytic carbon materials obtained at different temperatures. Inset is the XRD pattern of commercial graphite. (d) Raman spectra of electrolytic carbon materials obtained at different temperatures. The spectrum of commercial graphite is displayed for comparison; SEM images of the electrolytic carbon materials obtained at different temperatures (e) 675°C, (f) and (g) 725°C and (h) 775°C [64].

carbon has attracted a lot of attention from researchers as a potential alternative adsorbent for CO_2 capture applications on an industrial scale because to its low discarding difficulties, reduced cost of recycling adsorbent, and ease of regeneration. But I could not find any referenced research that details the specific sectors that make use of activated porous carbons. Scientists and researchers in the field of CO_2 adsorption, whether in academia or industry, should thus immediately put these materials to use in large-scale applications and study how well they absorb CO_2. There are several factors to consider when using carbon nanomaterials in large-scale industrial applications:

1. A more equitable approach would result from the development of cost-effective sorbents.
2. It should have good stability in flue gases over the long term.
3. The adsorbent needs to be highly recyclable with low regeneration energy requirements.
4. It is important to study the regeneration mechanism.
5. It is vital to examine the impact of additional variables, including different gas combinations, environmental factors, outside ions, and water vapour.

The use of Direct Air Capture (DAC) technology to remove CO_2 from the air is currently the subject of a heated controversy as a means to mitigate climate change. Combining DAC technology with low-cost, reusable carbonaceous porous adsorbents is a remarkable idea if we want to produce DAC technology that is economically friendly. With its CO_2 adsorption capabilities in a low-concentration medium, it can help mitigate climate change.

1.6 CONCLUSIONS

Innovations in science and technology are being created all the time. Science and technology have advanced to the point where they are good for both people and the planet. Thanks to its ability to control the characteristics of bulk matter on a microscopic scale, nanotechnology has attracted a lot of interest from academics. Many fields have found useful applications for nanomaterials, and 0D carbon nanomaterials are no exception. These include biomedicine, sensors, environmental monitoring, robotics, agriculture, and more. 0D carbon nanomaterials have distinct and adjustable features due to their incredibly tiny size, which provides a lot of potential applications. Tunable microporosities, hierarchical porosities, a large surface area, and other unique properties made this carbon nanomaterial an effective CO_2 collection agent. One way to improve the CO_2 capture efficiency is to modify the surface and functionalise the carbon nanomaterial. As an example, carbon nanoparticles have been imine-and amine-functionalised, subjected to acid treatment, formed as composites or hybrids, activated chemically or physically, impregnated with metal oxides, and mono- or co-doped with various atoms. In this chapter we described the difficulties of reducing carbon emissions, why nanomaterials are important for this goal, and how to use them for zero-carbon emission capture and storage, both theoretically and in practice.

REFERENCES

[1] Lin, J., Huang, Y. and Huang, P. 2018. Graphene-based nanomaterials in bioimaging. Biomedical Applications of Functionalized Nanomaterials. 247–287.

[2] Xu, M., Zhu, J., Wang, F., Xiong, Y., Wu, Y., Wang, Q., et al. 2016. Improved in vitro and in vivo biocompatibility of graphene oxide through surface modification: poly(acrylic acid)-functionalization is superior to PEGylation. ACS Nano. 10(3): 3267–3281.

[3] Liu, J.H., Cao, L., LeCroy, G.E., Wang, P., Meziani, M.J., Dong, Y., et al. 2015. Carbon "quantum" dots for fluorescence labeling of cells. ACS Applied Materials and Interfaces. 7(34): 19439–19445.

[4] Baek, S., Singh, R.K., Khanal, D., Patel, K.D., Lee, E.J., Leong, K.W., et al. 2015. Smart multifunctional drug delivery towards anticancer therapy harmonized in mesoporous nanoparticles. Nanoscale. 7(34): 14191–14216.

[5] Garcia-Alegria, E., Iliut, M., Stefanska, M., Silva, C., Heeg, S., Kimber, S.J., et al. 2016. Graphene Oxide promotes embryonic stem cell differentiation to haematopoietic lineage. Scientific Reports. 6(1): 25917.

[6] Soikkeli, M., Kurppa, K., Kainlauri, M., Arpiainen, S., Paananen, A., Gunnarsson, D., et al. 2016. Graphene biosensor programming with genetically engineered fusion protein monolayers. ACS Applied Materials and Interfaces. 8(12): 8257–8264.

[7] Shi, H., Wei, J., Qiang, L., Chen, X. and Meng, X. 2014. Fluorescent carbon dots for bioimaging and biosensing applications. Journal of Biomedical Nanotechnology. 10(10): 2677–2699.

[8] Bao, H., Pan, Y., Ping, Y., Sahoo, N.G., Wu, T., Li, L., et al. 2011. Chitosan-functionalized graphene oxide as a nanocarrier for drug and gene delivery. Small. 7(11): 1569–1578.

[9] Yang, K., Li, Y., Tan, X., Peng, R. and Liu, Z. 2013. Behavior and toxicity of graphene and its functionalized derivatives in biological systems. Small. 9(9–10): 1492–1503.

[10] Tian, B., Wang, C., Zhang, S., Feng, L. and Liu, Z. 2011. Photothermally enhanced photodynamic therapy delivered by nano-graphene oxide. ACS Nano. 5(9): 7000–7009.

[11] Ge, J., Lan, M., Zhou, B., Liu, W., Guo, L., Wang, H., et al. 2014. A graphene quantum dot photodynamic therapy agent with high singlet oxygen generation. Nature Communications. 5(1): 4596.

[12] Hong, E.J., Choi, D.G. and Shim, M.S. 2016. Targeted and effective photodynamic therapy for cancer using functionalized nanomaterials. Acta Pharmaceutica Sinica B. 6(4): 297–307.

[13] Du, D., Wang, K., Wen, Y., Li, Y. and Li, Y.Y. 2016. Photodynamic graphene quantum dot: reduction condition regulated photoactivity and size dependent efficacy. ACS Applied Materials and Interfaces. 8(5): 3287–3294.

[14] Pinto, A.M., Goncalves, I.C. and Magalhaes, F.D. 2013. Graphene-based materials biocompatibility: A review. Colloids and Surfaces B: Biointerfaces. 111: 188–202.

[15] Yang, P., Gai, S. and Lin, J. 2012. Functionalized mesoporous silica materials for controlled drug delivery. Chemical Society Reviews. 41(9): 3679–3698.

[16] Eda, G., Lin, Y.Y., Mattevi, C., Yamaguchi, H., Chen, H.A., Chen, I.S.C.W., et al. 2009. Blue photoluminescence from chemically derived graphene oxide. arXiv preprint arXiv:0909.2456.

[17] Zhu, S., Zhang, J., Tang, S., Qiao, C., Wang, L., Wang, H., et al. 2012. Surface chemistry routes to modulate the photoluminescence of graphene quantum dots: from fluorescence mechanism to up-conversion bioimaging applications. Advanced Functional Materials. 22(22): 4732–4740.

[18] Sciortino, A., Cannizzo, A. and Messina, F. 2018. Carbon nanodots: a review—from the current understanding of the fundamental photophysics to the full control of the optical response, C. 4(4): 67.

[19] Shen, J., Zhu, Y., Chen, C., Yang, X. and Li, C. 2011. Facile preparation and upconversion luminescence of graphene quantum dots. Chemical Communications. 47(9): 2580–2582.

[20] Gu, J., Zhang, X., Pang, A. and Yang, J. 2016. Facile synthesis and photoluminescence characteristics of blue-emitting nitrogen-doped graphene quantum dots. Nanotechnology. 27(16): 165704.

[21] Chen, S., Hong, Y., Liu, Y., Liu, J., Leung, C.W., Li, M., et al. 2013. Full-range intracellular pH sensing by an aggregation-induced emission-active two-channel ratiometric fluorogen. Journal of the American Chemical Society. 135(13): 4926–4929.

[22] Liu, M.A. and Reeves, P.R. 2019. Customizable cloning of whole polysaccharide gene clusters by yeast homologous recombination. Bacterial Polysaccharides: Methods and Protocols. 1–14.

[23] Zhang, R., Adsetts, J. R., Nie, Y., Sun, X. and Ding, Z. 2018. Electrochemiluminescence of nitrogen-and sulfur-doped graphene quantum dots. Carbon. 129: 45–53.

[24] Lu, Q., Wei, W., Zhou, Z., Zhou, Z., Zhang, Y. and Liu, S. 2014. Electrochemiluminescence resonance energy transfer between graphene quantum dots and gold nanoparticles for DNA damage detection. Analyst. 139(10): 2404–2410.

[25] Xu, Y., Liu, J., Gao, C. and Wang, E. 2014. Applications of carbon quantum dots in electrochemiluminescence: a mini review. Electrochemistry Communications. 48: 151–154.

[26] Wang, R., Wu, H., Chen, R. and Chi, Y. 2019. Strong electrochemiluminescence emission from oxidized multiwalled carbon nanotubes. Small. 15(48): 1901550.

[27] Patel, D.K., Kim, H.B., Dutta, S.D., Ganguly, K. and Lim, K.T. 2020. Carbon nanotubes-based nanomaterials and their agricultural and biotechnological applications. Materials. 13(7): 1679.

[28] Zhang, X., Huang, Q., Deng, F., Huang, H., Wan, Q., Liu, M., et al. 2017. Mussel-inspired fabrication of functional materials and their environmental applications: progress and prospects. Applied Materials Today. 7: 222–238.

[29] Devi, P., Rajput, P., Thakur, A., Kim, K.H. and Kumar, P. 2019. Recent advances in carbon quantum dot-based sensing of heavy metals in water. TrAC Trends in Analytical Chemistry. 114: 171–195.

[30] Sun, X. and Lei, Y. 2017. Fluorescent carbon dots and their sensing applications. TrAC Trends in Analytical Chemistry. 89: 163–180.

[31] Jaishankar, M., Tseten, T., Anbalagan, N., Mathew, B.B. and Beeregowda, K.N. 2014. Toxicity, mechanism and health effects of some heavy metals. Interdisciplinary Toxicology. 7(2): 60.

[32] Mintz, K. J., Guerrero, B. and Leblanc, R. M. 2018. Photoinduced electron transfer in carbon dots with long-wavelength photoluminescence. The Journal of Physical Chemistry C. 122(51): 29507–29515.

[33] Bhaisare, M.L., Gedda, G., Khan, M.S. and Wu, H.F. 2016. Fluorimetric detection of pathogenic bacteria using magnetic carbon dots. Analytica Chimica Acta. 920: 63–71.

[34] Zhong, D., Zhuo, Y., Feng, Y. and Yang, X. 2015. Employing carbon dots modified with vancomycin for assaying Gram-positive bacteria like *Staphylococcus aureus*. Biosensors and Bioelectronics. 74: 546–553.

[35] Yang, L., Deng, W., Cheng, C., Tan, Y., Xie, Q. and Yao, S. 2018. Fluorescent immunoassay for the detection of pathogenic bacteria at the single-cell level using carbon dots-encapsulated breakable organosilica nanocapsule as labels. ACS Applied Materials and Interfaces. 10(4): 3441–3448.

[36] Zheng, L., Qi, P. and Zhang, D. 2019. Identification of bacteria by a fluorescence sensor array based on three kinds of receptors functionalized carbon dots. Sensors and Actuators B: Chemical. 286: 206–213.

[37] Duran, N., Simoes, M.B., de Moraes, A., Favaro, W.J. and Seabra, A.B. 2016. Nanobiotechnology of carbon dots: A review. Journal of Biomedical Nanotechnology. 12(7): 1323–1347.

[38] Li, X., Shi, L., Li, L., Dong, C., Li, C.Z. and Shuang, S. 2019. Recent advances in carbon nanodots: properties and applications in cancer diagnosis and treatment. Journal of Analysis and Testing. 3: 37–49.

[39] Kam, N.W.S., O'Connell, M., Wisdom, J.A. and Dai, H. 2005. Carbon nanotubes as multifunctional biological transporters and near-infrared agents for selective cancer cell destruction. Proceedings of the National Academy of Sciences. 102(33): 11600–11605.

[40] Geng, B., Yang, D., Pan, D., Wang, L., Zheng, F., Shen, W., et al. 2018. NIR-responsive carbon dots for efficient photothermal cancer therapy at low power densities. Carbon. 134: 153–162.

[41] Yang, S.T., Wang, X., Wang, H., Lu, F., Luo, P.G., Cao, L., et al. 2009. Carbon dots as nontoxic and high-performance fluorescence imaging agents. The Journal of Physical Chemistry C. 113(42): 18110–18114.

[42] Ding, H., Du, F., Liu, P., Chen, Z. and Shen, J. 2015. DNA–carbon dots function as fluorescent vehicles for drug delivery. ACS Applied Materials and Interfaces. 7(12): 6889–6897.

[43] Khodakovskaya, M.V., Kim, B.S., Kim, J.N., Alimohammadi, M., Dervishi, E., Mustafa, T. and Cernigla, C.E. 2013. Carbon nanotubes as plant growth regulators: effects on tomato growth, reproductive system, and soil microbial community. Small. 9(1): 115–123.

[44] Tripathi, S. and Sarkar, S. 2015. Influence of water soluble carbon dots on the growth of wheat plant. Applied Nanoscience. 5: 609–616.

[45] Zheng, Y., Xie, G., Zhang, X., Chen, Z., Cai, Y., Yu, W., et al. 2017. Bioimaging application and growth-promoting behavior of carbon dots from pollen on hydroponically cultivated Rome lettuce. ACS Omega. 2(7): 3958–3965.

[46] Mehndiratta, P., Jain, A., Srivastava, S. and Gupta, N. 2013. Environmental pollution and nanotechnology. Environment and Pollution. 2(2): 49.

[47] González, A.S., Plaza, M.G., Pis, J.J., Rubiera, F. and Pevida, C. 2013. Post-combustion CO_2 capture adsorbents from spent coffee grounds. Energy Procedia. 37: 134–141.

[48] Zhao, Z., Wang, Y., Li, M. and Yang, R. 2015. High performance N-doped porous activated carbon based on chicken feather for supercapacitors and CO_2 capture. RSC Advances. 5(44): 34803–34811.

[49] Shafeeyan, M.S., Daud, W.M.A.W., Houshmand, A. and Shamiri, A. 2010. A review on surface modification of activated carbon for carbon dioxide adsorption. Journal of Analytical and Applied Pyrolysis. 89(2): 143–151.

[50] Plaza, M.G., Pevida, C., Arias, B., Casal, M.D., Martín, C.F., Fermoso, J., et al. 2009. Different approaches for the development of low-cost CO_2 adsorbents. Journal of Environmental Engineering. 135(6): 426–432.

[51] Zhang, Y., Maroto-Valer, M.M. and Tang, Z. 2004. Microporous activated carbons produced from unburned carbon in fly ash and their application for CO_2 capture. Fuel Chemistry Division Preprints. 49(1): 304–305.

[52] Mangun, C.L., Benak, K.R., Daley, M.A. and Economy, J. 1999. Oxidation of activated carbon fibers: effect on pore size, surface chemistry, and adsorption properties. Chemistry of Materials. 11(12): 3476–3483.

[53] Montes-Morán, M.A., Suárez, D., Menéndez, J.A. and Fuente, E. 2004. On the nature of basic sites on carbon surfaces: An overview. Carbon. 42(7): 1219–1225.

[54] Plaza, M.G., González, A.S., Pis, J.J., Rubiera, F. and Pevida, C. 2014. Production of microporous biochars by single-step oxidation: Effect of activation conditions on CO_2 capture. Applied Energy. 114: 551–562.

[55] Creamer, A.E., Gao, B. and Zhang, M. 2014. Carbon dioxide capture using biochar produced from sugarcane bagasse and hickory wood. Chemical Engineering Journal. 249: 174–179.

[56] Wang, R., Wang, P., Yan, X., Lang, J., Peng, C. and Xue, Q. 2012. Promising porous carbon derived from celtuce leaves with outstanding supercapacitance and CO_2 capture performance. ACS Applied Materials and Interfaces. 4(11): 5800–5806.

[57] Manabe, S. 2019. Role of greenhouse gas in climate change. Tellus A: Dynamic Meteorology and Oceanography. 71(1): 1620078.

[58] Davis, W.J. 2017. The relationship between atmospheric carbon dioxide concentration and global temperature for the last 425 million years. Climate. 5(4): 76.

[59] Global Monitoring Laboratory. https://www.esrl.noaa.gov/gmd/ccgg/trends/monthly.html (accessed Jan 11, 2021)

[60] Shin, M.D., Shukla, S., Chung, Y.H., Beiss, V., Chan, S.K., Ortega-Rivera, O.A., et al. 2020. COVID-19 vaccine development and a potential nanomaterial path forward. Nature Nanotechnology. 15(8): 646–655.

[61] Wang, C., Li, F., Qu, H., Wang, Y., Yi, X., Qiu, Y., et al. 2015. Fabrication of three dimensional carbon nanotube foam by direct conversion carbon dioxide and its application in supercapacitor. Electrochimica Acta. 158: 35–41.

[62] Kim, G.M., Lim, W.G., Kang, D., Park, J.H., Lee, H., Lee, J. and Lee, J.W. 2020. Transformation of carbon dioxide into carbon nanotubes for enhanced ion transport and energy storage. Nanoscale. 12(14): 7822–7833.

[63] Chen, Z., Deng, B., Du, K., Mao, X., Zhu, H., Xiao, W., et al. 2017. Flue-gas-derived sulfur-doped carbon with enhanced capacitance. Advanced Sustainable Systems. 1(6): 1700047.

[64] Chen, Z., Gu, Y., Hu, L., Xiao, W., Mao, X., Zhu, H., et al. 2017. Synthesis of nanostructured graphite via molten salt reduction of CO_2 and SO_2 at a relatively low temperature. Journal of Materials Chemistry A. 5(39): 20603–20607.

[65] Chen, Z., Gu, Y., Du, K., Wang, X., Xiao, W., Mao, X., et al. 2017. Enhanced electrocatalysis performance of amorphous electrolytic carbon from CO_2 for oxygen reduction by surface modification in molten salt. Electrochimica Acta. 253: 248–256.

[66] Liu, X., Ren, J., Licht, G., Wang, X. and Licht, S. 2019. Carbon nano-onions made directly from CO_2 by molten electrolysis for greenhouse gas mitigation. Advanced Sustainable Systems. 3(10): 1900056.

Chapter 2

Nanomaterials for Carbon Capture

Akanksha Andotra
Department of Nanoscience and Materials,
Central University of Jammu, Jammu-181143, Jammu and Kashmir India

2.1 INTRODUCTION

The topic of global warming has garnered a lot of attention and discussion in recent years [1]. Figure 2.1 shows that the main driver of global warming, as

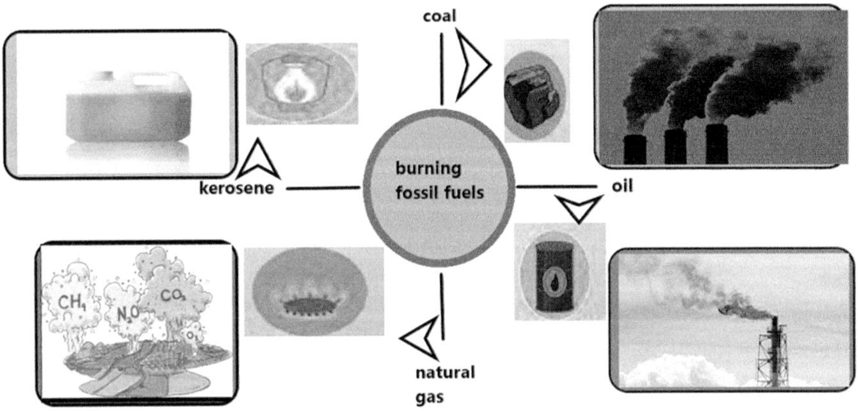

Figure 2.1 Shows some burning fuels result in CO_2 and other gases emission which cause global warming.

*For Correspondence: akanksha2194andotra@gmail.com

DOI: 10.1201/9781003492184-2

indicated by the excessive emissions of greenhouse gases like carbon dioxide (CO_2), is the burning of fossil fuels such as coal, oil, and natural gas. Commercial heating and cooling, energy generation, and human transportation are further contributors to CO_2 emissions. There has been substantial socioeconomic loss and extensive ecological harm as a result of climate change and global warming [4]. Anthropogenic increases in atmospheric concentrations of greenhouse gases, most notably carbon dioxide, are the primary drivers of climate change and global warming. Global agricultural productivity is declining as a result of climate change, which is characterized by less rainfall, changing seasons, and increasing temperatures [5].

Reducing atmospheric CO_2 levels and mitigating climate change are the goals of carbon capture and conversion. Carbon capture and storage technologies are estimated by the Centre for Climate and Energy Solutions to be able to capture over 90% of CO_2 emissions from power plants and industrial facilities and reduce 14% of the world's greenhouse gas emissions by the year 2050 [5]. We need better technology to capture carbon dioxide in an inexpensive and energy-efficient manner. Recent developments have also prompted studies into potential alternatives to sequestration, such as the direct transformation of carbon dioxide into useful goods. Global CO_2 emissions can be drastically reduced with the use of nanotechnology [6]. Currently, there are three primary technological pathways for capturing CO_2 both domestically and internationally: oxygen-enriched combustion, postcombustion capture, and pre combustion capture. Because of its excellent adaptability to the flue gas characteristics of current power plants and the requirement for large-scale transformation, post-combustion capture technology is considered to be the most practical way of reducing CO_2 emissions in the future. There are many promising new materials available for CO_2 capture from oxygen-enriched combustion processes, post-combustion, and pre-combustion. These materials include alkanolamines, silica-based materials, carbonaceous adsorbents like zeolites, basic metal compounds, porous materials, nanocomposites materials, ionic liquids, and porous materials [7]. Literature suggests that CO_2 capture through different technologies is costly [8].

2.2 TYPES OF NANOMATERIALS FOR CARBON CAPTURE

Nanoparticles: In CO_2 capturing-related technologies, nanomaterials and nanoparticles are currently playing a significant role. The goal of researchers' work with nanomaterials is to improve the performance of current technologies [9]. Because of their high surface areas and adjustable characteristics, nanoparticles are an essential part of carbon capture technologies, increasing their efficiency and selectivity. In order to examine the ability of nanoparticles to absorb CO_2 and attempt to enhance their thermal stability, Park and Lin synthesised organic hybrid materials (NOHMs). The impact of amine and ether groups on NOHM-assisted CO_2 capture was examined. There were also effects related to pressure and temperature. It was determined that NOHMs were more selective in capturing CO_2 than O_2 and N_2 [10]. Also, copper oxide nanoparticle-loaded Porous Carbons (PCs) with high efficiency led to CO_2 capture. It was determined that the enhanced carbons' ability to adsorb CO_2 is caused by copper oxide nanoparticles. Because CO_2 molecules are typically electron acceptors, copper oxide nanoparticles with donor electron properties exhibit strong interaction and efficient capture as shown in Figure 2.2. Experiments using different nanoparticle synthesis showed promising results for CO_2 capture at high temperatures. However, they still have a lot of drawbacks, and still need to be financially feasible. Therefore, in order to use nanoparticles for CO_2 capture, more attention needs to be paid to their affordable synthesis [11].

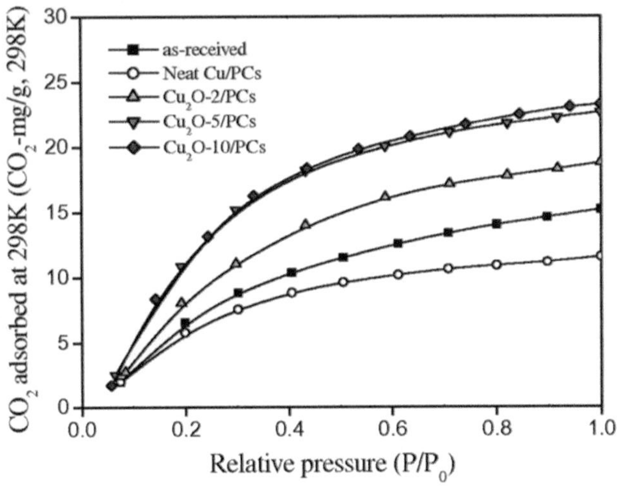

Figure 2.2 CO_2 adsorption behaviours of as-received, neat Cu/PCs, and Cu_2O/PCs as a function of the oxidation time (e.g., Cu_2O-2/PC, Cu_2O-5/PC etc.) [12].

Nanotube: Since carbon nanotubes (CNTs) are effective at absorbing carbon dioxide (CO_2), their numerous properties make them valuable for carbon capture as shown in Figure 2.3. CNTs are cylinders of pure carbon with a length ranging from 100 nanometres to millimetres. They have a high surface area, pore size-tunable, and chemically stable. They also possess good mechanical strength, light weight, and high porosity. CNTs can be renewed and reused, have quick adsorption and desorption kinetics, and work with a variety of capture techniques [12]. They are effective at absorbing CO_2 because of their large surface area, adjustable pore size, and chemical stability. In addition, CNTs have high mechanical strength, are regenerable and reusable, and show fast adsorption and desorption kinetics [13, 14].

2.2.1 Nanocomposites

Nanocomposites based on multi-component polymers provide an efficient and cost-effective method of directly capturing and storing carbon dioxide as

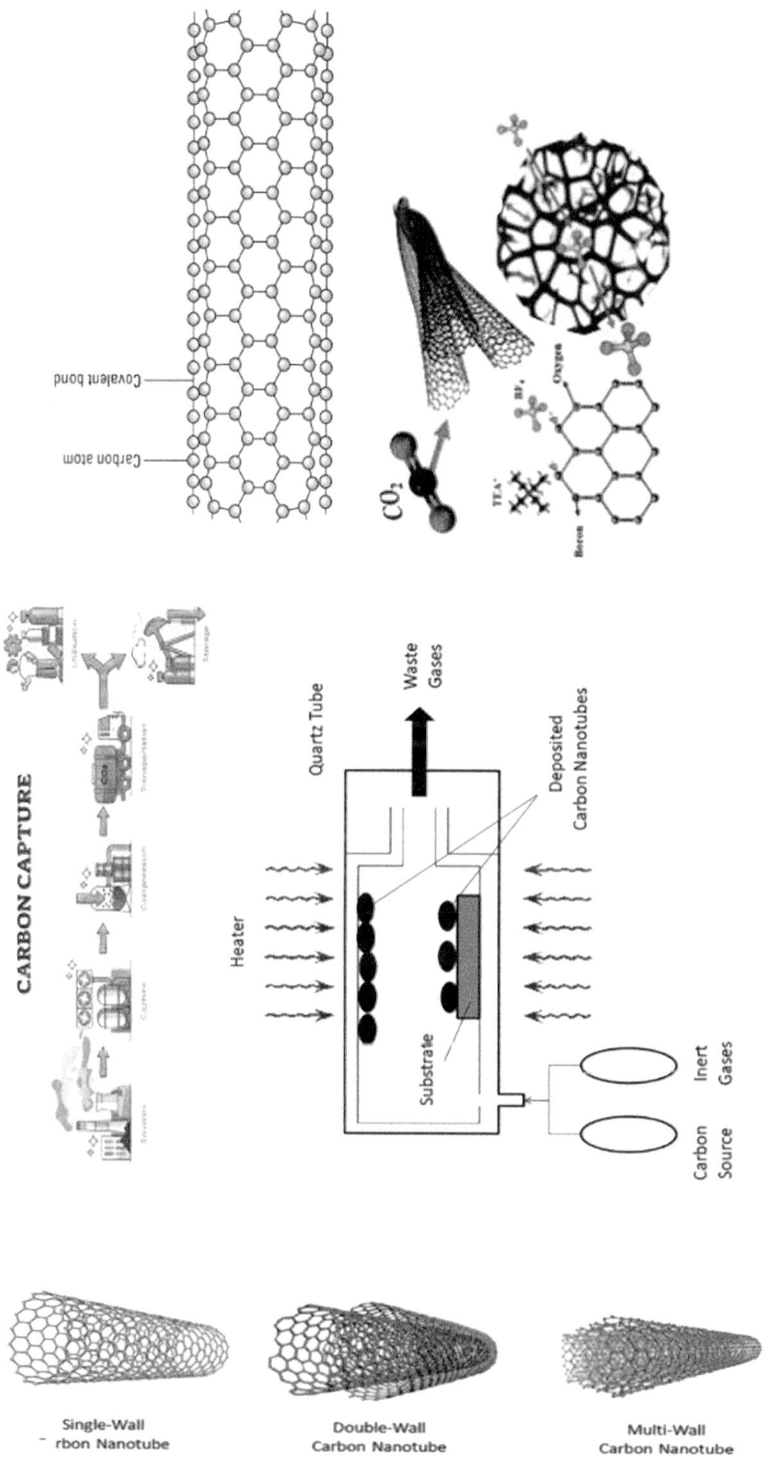

Figure 2.3 Shows different types of nanotubes and carbon capture process.

carbonate minerals that do not harm the environment. For the purpose of synthesizing these nanocomposites, an organic polymer is mixed with many metal oxide nano-powders in a uniform distribution, as shown schematically in [15]. Another option for improving CO_2 capture is to spray the finished product as a nanocomposite membrane over any surface. This is an exciting new area of research that could provide low-cost, large-scale carbon collection as well as safe environmental storage [16]. Increases in the capture capacity or the number of CO_2 active sites of sorbents are the focus of much research into amine-functionalized nanocomposites. An innovative CO_2 collection platform has been developed by researchers, which makes use of oligomeric amines mounted atop mesoporous hollow particles that have been custom-made. Amine incorporation and CO_2 collection ability have both been enhanced thus far [17].

2.2.2 Metal-Organic Frameworks (MOFs)

One of the most urgent issues facing our planet right now is the problem of carbon dioxide. Every phase of the carbon cycle, namely capture, regeneration, and conversion, has specific material needs. The potential and efficacy of Metal–Organic Frameworks (MOFs) in tackling this issue have been exhibited by recent research as discussed in Figure 2.4.

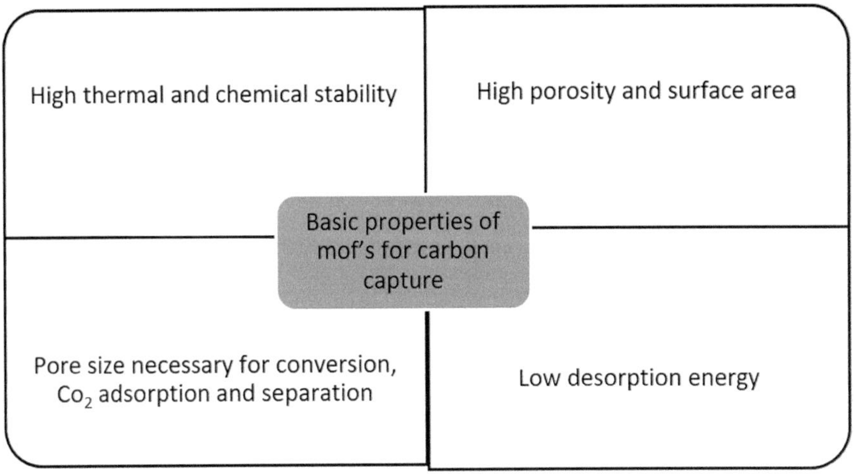

Figure 2.4 Shows basic properties of carbon capture.

The highest capture capacities, the most successful regeneration and separation procedures, and the most successful catalytic conversions have all been achieved by MOFs with specific structural and chemical characteristic. With an unprecedented degree of precision, the interior of MOFs can be engineered to contain coordinative unsaturated metal sites, particular heteroatoms, covalent functionalisation, other building unit interactions, hydrophobicity, porosity, defects, and embedded nanoscale metal catalysts—all of which are essential for the development of MOFs with improved performance [18].

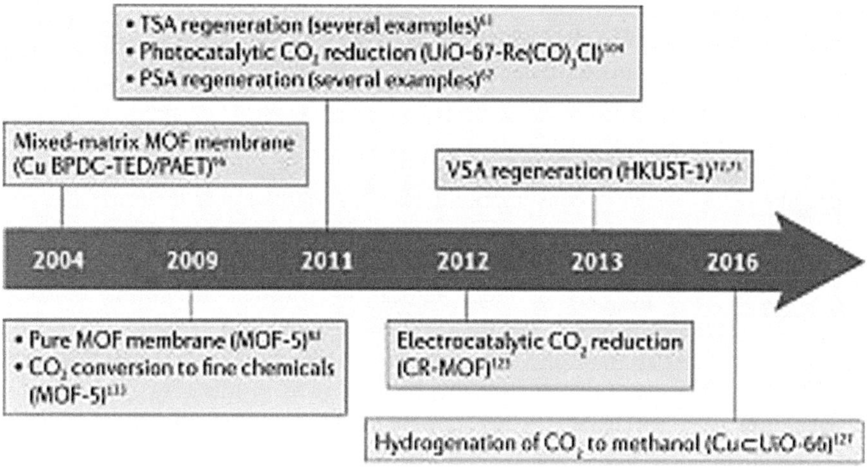

Figure 2.5 Shows the chronology of key achievements in the field of MOF's and CO_2 [18].

Expanding the strength of this interaction increases the material's capacity to absorb CO_2, especially at low loading pressure. In general, this interaction between the MOFs and CO_2 adsorbate molecules is crucial for the process of CO_2 capture. Thus, MOFs' chemical adjustability and compatibility can be leveraged to control the CO_2 molecules' affinity as shown in Figure 2.5 [18, 19]. Because MOFs can complete both stages of the process, they are a more energy-efficient carbon capture and conversion technology than alternative approaches. Another method for creating MOF composites is to cultivate MOFs on the outside or inside the pores of other materials. Because their surfaces can be functionalised and/or because they often have a variety of reactive functional groups, some 2D materials, including graphene, graphene oxide, layered double hydroxides, and amino clays, can be used as substrates for the growth of MOF composites [19]. MOFs have few industrial applications and are not widely manufactured due to the high cost of research and synthesis [20, 21]. As the debate over rising CO_2 emissions becomes more heated, MOFs remain a viable and energy-efficient solution [22]. The wide range of MOFs' characteristics (chemistry, geometry, percent porosity, pore size, etc.) and the different reaction condition combinations that can affect capture selectivity and conversion, however, results in a large MOF design space. It is difficult to determine the impact of each MOF design feature through experimental examination because of these variables [21]. Using MOFs has become more difficult because some of them eventually run out of power or deteriorate and need to be replaced or regenerated [22].

2.2.2.1 Porous Carbon Materials

High efficiency CO_2 capture of Porous Carbons (PCs) loaded with copper oxide nanoparticles was reported by another study that was recently conducted. The enhanced CO_2 adsorption process of the carbons was found to be caused by copper oxide nanoparticles. Strong interaction and effective capture result from the electron donor characteristics of copper oxide nanoparticles and the electron acceptor nature of CO_2 molecules in general Kim BJ, Cho KS, and Park SJ, copper-oxide decorated porous carbons for carbon dioxide adsorption behaviours. Porous carbons are a promising new approach to carbon capture, offering significant benefits in terms of cost, availability, large surface area, ease of design, and low energy consumption [23]. This method uses materials that are renewable to address the sustainability issue. Carbon sorbents are made from biomass or products derived from biomass, like cellulose, in order to capture CO_2 [24, 25]. The porosity of these substances controls the uptake of CO_2. Using carbonaceous materials has the benefit of allowing the design of porosity through the modification of operating parameters like activation temperature and activating agent dosage. CO_2 uptake can be optimised in this way.

Additional research on porous carbons may pave the way for low-cost, environmentally friendly carbon capture technology [26, 27].

2.2.2.2 Graphene-Based Materials

Graphene is a two-dimensional substance with a very large specific surface area. Curiously, adsorbents that are highly selective for CO_2 and have a high capacity to absorb CO_2 can be synthesized by purposefully deleting atoms from the hexagonal lattice, resulting in pores with radii large enough to admit CO_2. Graphene is more desirable than conventional CO_2 adsorbents because of these characteristics. These properties of graphene render it more appealing than traditional CO_2 adsorbents [28]. Appropriate for gas separation procedures, graphene oxide membranes have high permeability and selectivity. Researchers can modify graphene oxide's affinity for CO_2 molecules to achieve effective carbon capture from gas mixtures by adjusting the porosity and functional groups on its surface. With their vast surface areas and profusion of surface functional groups, graphene, and its derivatives, such as graphene nanosheets and graphene quantum dots, are excellent absorbent materials. These characteristics enable robust interactions with CO_2 molecules, resulting in elevated capacities for adsorption. Moreover, the mass transfer kinetics of graphene-based materials are improved by their porous structure, which enables the quick extraction of CO_2 from gas streams [29]. Graphene can be added to composite structures along with other materials to improve the efficiency of carbon capture. To increase the selectivity and capacities of CO_2 adsorption, graphene can be added to mesoporous silica, porous polymers, and metal-organic frameworks (MOFs). In comparison to individual materials, these graphene-based composites achieve superior carbon capture performance by utilising the synergistic effects of multiple components. Electrodes based on graphene have been investigated for electrochemical processes related to carbon capture. CO_2 molecules can be selectively captured and transformed into products with added value, like carbonates or hydrocarbons, by passing a voltage across graphene-based electrodes submerged in an electrolyte solution containing CO_2. This strategy has the potential to efficiently capture carbon while also yielding valuable chemicals or fuels. Although graphene-based materials show promise in carbon capture, large-scale implementation still faces significant challenges related to scalability and cost-effectiveness. Research has been conducted to create scalable fabrication processes and economical synthesis methods so that graphene-based carbon capture technologies can be made into a profitable industry [30].

2.3 ADSORPTION PROCESS IN NONMATERIAL FOR CARBON CAPTURE

Adsorption describes the process by which molecules cling to the surface of a solid material (the adsorbent) by means of physical forces. The technique of carbon capture via adsorption involves removing carbon dioxide (CO_2) from either industrial processes or the atmosphere. To achieve this goal, a wide range of adsorbent materials with specific properties and applications can be employed.

Different adsorption mechanisms allow for the categorization of the adsorption process into physical and chemical adsorption, respectively [31]. In physical adsorption, stable adsorption is formed without the formation of chemical bonds by the combination of adsorbate molecules and the adsorbent surface through microscopic forces (Coulomb force and Van der Waals force). A layer of adsorption will form on the surface of the adsorbent when the adsorbed gas comes into contact with it because the adsorbent's surface attracts the adsorbed gas. The adsorption phase is the name given to this layer of adsorption. Adsorption phase density is substantially higher than gas density. The process by which gas molecules and the adsorbent surface combine chemically to form bonds is known as chemical adsorption [32]. MOFs are crystalline substances made up of organic ligands connecting metal ions or clusters. They are promising for CO_2 adsorption because of their large surface areas and adjustable structures. Van der Waals forces and other weak interactions cause CO_2 molecules to be adsorbed onto the open metal sites or inside the pores of the MOF structure. By altering the chemical composition of the MOF, the selectivity and capacity of CO_2 adsorption can be customised [33]. Given that it can save a significant amount of energy when compared to the conventional amine-based absorption capture method, the adsorption-based CO_2 postcombustion capture is regarded as one of the most promising methods in this framework [33]. On the other hand, to separate CO_2 using regeneration techniques, numerous adsorption processes are employed. Included in this are (1) Temperature Swing Adsorption (TSA), (2) Electric Swing Adsorption (ESA), (3) Vacuum and pressure Swing Adsorption (VSA and PSA), (4) Simulated Moving Beds (SMB), and (5) purge displacement. More specifically, the CO_2 adsorption capacity is dependent on the heat of adsorption under low pressure conditions and the specific surface area under high pressure conditions. Increasing the number of linkers connected to each node can improve the thermal stability of MOFs because the bond between the node and linker is stronger. In this regard, substituting commonly used linkers, such as the carboxylate group and divalent cations, with oxyanion-terminated trivalent linker cations, such as Al (3), Zr (3), and Ti (3), which are characterised by high bond enthalpy and bond dissociation energy, also improves the thermal stability. Developing new, state-of-the-art adsorbent materials has taken up the majority of research efforts to date. The main goal is to increase adsorption capacity and lifetime while decreasing adsorption heat, which lowers the energy needed for sorbent regeneration as described in Figure 2.6 [34].

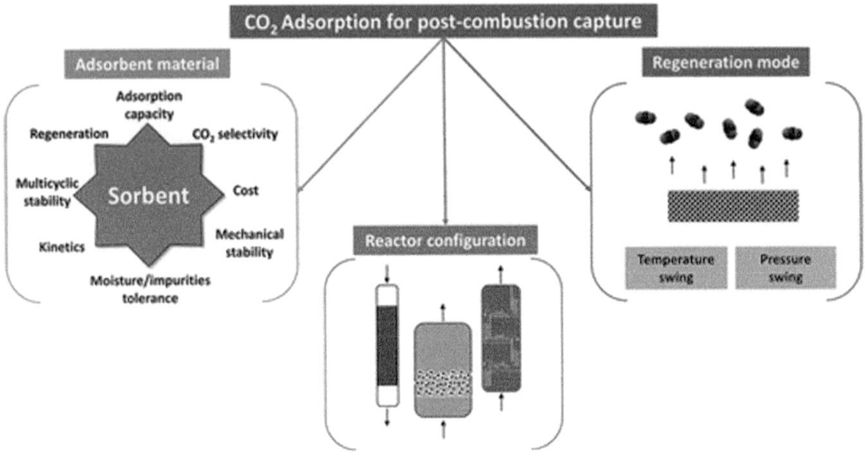

Figure 2.6 Represents CO_2 adsorption for post-combustion capture [34].

The gas-solid contacting system, which affects the sorbent utilisation efficiency, and the regeneration strategies, which determine the majority of the global CO_2 capture costs, are two other factors that significantly impact the competitiveness of the CO_2 capture based on the adsorption route, aside from the sorbents as shown in Figure 2.7.

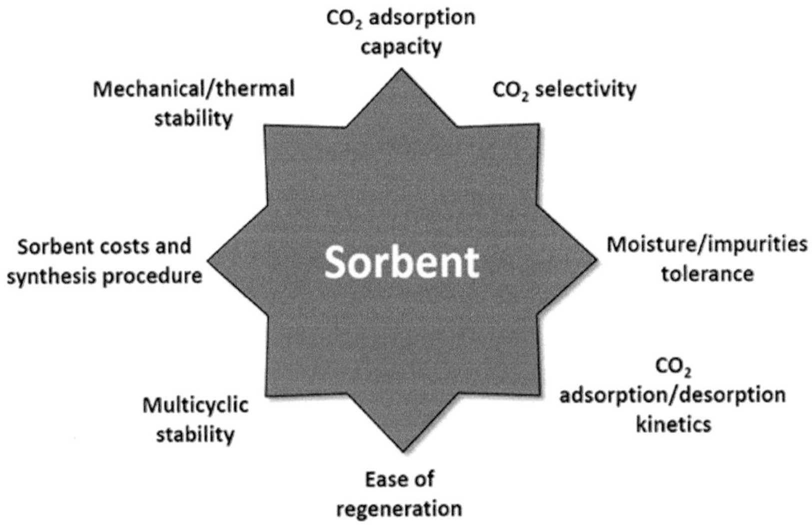

Figure 2.7 Schematic representation of properties of the required adsorbents [34].

2.4 ABSORPTION PROCESS IN NONMATERIAL FOR CARBON CAPTURE

The process of absorption involves a genuine transfer of mass from the gas to the absorbent or absorbate, whether the gas is dissolved chemically or physically. Both physical and chemical absorption bind carbon dioxide to the absorbent in distinct ways. No less than that, both methods can replenish the absorbed carbon dioxide. According to Henry's Law, the partial pressure of a gas is proportionate to the concentration of gas in a solution, and this law forms the basis of physical absorption. Since the temperature and pressure are critical variables, it is necessary to determine the optimal conditions for the physical process. Since the partial pressure of carbon dioxide is high before burning, physical absorption is usually employed to attain high efficiency in carbon capture. Disadvantages of physical absorption include the high energy required to recover CO_2 from the absorbent and the possibility of liquid loss due to the gas's high volatility. On the other hand, the chemical reaction between carbon dioxide and the absorbent material occurs during chemical absorption. The absorbing species influences the products and the mechanism. There are similar drawbacks for both chemical and physical absorption. In order to absorb CO_2 or convert it to mineral carbonates through wet chemical conversion, one crucial step in the process is the hydration of CO_2. For the separation of CO_2 absorption, amines and carbonates are frequently employed. Carbonic acid reacts with amine solutions faster than with $CO_{2(aq.)}$. The rate of CO_2 separation and mineralisation can thus be improved by increasing the rate of CO_2 hydration to carbonic acid. Another method of capturing carbon dioxide is absorption, which usually entails applying a solvent to a gas stream in order to selectively absorb CO_2 while permitting other gases, such as nitrogen, oxygen, and water vapour, to pass through. An amine-based solvent called carbamate is created when it reacts with CO_2 in the absorption process. If methods for synthesising nanoparticles are made commercially viable, ZnO nanofluid will be a promising option for absorbing CO_2. ZnO nanofluid achieved the highest CO_2 absorption capacity in the work. Table 2.1 presents a comparison of CO_2 absorption capacities among different nanomaterials reported in recent studies.

Table 2.1 Table gives comparison of CO_2 absorption capacity

S. No.	Properties	Activated Carbon	Zeolites (CNTs)	TiO_2	Si/Ca	CaO-$CaCO_3$	Ref.
1	Surface area (m^2/g)	1284	788	43	78	10.40	[35]
2	Temperature (°C)	25	20	30	10	20	[35]
3	Pressure (kPa)	100	100	20	150	–	[35]
4	CO_2 adsorption capacity (Mol/kg)	2.23	1.44	–	–	–	[35]

Following its breakdown, carbamate releases free CO_2, which can be gathered, compressed, and transported for use or disposal. The ability to be applied at scale makes absorption valuable in large scale industries, such as the production of electric power. This is just one of its many noteworthy advantages [35]. It is discovered that, when it comes to carbon capture at atmospheric pressure, physical absorption is most economical when the CO_2 content is greater than 59.3%, hybrid absorption is most efficient when the CO_2 composition is medium (30.4–59.3%), and chemical absorption performs best at low CO_2 concentrations (<30.4%). The hybrid physical–chemical absorption process is the more cost effective method for handling high pressure CO_2-containing gases (such as those produced during precombustion carbon capture) [36, 37]. In order to separate CO_2, absorption processes—such as chemical or physical ones—are extensively employed in the chemical, coal-fired, natural gas, and petroleum industries. Chemical absorption, which includes aqueous alkanolamine-based scrubbers, is the most developed capture technology and has been used since the 1930s [38].

2.5 ROLE OF SURFACE AREA, POROSITY, AND FUNCTIONALISATION IN ENHANCING CARBON CAPTURE EFFICIENCY

Surface area: Enhancing the efficiency of carbon capture materials requires consideration of surface area, porosity, and functionalisation. Greater surface area materials offer more places for CO_2 molecules to be adsorbed. This improves the material's overall ability to absorb carbon dioxide from the atmosphere or from emissions in the industry. High surface area materials, such as activated carbon, metal-organic frameworks (MOFs), and some zeolites, are known to improve carbon capture efficiency [39].

Porosity: Effective CO_2 molecule diffusion and adsorption are made possible by the interconnected pore networks found in porous materials. Because CO_2 molecules and pore surfaces interact strongly, microporous materials—those with pore sizes smaller than 2 nm—are especially useful for capturing CO_2. By offering more adsorption sites and promoting mass transfer, mesoporous and microporous materials also improve CO_2 capture efficiency. Porous Organic Polymers (POPs) are among the most promising candidates for carbon capture because of their exceptionally high porosity, structural diversity, and physicochemical stability. As a result, extensive research has been done on these advanced porous materials [40]. A critical aspect of CO_2 capture is pore/volume filling, which is correlated with an adsorbent's surface area and pore volume. Greater CO_2 adsorption is produced by larger surface areas and pore volumes because they provide more active sites for CO_2 molecules to interact. Mesoporous structures permit more adsorption at high pressures, whereas microporous structures, like micropores, boost their adsorption capacity at low pressures. The pore filling/volume of an adsorbent also affects its dynamic adsorption-desorption properties. Adsorption and desorption cycles occur more quickly when kinetics are accelerated and mass transfer is facilitated by well-designed and optimised pore structures [40, 41].

Functionalisation: It is the process of altering a material's surface to add specific chemical groups or functionalities that can improve its capacity to absorb CO_2.

A material's selectivity and capacity for CO_2 capture can be increased by reacting with CO_2 to form stable chemical bonds with functional groups like amines, aminoxides, and alkylamines. Additionally, functionalisation can increase the material's renderability and stability, which will increase its suitability for long term carbon capture applications [39, 42, 43].

2.6 CONCLUSION

Nanomaterials have good CO_2 capture capacities, exceptional durability in sorption/desorption cycles, and low energy requirements, making them potential sorbents to overcome a number of the process's limitations. Certain precursor materials are frequently needed to customise and regulate the size and dimensions of nanomaterials, which raises the cost and complicates the synthesis process. Future studies should concentrate on developing an easy to use, low cost production process for these nanomaterials, aside from screening and determining the optimal nonmaterial absorbent [39]. In conclusion, because of their special structural and chemical characteristics, nonmaterial present promising paths for the advancement of carbon capture technologies. For the purpose of absorbing carbon dioxide from diverse sources, their large surface area, adjustable pore sizes, and effective adsorption capabilities make them desirable options. To fully realise their potential in mitigating climate change through carbon capture and storage initiatives, more research is necessary to address issues like scalability, stability, and cost effectiveness. Sustained funding and interdisciplinary cooperation will be essential for fostering innovation and realising carbon capture solutions based on nonmaterial. Many possible CO_2 utilisation pathways were analysed and explored. The CO_2 can be used directly or as a raw material to make chemical products. It should be noted, nevertheless, that the amount of CO_2 that must be removed for climate stabilisation to have any meaningful impact differs significantly (by orders of magnitude) from the potential market for CO_2 as feedstock.

REFERENCES

[1] Olah, G.A., Goeppert, A. and Surya Prakash, G.K. 2009. Chemical recycling of carbon dioxide to methanol and dimethyl ether: from greenhouse gas to renewable, environmentally carbon neutral fuels and synthetic hydrocarbons. Journal of Organic Chemistry. 74(2): 487–498.

[2] Goulder, L.H. and Pizer, W.A. 2011. The Economics of Climate Change, National Bureau of Economic Research Books.

[3] Specht, E., Redemann, T. and Lorenz, N. 2016. Simplified mathematical model for calculating global warming through anthropogenic CO_2. International Journal of Thermal Sciences. 102.

[4] Center for Climate and Energy Solutions. 2021. Carbon Capture. https://www.c2es.org/content/carboncapture/

[5] Hyseni, S. 2017. Carbon capture and storage as a method to mitigate climate change. Inquiries Journal. 9(3). http://www.inquiriesjournal.com/articles/1551/carbon-capture-and-storage-as-a-method-to-mitigate-climate-change

[6] Szulczewski, M.L., MacMinn, C.W., H.J., Herzog and Juanes, R. 2012. Lifetime of carbon capture and storage as a climate-change mitigation technology. Proceedings of the National Academy of Sciences of the United State of America. 109(14): 5185–5189. https://doi.org/10.1073/pnas.1115347109

[7] International Panel on Climate Change (IPCC). Carbon dioxide capture and storage, IPCC.

[8] Rubin, E.S., Davison, J.E. and Herzog, H.J. 2015. The cost of CO_2 capture and storage. International Journal of Greenhouse Gas Control. 40: 378–400. https://doi.org/10.1016/j.ijggc.2015.05.018

[9] Chen, Y., Li, C.W. and Kanan, M.W. 2012. Journal of the American Chemical Society. 134(49): 19969–19972. https://doi.org/10.1021/ja309317u

[10] Lin, K.Y. and Park, A.H. 2011. Effects of bonding types and functional groups on CO_2 capture using novel multiphase systems of liquid-like nanoparticle organic hybrid materials. Environmental Science and Technology. 45(15): 6633–6639. https://doi.org/10.1021/es200146g

[11] Kumar, R., Mangalapuri, R., Ahmadi, M.H., Vo, D.V.N., Solanki, R. and Kumar, P. 2020. The role of nanotechnology on post-combustion CO_2 absorption in process industries. International Journal of Low-Carbon Technologies. 15(3): 361–367. https://doi.org/10.1093/ijlct/ctaa006

[12] Verma, S.K., Tripathi, P. and Bhatnagar, A. 2023. Carbon nanotubes for CO_2 capture and conversion. *In*: Nanomaterials for Carbon Dioxide Capture and Conversion Technologies. 245–260. Elsevier.

[13] Langie, K.M.G., Tak, K., Kim, C., Lee, H.W., Park, K., Kim, D., et al. 2022. Toward economical application of carbon capture and utilization technology with near-zero carbon emission. Nature Communications. 13(1): 1–10. https://doi.org/10.1038/s41467-022-35239-9

[14] Ren, J. and Licht, S. 2016. Tracking airborne CO_2 mitigation and low cost transformation into valuable carbon nanotubes. Scientific Reports. 6(1): 1–11. https://doi.org/10.1038/srep27760

[15] Mishra, A.K. and Ramaprabhu, S. 2012. Palladium nanoparticles decorated graphite nanoplatelets for room temperature carbon dioxide adsorption. Chemical Engineering Journal. 187: 10–15.

[16] Biswas, A., Tokoly, T., Wang, T., Ramidi, P., Ghosh, A., Dervishi, E., et al. 2011. Design and synthesis of sprayable nanocomposite coatings for carbon capture and direct conversion into environmentally safe stable carbonates. Chemical Physics Letters. 508(4–6): 276–280.

[17] Qi, G., Wang, Y., Estevez, L., Duan, X., Anako, N., Park, A.-H.A., et al. 2011. High efficiency nanocomposite sorbents for CO_2 capture based on amine-functionalized mesoporous capsules. https://doi.org/10.1039/C0EE00213E

[18] Trickett, C.A., Helal, A., Al-Maythalony, B.A., Yamani, Z.H., Cordova, K.E. and Yaghi, O.M. 2017. The chemistry of metal–organic frameworks for CO_2 capture, regeneration and conversion. Nature Reviews Materials. 2: 17045. https://doi.org/10.1038/natrevmats.2017.45

[19] Ding, M., Flaig, R.W., Jiang, H.L. and Yaghi, O.M. 2019. Carbon capture and conversion using metal-organic frameworks and MOF-based materials. Chemical Society Reviews. 48: 2783–2828. https://doi.org/10.1039/C8CS00885H

[20] Hendon, C.H., Rieth, A.J., Korzyński, M.D. and Dincă, M. 2017. Grand challenges and future opportunities for metal–organic frameworks. ACS Central Science. 3(6): 554–563. https://pubs.acs.org/doi/10.1021/acscentsci.7b00197

[21] Niklas. 2020. Price vs. value of metal-organic frameworks. NovoMOF Blog. https://blog.novomof.com/blog/price-vs-value-ofmetal-organic-frameworks

[22] Demir, H., Aksu, G.O., Gulbalkan, H.C. and Keskin, S. 2022. MOF membranes for CO_2 capture: Past, present and future. Carbon Capture Science and Technology. 2: 100026. https://doi.org/10.1016/j.ccst.2021.100026

[23] Kim, B.J., Cho, K.S. and Park, S.J. 2010. Copper-oxide decorated porous carbons for Carbon Dioxide adsorption behaviours. Journal of Colloid and Interface Science. 342(2): 575–578.

[24] Sevilla, M. and Fuertes, A.B. 2011. Sustainable porous carbons with a superior performance for CO_2 capture. Energy and Environmental Science. 4: 1765–1771.

[25] Lively, R., Koros, W. and Johnson, J. 2012. Enhanced cryogenic CO_2 capture using dynamically operated low-cost fibre beds. Chemical Engineering Science. 71: 97–103.

[26] Pacheco, D.M., Johnson, J.R. and Koros, W.J. 2012. Aminosilane-functionalized cellulosic polymer for increased Carbon Dioxide sorption. Industrial and Engineering Chemistry Research. 51(1): 503–514.

[27] Gebald, C., Wurzbacher, J., Tingaut, P., Zimmermann, T. and Steinfeld, A. 2011. Amine-based Nano fibrillated cellulose as adsorbent for CO_2 capture from air. Environmental Science and Technology. 45(20): 9101–9108.

[28] Jarvis, K.A., Wang, C.-C., Manthiram, A. and Ferreira, P.J. 2014. The role of composition in the atomic structure, oxygen loss, and capacity of layered Li–Mn–Ni oxide cathodes. Journal of Materials Chemistry A. 2, 1353–1362. https://doi.org/10.1039/C3TA12440A

[29] Arango Hoyos, B.E., Osorio, H.F., Valencia Gómez, E.K., Guerrero Sánchez, J., Del Canto Palominos, A.P., Larrain, F.A., et al. 2023. Exploring the capture and desorption of CO_2 on graphene oxide foams supported by computational calculations. Science Reports. 13: 14476. https://doi.org/10.1038/s41598-025-88643-8

[30] Balasubramanian, R. and Chowdhury, S. 2015. Recent advances and progress in the development of graphene-based adsorbents for CO_2 capture. Journal of Materials Chemistry. https://doi.org/10.1039/C5TA04822B

[31] Abd, A.A., Naji, S.Z., Hashim, A.S. and Othman, M.R. 2020. Carbon dioxide removal through physical adsorption using carbonaceous and non-carbonaceous adsorbents: A review. Journal of Environmental Chemical Engineering. 8: 104142. https://doi.org/10.1016/j.jece.2020.104142

[32] Li, Z., Jin, J., Yang, F., Song, N. and Yin, Y. 2020. Coupling Magnetic and Plasmonic Anisotropy in Hybrid Nano rods for Mechanochromic Responses. Nature Communications. 11: 2883–2894. https://doi.org/10.1038/s41467-020-16678-8

[33] Younas, M., Rezakazemi, M., Daud, M., Wazir, M.B., Ahmad, S., Ullah, N., et al. 2020. Recent progress and remaining challenges in post-combustion CO_2 capture using metal–organic frameworks (MOFs). Progress in Energy and Combustion Science. 80. 100849. https://doi.org/10.1016/j.pecs.2020.100849

[34] Raganati, F., Miccio, F. and Ammendola, P. 2021. Adsorption of carbon dioxide for post-combustion capture: A review. Energy and Fuels. 35(16): 12845–12868. https://doi.org/10.1021/acs.energyfuels.1c01618

[35] Kumar, R., Mangalapuri, R., Ahmadi, M.H., Vo, D.V.N., Solanki, R. and Kumar, P. 2020. The role of nanotechnology on post-combustion CO_2 absorption in process industries. International Journal of Low-Carbon Technologies. 15(3): 361–367.

[36] Zhang, X., Song, Z., Gani, R. and Zhou, T. 2020. Comparative economic analysis of physical, chemical, and hybrid absorption processes for carbon capture. Industrial and Engineering Chemistry Research. 59(5): 2005–2012. https://doi.org/10.1021/acs.iecr.9b05510

[37] Ai, S., Dang, B., Lv, B., Zhou, Z. amd Jing, G. 2019. Absorption characteristics and kinetics of CO_2 capture into N-methyldiethanolamine aqueous solution catalyzed by the immobilized carbonic anhydrase. Biocatalysis and Biotransformation. 37(5): 331–340. https://doi.org/10.1080/10242422.2018.1551377

[38] Choi, W.-J., Seo, J.-B., Jang, S.-Y., Jung, J.-H. and Oh, K.-J. 2009. Removal characteristics of CO_2 using aqueous MEA/AMP solutions in the absorption and regeneration process. Journal of Environmental Sciences. 21: 907–913.

[39] Chi, J., Yin, Y., Zhang, W., Putnis, C.V. and Wang, L. 2021. Nanoscale imaging of the simultaneous occlusion of nanoplastics and glyphosate within soil minerals. Environmental Science: Nano. 8(10): 2855–2865.

[40] Singh, R., Wang, L., Ostrikov, K.K. and Huang, J. 2024. Designing carbon-based porous materials for carbon dioxide capture. Advanced Materials Interfaces. 11: 2202290. https://doi.org/10.1002/admi.202202290

[41] Zhang, Z., Cano, Z.P., Luo, D., Dou, H., Yu, A. and Chen, Z. 2019. Rational design of tailored porous carbon-based materials for CO_2 capture. Journal of Materials Chemistry. 7: 20985.

[42] International Journal of Low-Carbon Technologies. 2020. 15(3): 361–367. https://doi.org/10.1093/ijlct/ctaa002

[43] Wang, J., Huang, L., Yang, R., Zhang, Z., Wu, J., Gao, Y., et al. 2014. Recent advances in solid sorbents for CO_2 capture and new development trends. Energy and Environmental Science. https://doi.org/10.1039/C4EE01647E

Chapter 3

Nanomaterials for Carbon Detection

Nair Chithra Harinarayanan*[1] and Menassery Lakshmi[2]

[1]Department of Food Science and Technology,
Faculty of Ocean Science and Technology,
Kerala University of Fisheries and Ocean Studies, Kochi-682506, Kerala, India

[2]Department of Chemistry, Al-Azhar College of Arts and Science,
Thodupuzha-685605, Kerala, India

3.1 INTRODUCTION

It is undeniable that environmental pollution is a major issue in society today. As a result of technological advancements, new methods for remediating air, water, and soil pollutants are constantly being developed [1]. It occurs most often due to emissions from industry and automobiles that contribute to environmental pollution [2]. These contaminants are toxic gases, fertiliser, herbicides, oil spills, sewage, industrial effluents, and other organic compounds [3, 4]. Carbon footprints have become increasingly important with growing environmental awareness and the urgency of action against climate change. As an indicator that quantifies global warming's impacts on the world's environment, this tool helps identify regions that can reduce emissions. By increasing emissions, greenhouse gases become more concentrated, forming a thicker "blanket" that prevents some heat from escaping, which results in climate change and an increase in global temperatures. Activities such as the burning of fossil fuels, transportation, and other industrial processes lead to the release of gases such as Methane (CH_4) and Carbon dioxide (CO_2), and hence, it is important to address these issues for a more sustainable

*For Correspondence: chithrahari1999@gmail.com

DOI: 10.1201/9781003492184-3

and safer future. Environmental remediation is thus employed using a number of approaches but is challenging due to the complexity of the compounds, reactivity, and volatility and hence, studies have been carried out in using nanomaterials for this purpose.

As nanotechnology continues to develop and use environmentally friendly materials, it reduces energy use during manufacturing and production pathways [5]. It possesses the potential to produce cleaner, greener technologies that are more beneficial to the environment and health. It is also important to note that, by intentionally altering the nanomaterials' physical properties (such as size, chemical composition, porosity, and morphology), additional beneficial features can be conferred that directly affect the remediation performance of the material. The modification of its physical and chemical parameters will make it significant for addressing environmental remediation. When different methods are combined, they can potentially yield better results by harnessing the unique properties of each component. These methods, known as 'hybrids' or 'composites', can be more efficient than those that rely on a single nanoplatform [6]. Therefore, nanomaterials (e.g., metallic nanoparticles, carbon nanotubes, etc.) provide many advantages, including quickly diagnosing environmental conditions, detecting wastewater pollutants, monitoring them, and treating them in real-time [7]. Polymer materials reinforced with nanoparticles are being developed by several companies for automotive applications [8]. Using nanocomposites extensively could save 1.5 billion litres of gasoline annually and more than 5 billion kilograms of carbon dioxide-related pollution [9]. Nanotechnology has enabled the manufacture of lightweight and durable nanocomposites, which can replace heavy metal components, reducing equipment and auto parts weight and energy consumption, and hence, air pollution [10]. Molecular-level sensors enhance human health and environmental sustainability by detecting pollutants quickly and accurately [11]. Sensors in the nanoscale can be chemical, physical, or biological sensors that provide qualitative or quantitative measurements of changes in the nanoscale. The continuous monitoring of air pollution is a basic requirement for environmental pollution control [12]. As home, office, and factory air quality is monitored, as well as safety in mining and energy, detecting dangerous gases is increasingly necessary [13]. The measurement of greenhouse gases is also widely carried out using gas sensors. There are several limitations to these conventional sensors, including their high cost, weight, and size, as well as their limited durability [16]. Hence, a significant improvement in air pollution control has been made with the use of nanosensors [14]. Sensors based on different principles have been developed, including field-effect transistors, electrical resistance, and optical absorption [15–25]. Long term suspension of these nanosensors in the air is possible [26]. Silicon particles can communicate wirelessly with a central base, sending the collected information. Wearable nanosensors for monitoring air pollution are presented in this paper as an example of cutting-edge advancements in this field. A review of the materials, device architectures, and sensing approaches presented in the following sections will include discussions and presentations of gas-sensing parameters, including sensor response, detection limits, sensitivity, recovery times, selectivity, and stability.

3.2 TYPES OF CARBON IN THE ENVIRONMENT

Pollutants from increasing urbanisation and industrialisation mainly consist of toxic gases such as CO, CO_2, dioxins, Volatile Organic Carbons (VOCs), and so on, which leads to the greenhouse effect, acid rains, and ozone depletion. Also, industrial workers, when exposed to these harmful gases have deteriorating effects on their health, which makes it important to develop a safety gas alert system to detect and quantify the toxic gases and chemicals.

3.2.1 Carbon Monoxide (CO)

World Health Organisation (WHO) has identified it as one of the six primary pollutants and is an odorless, colourless gas that has a density less than that of the air. This pollutant affects all the natural resources such as water, land, earth, and all living organisms. CO is mainly produced as a result of incomplete combustion of fossil fuels, industrial waste, water heaters, automobile emissions, fires, open flames, and other biological activities [27]. Mild exposure to this gas may cause headaches, nausea, dizziness, and breathing problems, and prolonged exposure may prove to be fatal [28–30]. CO has a good ability to bind up with haemoglobin, which in very large amounts is present in blood, and makes it incapable of transporting oxygen to the tissues leading to hypoxia in the body [31]. Since abnormal CO concentrations can mediate various diseases, early warning and real time detection of CO concentrations are essential for human and social development [32].

3.2.2 Carbon Dioxide (CO_2)

Many of the industrial and biological processes require CO_2 as a vital component. This gas is generally released during the burning and extraction of fossil fuels, deforestation, cement production, transportation, agriculture, and even volcanic eruptions. High levels of CO_2 can cause loss of concentration and tiredness, drowsiness, and even death. The buildup of CO_2 gas has been identified as one of the major contributors to climate change and is even corresponded to a rise in extreme weather events, such as wildfires, droughts, storms, etc [33]. The increase in CO_2 concentration in the atmosphere is causing a rise in temperature and ocean acidification, making it a key factor in climate change [34]. Hence, sensing and monitoring of this gas is necessary to gain knowledge about CO_2-affected mechanisms and control them.

3.2.3 Volatile Organic Compounds (VOCs)

Atmospheric reactions can result in the formation of numerous chemicals, such as nitric acid, polyaromatic compounds, soot formation, and volatile compounds, along with nitrogen and sulphur oxides. The detection of these substances is much more complex and difficult. This class mainly consists of aromatic hydrocarbons,

Gas detection is one of the many applications that CPs can be used for because of their unique optoelectronic characteristics. A lot of attention has been paid to flexible gas sensors made from CP-sensing materials, due to their many advantages, including low cost, flexibility, operation at room temperature, environmental stability, functionalised operations, and tunable conductivity [50]. A capacitive nanohybrid sensor for the detection of CO_2 was created by combining polypyrrole (Ppy) and copper phthalocyanine (CuPc). The response and sensitivity of this sensor were higher than Ppy and CuPc hybrids or PPy and can be attributed to the highly interconnected network of Ppy nanofibres and the synergistic effects of the individual components, enhancing gas diffusion easily through their porous structures [51, 52]. Another sensor made of Ppy-modified TiO_2 and Cu-interdigitated electrodes demonstrated fast response and recovery times to CO gas of 1–320 ppm at room temperature [53].

3.3.3 2D Nanostructures

The exceptional material and electrical properties, as well as their strong photo response, have made 2D nanostructures very interesting. The remarkable physical and chemical characteristics of 2D nanostructures have been demonstrated in numerous studies, as have their many uses in the fields of food safety, healthcare, energy, and defence. Being an electrically conductive and electron-mobile material with a large surface area, these two-dimensional nanostructures—graphene, poly-types of transition-metal dichalcogenides (TMDC or TMD) materials and graphitic carbon nitride ($g\text{-}C_3N_4$), oxides of group B elements like MoO_3, MnO_2, and WO_3 can have extraordinary electrical, chemical, physical, and optical properties, making it highly beneficial for sensing applications such as being used in colourimetric biosensors, Field-Effect Transistors (FET), and Surface Enhance Raman Spectroscopy (SERS) [54]. Materials can now be integrated into lab-on-a-chip systems that evaluate a multitude of analytes in gaseous form, due to recent developments in device integration, device. and material processing, and microfabrication [55]. Because of their special electronic properties, high sensitivity, quick response, good selectivity, and low temperature of operation, layered 2D nanomaterials—also known as layered metal disulphides—have been applied in the gas sensing industry.

3.3.4 Metal Oxide Semiconductors (MOS)

As a result of their excellent response, simplicity, low cost, ease of manufacturing, compatibility with microelectronic manufacturing, and long term stability, a wide range of gas sensing devices utilise MOS materials [56]. Metal oxide films have been shown to change resistance at a specific temperature when exposed to toluene, CO_2, and propane gases [57]. Semiconductors have shown this chemiresistive property, making them ideal candidates for gas sensing. Sensors based on semiconductors detect analyte molecules by measuring changes in electrical parameters following gas adsorption or desorption on their surfaces. Chemical gas sensors are designed to detect gaseous compounds in the environment based on the composition and structure of analytes as well as the morphology of semiconductor materials, so these sensors are the right choice for environmental gas detection [58–61]. MOS gas sensors based on nanostructured ZnO, which represent n-type semiconductors, are among the most promising candidates for gas sensing, especially for wearable devices, due to their non-toxicity, cost effectiveness, and fast and stable reaction time. 1D ZnO nanostructures-based sensors are able to detect gases like C_3H_8, NO_2, H_2, NH_3, and CH_4, though their sensitivity varies greatly depending on operating conditions [62, 63]. In gas molecules, adsorption and desorption occur into chemiresistive structures. In addition, the diameter of the electrodes, in proportion to the Debye length, affects their electrical properties. Also, recent advances in two-dimensional nanomaterials like metal-organic frameworks (MOFs), TMDs, and graphene, can have thicknesses ranging from a few atomic layers to tens of nanometres, have made them suitable structures for the development of functional devices due to their unique physical (large surface area), chemical, and electronic properties [64–66].

3.3.5 Nanohybrids

Because of their affordability, viability, ease of maintenance, simple construction, gas detection capability and versatility, nanohybrid sensors are being used more often to detect minute amounts of dangerous, toxic gases [67]. Nanohybrid sensors have improved optical and electrical potential, porosity, conductivity, and catalytic activity and have a low LOD for gases even at ppm concentrations [68–70]. A new class of hybrid nanomaterials has been introduced through the conjugation of inorganic and organic nanomaterials [71]. The interactions of some of these depend on the size, shape, and distribution of the nanoparticles as well as the dispersion form and strength of the chemical bonds present at the interface. Different chemical, physical, and electrochemical techniques are generally used to synthesise inorganic/polymeric nanohybrids to detect the presence of toxic gases such as CO, NH_3, CO_2, NO_2, H_2, liquified petroleum gas, and H_2S [72–74]. Nanohybrids provide flexible physical and chemical functionalities that are advantageous in a number of fields, as opposed to single-component nanomaterials [75]. The conductivity of certain polymers, such as Ppy,PT, Poly (Phenylene Vinylene) (PPV), PANI, and PEDOT, changes when exposed to gas molecules, making them potential candidates for use in gas sensing [76]. The advantages of using these polymers include large surface area, outstanding stability, high functionality, ease of synthesis, low cost, quick reaction time, and spectacular recovery. Due to their high sensitivity, affordability, and capacity to identify a broad variety of hazardous gases, semiconductor-based gas sensors like ZnO, SnO_2, NiO, and CuO have also garnered a lot of interest [77].

3.3.6 Paper-Based

Paper is gaining popularity in electronic devices because it is a lightweight, flexible, affordable, customisable, biodegradable, and renewable material that can help cut down on the amount of electronic waste produced for future generations. Hence, to create flexible and wearable gas sensors, a PB substrate can also be used as the foundation. Thick film deposition and transfer on PB substrates are incompatible due to their higher surface roughness and hole content when compared to other substrates. PB gas sensors are made of sensing materials such as MOS, organic materials, and carbon compounds and derivatives, the latter of which are especially useful. The high hygroscopicity of PB substrates makes them inappropriate for VOC detection. This can be attributed to the fact that absorbed humidity has a greater impact on conductivity than target VOC presence [78]. The tracking of respiration rates and patterns has been demonstrated with PB moisture sensors [79]. The basis for these sensors is the paper's natural ability to absorb moisture from its environment. Based on the swelling and bending properties of the PB cantilever, a straightforward, light, and inexpensive hydrocarbon gas sensor has been proposed, using hydrocarbons such as toluene, xylene, hexane, ethylbenzene, and benzene with weighing paper serving as the substrate for the sensory cantilever [80]. This sensor exhibited a remarkable linear response to the hydrocarbon concentration.

3.3.7 Textiles

Since the integration of clothing with wearable gaseous pollutant sensors would be relatively simple, textiles are a good substrate for this kind of technology. Textiles are appealing for a variety of reasons. They are washable, breathable, and long-lasting. Silk and cotton are the most widely used textile fabrics because of their availability [81, 82]. The remarkable mechanical strength, excellent biocompatibility, low cost, high absorption of moisture, high flexibility, and biodegradability of cotton fabrics have attracted a lot of interest. Furthermore, when it comes to atmospheric conditions, encapsulation is crucial in wearable sensor technologies. Electronic sensor encapsulation provides a waterproof layer and serves as a skin-shielding barrier [83].

3.4 SENSOR TYPES

Though conventional detection techniques such as gas spectrometry and other electrochemical devices are reliable and stable, they have certain limitations such as large size, high power consumption, cost, and unportable [84, 85]. Hence, as a result of the growing demand for portable gas sensors with high accuracy and low power consumption, capacitive-based, field effect-based transistors, solid-state electrochemical, optical, cantilever, and colourimetric gas sensors were developed and have gained a great deal of attention [84].

3.4.1 Optical

Fast response times, low drift, and exceptional gas specificity and sensitivity without affecting the gas sample are among the benefits of optically based gas sensors. By comparing their unique gas optical fingerprints, optical gas sensors can distinguish a wide range of gas species, and offer exceptional chemical and physical stability over chemiresistive metal oxide sensors[196]. These include refractometric fibre optic sensors, non-dispersive infrared sensors, colourimetric sensing, and surface plasmon resonance sensors, which are discussed below.

3.4.1.1 Refractometric Fibre Optic Sensors

A wide range of applications involving fibre-optic CO_2 detection are available in the aquaculture industry and environmental monitoring services [85]. With Surface Plasmon Resonance fibre (SPR) optics, the core acts as a prism and the same mechanism for plasmon excitation is used as with prism-based fibre optics. Claddings are etched and plasmonic nanomaterials are deposited at specific intervals [86]. Fibre claddings are usually etched with hydrofluoric acid (HF) so that the core of the fibre can be exposed. An etched place is then coated with porous nanostructured films or a sol. It is possible for nanomaterials to be more sensitive due to their increased surface-to-volume ratio and high porosity. Based on changes in Refractive Index (RI), fibre optic gas sensors calculate gas concentrations (Figure 3.1) [87]. A shift in optical resonance frequency is observed on detecting gas concentration, such as CO_2 and RI changes. Many studies have been conducted on CO_2 interaction with nanomaterials coated over fibres with or without cladding, resulting in a variation in their refractive index. The clad-etched Fibre Bragg Grating (FBG) was coated with polyallylamine-amino-carbon nanotubes to detect CO_2 gas concentrations inside a wide range of concentrations

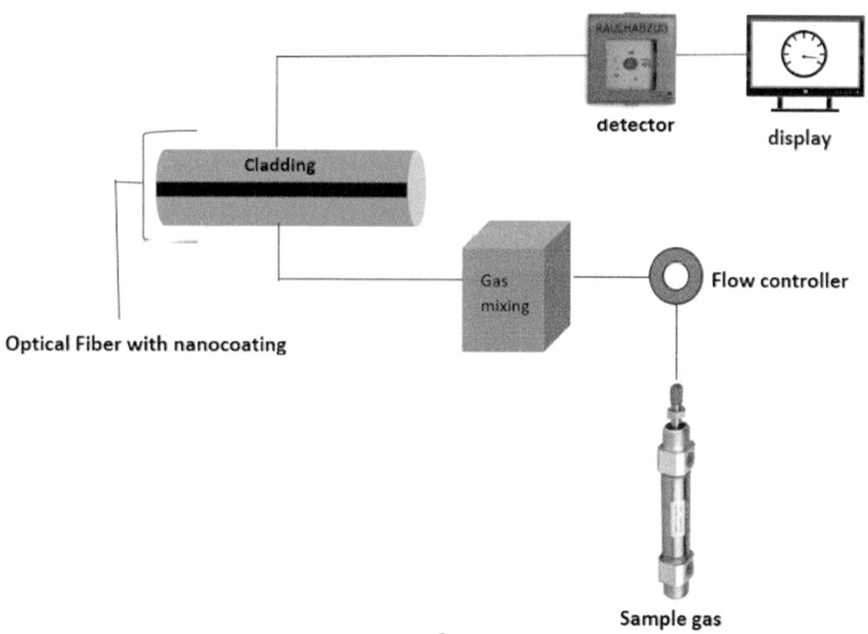

Figure 3.1 Refractometric fibre optic sensors.

(1000–4000 ppm) at room temperature [88]. With a Limit Of Detection (LOD) of 75 ppm, the Bragg wavelength shift exhibited a linear reproducible, and reversible response. By coating Cu-BTC (Cu-benzene-1,3,5-tricarboxylate) over single-mode optical fibre without cladding, a metal-organic framework (MOF) a nanoporous structure was developed. The CO_2 concentration was reported as 20 ppm and the Figure Of Merit (FOM) as 100 ppm [89].

3.4.1.2 Surface Plasmon Resonance Sensors

During Surface Plasmon Resonance (SPR), incident light causes resonant oscillations of electrons at the interface between two mediums when incident light strikes at a particular angle of incidence [190]. Adsorption and desorption of carbon gases, such as CO_2 are determined by the high sensitivity of the angle of incidence to changes in refractive index. Sensor nanomaterials exhibit a change in refractive index with changes in CO_2 concentration. This phenomenon is referred to as Kretschmann configuration. The surface plasmon capability of materials with small imaginary dielectric constants is observed in only a few materials like Cu, Ag, Al, and Au [191]. Illumination of a plasmonic nanomaterial is performed using a light source in this method (Figure 3.2). The surface electrons of the plasmonic nanomaterial absorb a portion of the light at a unique angle called the angle of resonance, while some of the light is reflected off the surface [192], 193]. As a result, they become surface plasmons, which are electronic resonances. An effect of surface plasmon resonance is an alteration of the dielectric constant surrounding a nanomaterial. Surface plasmons lose intensity because of their sensitivity. It is possible to determine the surface of the sensor using the shape and location of the dips on the SPR reflection intensity curve. It is widely used in pharmaceutical and biosensor applications because of its non-toxicity, low cost, and high sensitivity [194, 195].

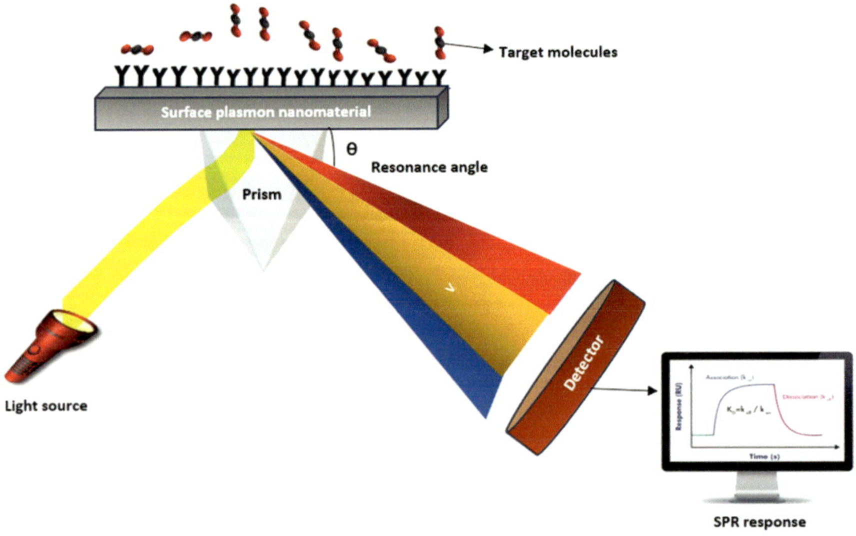

Figure 3.2 Surface plasmon resonance sensor [189].

3.4.1.3 Non-Dispersive Infrared Sensors

A very well-known and well-established technique for detecting gases is Non-Dispersive Infrared Detection (NDIR) which is commonly used in monitoring indoor air quality and in automobile industries as well [90, 91]. The gas sample is illuminated by an infrared (IR) source in a chamber with an infrared detector and an optical filter. Optical filters attenuate short wavelengths and transmit longer wavelengths as light passes through them. Almost all toxic gases accumulate in the mid-IR (2.5 to 14.4 μm) and display a strong absorption, about 100 times greater than in near-IR [92]. The infrared spectroscopy of carbon dioxide is easy because it is an infrared active molecule, absorbing energy at a wavelength of 4.24 μm [93]. According to the Beer-Lambert Law, the IR intensity at the detector will diminish exponentially if CO_2 is absent or below LOD [94, 95]. Nanomaterials are generally used to improve either photodetectors or IR sources [96, 97]. A study was conducted on the replacement of standard thermal emitters with CMOS nanoplasmonic tungsten crystals. As a result of low-wave plasmonic lattice resonance, as well as elevated light-to-plasmon coupling, the signal-to-noise ratio and sensitivity for CO_2 increased by 400% [98]. For mass-producing small and low-cost devices and systems, mid-wave infrared (MWIR) absorption spectroscopy is an additional gas-sensing technique that is worth mentioning [99]. For on-chip measurement of gases using absorption spectroscopy, MWIR graphene photodetectors were recently proposed for integration into Silicon-On-Sapphire (SOS) slot waveguides. The suggested device may be able to detect CO_2, N_2O, and CH_4 at lower than ppb levels (0.14, 0.52, and 0.72 ppb), according to simulations [99]. A study also demonstrated that Pt-on-Si needles integrated with an IR emitter emit IR 2.6 times higher than uncoated Si emitters at the same membrane temperature with no wavelength-dependent interference patterns [96]. The optoelectronic mode of operation of this sensor has several intrinsic disadvantages, including cost, scalability, power consumption, and device complexity. Despite its high accuracy, quick measurements, and strong long-term stability, some of these shortcomings have not been addressed even after decades of optimisation [100, 101].

3.4.1.4 Colourimetric Sensing

The use of colourimetric sensors has been widely used in biological applications as a noninvasive, semi-quantitative, or qualitative sensor [102]. A report describes their use in evaluating soils in carbon capture and storage facilities as well as for monitoring air quality, food quality, detection of hazardous chemicals, etc [103]. Sensors detect colour changes caused by wavelength changes in dyes based on visual observation. Colour-based gas sensors can be used to identify gaseous analytes and their concentrations employing chemical reactions that cause colour changes in the sensor. Washable textile gas sensors based on thread are becoming more popular due to their low power requirements, low LOD, multiplexity, selectivity, and relatively low cost [104]. Colourimetric sensors, which are used to detect target molecules like dyes, polymers, and other foetal chemical complexes, are based on the ring opening and closing interactions of organic compounds

[105, 106]. Many dyes and organic compounds with aromatic structures that have many conjugated π-systems can be used as colourimetric sensors, based on their ring opening and closing processes that change the number of conjugated π-bindings. The absorbance spectra of a material shift when the number of conjugated π-systems changes, or when a conjugated system is lengthened/ extended. This results in the capture of specific photons, which causes the colour of the material to change measurably and, ultimately, the identification of a target analyte. Additionally, colourimetric sensors may be based on functional groups. Organic compounds can benefit from specific moieties, enriching their characteristics and increasing their reactivity [105]. The hydroxyl groups in organic compounds, for example, can oxidise rapidly. This reaction results in a detectable change in the colour of the material when it reacts with the target gas. The exchange of ligands is exploited in a different colourimetric sensor mechanism. The colour of transition metal complexes is determined by electronic transitions in d-orbitals. Phase transition is the foundation of another popular strategy. Different inorganic compounds can undergo catalytic reactions with gases to cause phase transitions and, consequently, colour changes. One instance is the production of iron oxide when it is in air [105]. It is also common to perform spectroscopic measurements with Quantum Cascade Laser (QCL) technology in which the mid-IR semiconductor lasers are stacked and focused through multiple lenses, producing atomic level thickness [107]. However, because of their large, expensive embodiments, and the alignment of their optics, chambers, and vacuum pumps, they are not suitable for wearable applications. To address this, they have been combined with other technologies, such as tunable IR laser differential absorption spectroscopy (TIL-DAS), photoacoustic spectroscopy (PAS), quartz enhancement, and Wavelength Modulation Spectroscopy (WMS) to decrease their overall size and increase portability [107]. The simultaneous detection of CH_4, N_2O, and H_2O vapour using off-beam quartz-enhanced photoacoustic spectroscopy (OB-QEPAS) led to the development of an external-cavity QCL (EC-QCL) and received a LOD of 98 ppbv for CH_4. On implementing a real-time Kalman filter to enhance measurement accuracy by approximately four times while maintaining the same temporal resolution, the values for CH_4 shifted to 60 ppbv [108]. Colourimetric gas sensors interfered with humidity and required hydrophobic substrates (polyvinylidene fluoride, polyethylene terephthalate films, etc) to counter the interference [109–113]. However, their main disadvantage is that they are restricted to ionic compound detection, making them less suitable for accurately detecting hazardous and toxic gases.

3.4.2 Transistors

These sensors use a transistor—most notably a Metal Oxide Semiconductor Field Effect Transistor (MOSFET) to accomplish sensing. There are two approaches used for sensing:

(a) through the gate; and
(b) straight through the semiconducting channel

In addition, transistor-based sensors have the potential to overcome constraints associated with other technologies in terms of sensitivity, dynamic range, size, and power consumption [114]. However, despite being sensitive at low ppm levels, semiconductor-based gas sensors are hampered by fluctuating humidity levels, which may lead to drift and cross-reaction with other gases [115–117].

3.4.2.1 Channel Sensing

One way to directly alter the drain-source current for a given gate potential is by getting the transistor channel sensorised, which enables a transduction mechanism. The source and drain metal contacts to a specific need to be adequately passivated to guarantee that the measured response is solely the result of the channel. Currently, researchers have developed a new transistor architecture and material, such as n-doped graphene doped with ethylene amines, to be used in a transduction system. To improve the mobility, sensitivity, and specificity of the carrier in the channel current to oxidising gases, one method that was employed was the development of n-doped graphene doped with ethylene amines. The amount of amine functional groups found in various ethylene amines has a substantial impact on the decrease in channel current when exposing the oxidising gases at a fixed 0 V gate voltage [118]. When it comes to flexible transistors on thin, malleable substrates, InGaZnO (IGZO) is a promising material that has been used for gas sensing [119]. Further proposals for gas-sensing FETs (OFTs) (Figure 3.3) involve exposing the organic semiconductor to a particular gas and measuring the resulting change in the drain current. Several theories have been put forth to explain gas sensing with these kinds of devices, including doping the organic semiconductor, altering its dipole moment, or changing the organic film's texture [120]. The fact that organic field effect transistors can be solution-processed at low cost and can be operated in real-time conditions is one of their advantages (e.g., Spin-coating) and produced on pliable plastic substrates. Humidity can create electron traps at the conduction band's edge in n-type organic semiconductors. Simultaneously, it functions as a dopant in p-type materials, raising the device's off current [120]. To solve this problem, OFTs are typically encapsulated. Unfortunately, since the channel must be exposed and will therefore be exposed to humidity, this is not feasible in gas-sensing applications. This was recently resolved by using 4-amino benzonitrile (ABN) and tetracyanoquinodimethane (TCNQ) as molecular additives in an indacenodithiophene-co-benzothiadiazole (IDT-BT) semiconductor, which prevents moisture adsorption [120].

3.4.2.2 Gate Sensing

A MOSFET gate is often used for detection. Using the IV characteristics of a gas-sensing transistor, a shift in the gate potential that is pertinent can be directly transitioned to a proportional drain current. Operational amplifiers or instruments can use these transistors as one of differential input transistors. Many gas sensors work by measuring changes in resistance resulting from redox reactions between transition metal oxides and gas oxides [114]. The detected gas, rather than reacting directly with the metal oxide, enhances the stability for long

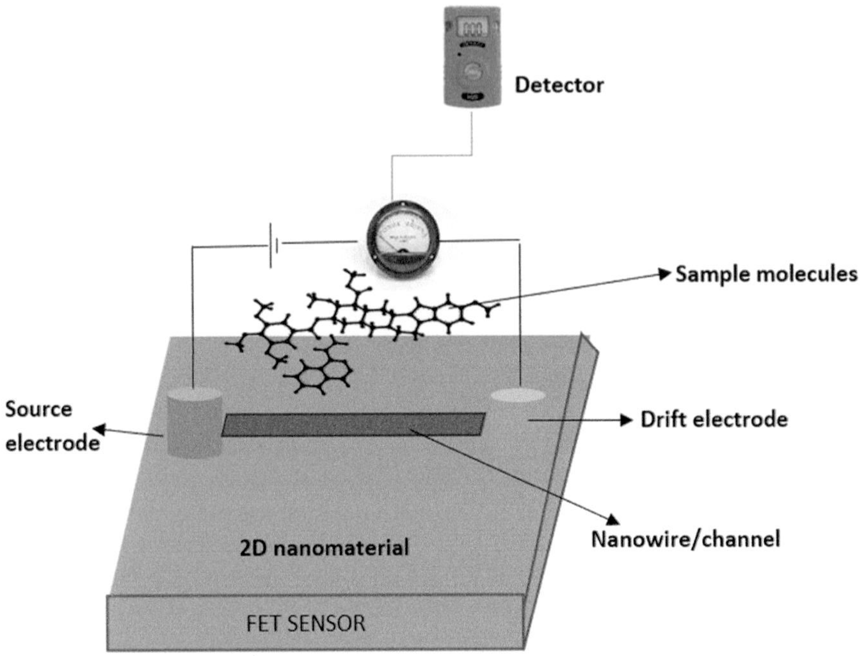

Figure 3.3 Field effect transistor.

term usage by reacting with the surrounding oxygen. Selectivity and sensitivity can only be improved with the use of a sensing film. Chemiresistive-type NOx gas sensors have been proposed using a variety of 2D materials. They consist of black phosphorous, graphene, and TMDs [114]. Many methods can be used to accomplish gate-based FET sensing. This is the earliest technique, based on the catalytic metal-gate FET. The detection technique is based on changes in the dipole moment that form at the interface between the metal and gate insulator. Thus, a dipole layer originates by permitting the gas molecules to migrate through the metal and be involved in a catalytic reaction at the gate. Through this, the device's channel current is modulated along with the threshold voltage. However, the target gas's molecular size restricts how broadly this approach can be used. Suspended-Gate FETs (SGFETs) were proposed to address this. Any molecular size of gas can be detected in this structure because of the air gap it uses [121].

A portion of the gate in a CCFET is in contact with the gate oxide, in contrast to an SGFET. Basically, there is a capacitor connected to the FET's gate electrode, one of whose plates has a floating potential and a sensing layer covering it. As the source electrode, the other has a clearly defined potential. The channel current is altered when gas molecules adsorb on the floating gate, modifying its potential. The last and most modern method that incorporates many of the previously mentioned methods is the Horizontal Floating-Gate FET (HFGFET), which is also likely the most popular [122]. The horizontal fabrication of a floating gate and a control occurs in the same location. The floating gate is insulated, and the sensing layer is directly connected to the control gate. Because the sensing layer is positioned above them both, there is a greater coupling ratio between the two layers. Since no air gap is required, this eliminates the manufacturing challenges that the SGFET and CCFET faced and allows for the sensing of gases of any molecular size. An investigation was carried out to detect HCHO using Y_2O_3 as a dielectric layer and CNT film as the semiconducting channel. At room temperature, it resulted in a 20 ppb LOD between 20 and 800 ppb at a Pd/Au catalytic metal gate. The amplification of the device achieves is responsible for the high sensitivity at room temperature [121]. Chemical reactions occur because the Pd/Au gate and the gas have different chemical potentials. A dipole layer will form between the gate and the dielectric due to the gate's dissociation of chemisorbed HCHO molecules, which produce hydrogen. The channel current rises as a result of widening the conductive channel. Au being the chemical sensitiser, the synergistic impact of the bimetallic structure promotes the chemical reaction of HCHO, which increases the sensitivity. The stacked Pd-Au gate is essential for the performance of the device [121].

3.4.3 Electrochemical

Electrochemical sensors are a great tool for detecting environmental contaminants and are used in many different applications such as biological analysis, environmental monitoring, clinical diagnostics, and food safety and quality. These sensors offer several advantages over traditional instrumentation approaches, which call for complex protocols and specialised equipment. The benefits include ppm- or ppb-level sensitivity, ease of use, low cost, high miniaturisation, and energy usage, and basic instrumentation [123]. High-conductivity electrode materials are required for this kind of sensor. They need to, also possess outstanding catalytic activity and an electrochemical potential range. The analyte and electrode react electrochemically at the interface, which produces a detectable electric signal proportional to the analyte concentration. The most commonly used gas detection methods are amperometry and voltammetry. In most cases, REs are Ag/AgCl electrodes. Counter Electrodes (CEs) normally have 10 times the surface area of the WorkingElectrodes (WEs) when composed of Pt, in order to avoid significant interfacial impedance effects on measurements. As a first step, CV can be combined with amperometry to determine redox potential, which can subsequently be used as a simple and main detection method. In case CE and RE are placed very closely together through miniaturisation, the current and ohmic drop between them can also be reduced to more than a few ohms, allowing two electrodes, WE and CE, to be used [107, 124]. There is a significant difference between the redox potentials of various gases. The negative reduction potential of O_2 and CO_2 is less than N_2O, while N_2 has a higher negative reduction potential [107]. Electrochemical fuel cells are often used for amperometric gas sensors (Figure 3.4). The WE and CE typically catalyse oxygen reduction and oxidation reactions of the target gases, respectively, by highly loaded Pt catalysts [125]. In the presence of increased potential, as the reaction rate attains a crucial stage, the target gas has attained its maximum diffusion flow, forming a plateau.

Another technique is to use potentiometry for electrochemical gas detection, in which the polarisation potential between two electrodes is measured. An indifferent electrode (RE), whose potential should be constant and unaffected by electrochemical changes, and a WE is used. There is no flow of current in potentiometry. The lambda sensor, a potentiometric gas sensor that detects oxygen in exhaust gases, is an example of a common potentiometric gas sensor. A mixed potential gas sensor (such as Electrochemical Impedance Spectroscopy (EIS)) uses a sinusoidal voltage with varying frequency to excite the electrochemical cell and detects a non-Nernstian response when the gas mixture is not in equilibrium. They can characterise gas mixtures containing CO, H, CO_2, and other hydrocarbons [126]. An electrochemical gas sensor based on solid electrolytes consists of two or three electrodes separated spatially. To close the circuit, a conductive medium must be present at each electrode for current to flow or polarisation potentials to be generated. A liquid or solid electrolyte is typically used to achieve this, with the latter being preferred for wearable and portable devices. This includes 85% mass zirconia, fluorites such as ZrO_2 stabilised with Y, Ca, Sc, Yb oxides or their mixtures, $SrZrO_3$, $LaYO_3$, Cu_2S, AgS, yttria-doped $BaCeO_3$, lanthanum gallate ($LaGaO_3$)-based materials, which demonstrates ionic conductivity [126]. Gas selectivity and specificity can be improved by a number of approaches. Two electrochemical cells are used in this process. An initially used reduction potential is lower than the reduction potential of the target gas, which is used for the second cell. Other species in the mixture can also be reduced by this potential.

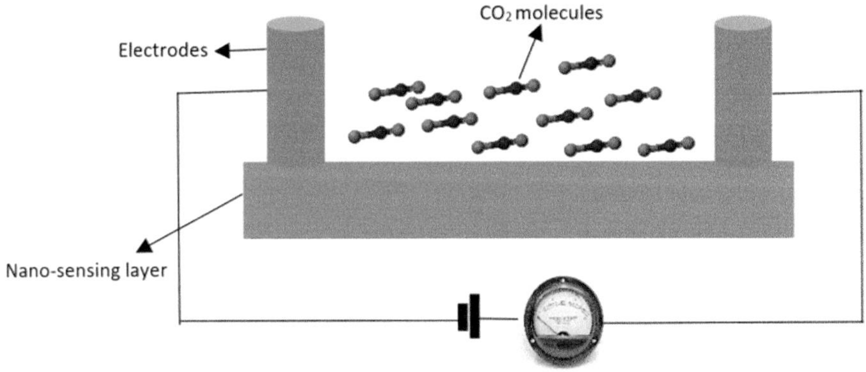

Figure 3.4 Electrochemical gas sensors [189].

3.4.4 Chemiresistors

It is most common to use chemical resistance sensing as a sensing approach. Due to the relatively inexpensive nature of portable devices, chemiresistors have been used more often in recent decades. A chemiresistor typically consists of two pairs of electrodes linked by a sensing film, which could be made of metals or semiconductors. Chemiresistive structures are typically used for printed gas sensors to detect trace gas amounts since they respond to gas analytes by changing channel conductance. The sensing response is often boosted with interdigitated electrodes (IDEs). An electrode pair can be applied with a known constant potential or current. As a result, the film's resistance is determined through Ohm's law by measuring the voltage or current generated by a gaseous analyte [127]. SnO_2, WO_3, ZnO, TiO_2, and In_2O_3 are among the MOS materials commonly used in chemiresistive gas sensors. Pulsed Laser Deposition (PLD), Chemical Vapour Deposition (CVD), radio frequency sputtering, hydrothermal growth, sol-gel deposition, co-precipitation, thermal evaporation, and direct current sputtering are just a few of the manufacturing methods that can be used to deposit these materials. When MOSs are exposed to a target gas, their resistance varies due to the oxidative reactions with the negatively charged oxygen deposited on them [107]. An n-type or a p-type MOS will determine whether the polarity of the change is positive or negative. Reduced electrical resistance is achieved in n-type materials by releasing electrons back into their conduction bands. In p-type material, a hole accumulation region is formed by adsorption. The resistance will then increase when exposed to reducing analytes. Alternatively, oxidative agents result in the oxidation of the material [107]. When an oxidising agent is added to n-type materials, their resistance increases, while when a reducing gas is added, their resistance decreases; when oxidising agents are added, p-type materials will have a decreased resistance, while when reducing agents are added, they will have a higher resistance. Creating a heterojunction interface using more than one metal oxide is another approach, e.g. graphene-CeO_2 for detection of CO. CO detection using CVD-grown SnO_2 NW stripe chemoresistors with a 500-fold increase in gas response [128]. The nanomaterial solutions of graphene on the silver electrodes (7 mm × 13.4 mm with a width of 210 μm) made graphene-based sensors capable of sensing a range of gases in the ppb range (NH_3, $(CH_3)_2CO$, C_2H_6O, $(CH_3)_2CHOH$, and NaOCl) [129].

3.4.5 Gas Sensors with Self-Powered Triboelectric Nanogenerators

In recent years, self-powered gas sensors using triboelectric nanogenerators (TENG) have gained increasing interest since they can be miniaturised and are portable, do not require external power sources, and require little maintenance [130, 131]. Energy is generated from these gas sensors while they also serve as sensors. A common material characteristic of triboelectricity is the temporary electrostatic charge that occurs in most materials. Therefore, many material options can be used to produce triboelectricity. In simple terms, triboelectricity is the generation of static electrical charges by friction. Sensing materials commonly used in TENG fabrication are conducting polymers and metal oxide semiconductors [132, 133]. As a versatile detection system for monitoring gas and movement, self powered flexible MXene sensors are developed using triboelectric-electromagnetic nanogenerators. To develop wearable environmental gas measurements TENGs with low power consumption, it is necessary to combine high-performance and miniaturised circuitry. A self powered TENG sensor that can identify formaldehyde at an ambient temperature has been developed, which highlights the need for minuscular gas sensors [134]. The Ag electrode seed layer was integrated into the active sensing layer by altering the surface of the augmented triboelectric layer. Comparing this sensor to the most advanced formaldehyde sensors, it demonstrates higher selectivity, an incredibly quick response time, sub-ppm detection, and superior selectivity.

3.4.6 Environmental Catalysts

Among the most important pollutants of water are Halogenated Organic Compounds (HOCs). In addition to pharmaceuticals, these organic molecules have extensive applications as additives and solvents. It is imperative that these toxic compounds are decomposed completely from water and wastewater so that health problems like cancer can be avoided [135]. There are no conventional methods for treating such water problems. Using palladium nano catalysts to selectively degrade HOCs is one innovative way to detoxify water. A water treatment plant can easily remove stable HOCs through biodegradation by converting them into organic compounds [136]. As an added benefit, they can be recovered through magnetic separation technology from desired water [137]. The most prominent air pollutants are CO, hydrocarbons, and NO. This pollution can be reduced with catalytic converters. Metal catalysts are currently used in catalytic converters, and the ratio of air to fuel must be stoichiometric. In order to address this problem, high-efficiency, low cost catalysts are highly sought [138]. A simple kinetics makes carbon monoxide an excellent model for studying the oxidation reaction. It is known that $CuCeO_2$ and $AuCeO_2$ catalysts for CO oxidation are highly active and stable as compared to copper or sodium oxide catalysts. The Au-CeO_2 or Cu-CeO_2 catalysts convert CO completely at specific temperatures (20°C and 80°C) at a spatial velocity [139].

3.4.7 Green Production

As a term, "green production" refers to a set of methods and technologies that are used to reduce waste, raw material consumption, and energy consumption [140, 141]. Developing environmentally friendly industrial processes is an integral part of green production. Water-based processes can be designed and replaced with organic solvent-based processes, and hazardous substances can also be reduced. Green chemicals can be developed that are harmless to the environment, and less energy is used in processes [142]. In textiles, cleaning, and oil extraction industries, toxic and carcinogenic compounds are usually used [143]. Nanosized microemulsions can be used for nanoscale extraction of these specific molecules. A micro emulsion synthesised by researchers makes it easier to separate oil from seawater by reducing the surface tension of organic matter [144].

Table 3.1 shows different types of nanomaterials used for the detection of CO and CO_2, and other types of nanosensors used in detection of environmental contaminants (Table 3.2).

Table 3.1 Different types of nanosensors used in CO_2 and CO detection

Materials	Application	Concentration/ Dynamic Range	Response Time	Recovery Time	Ref.
SnO_2 QD	CO	1000 ppm	–	–	[145]
ZnO NPs	CO	80 ppm	21	70 s`	[146]
Au-decorated ZnO nanorods (refractometric)	CO_2	0–2000 sccm	50	110 s	[147]
YPO_4 nanobelts (impedance)	CO_2	200–800 ppm	136	–	[148]
SnO_2: Ag	CO	1000	3	–	[149]
LaOCl-doped SnO_2 nanofibres (resistance)	CO_2	100–2000 ppm	24	92 s	[18]
Ru nanobeads doped HPTS in ormosil matrix (colourimetric)	CO_2	0–100%	30	<60 s	[150]
Poly(ionicliquid) alumina composite	CO_2	300–3200 ppm	420	2400 s	[151]
Pd:SnO_2 NF	CO	100 ppm	23.7	3.7	[152]
Inorganic silica nanoparticles (capacitance)	CO_2	400–2000	>60	>150	[93]
CuO nanoparticles (Surface charge)	CO_2	400–4000	720 @ 0% R.H 500 @ 20% R.H		[153]
Silica Nanoflower (colourimetric)	CO_2	400–70000 ppm	Instantaneous	–	[154]
Au/V: SnO_2	CO	1000 ppm	–	–	[155]

Table 3.2 Nanomaterials and their applications in detection of environmental contaminants

Material	Application	Ref.
Metal doped TiO_2	Water contaminants- endotoxin, *E. coli,* 2-chlorophenol	[156–158]
Amine modified aluminosilicates and porous silica	Gaseous CO_2, ketones, and aldehydes	[159–167, 197]
TiO_2 graphene	Gaseous Benzene	[168, 169]
Amine-modified xerogels	CO_2, H_2S	[170]
Ag ions/Ag NPs	Water disinfectant- *E. coli*	[171–175]
Iron based	Water-chlorinated organic solvents	[176, 177]
TiO_2 NPs	Soil, water disinfectant- *E. coli,* Polycyclic aromatic hydrocarbons (PAHs), phenanthrene	[178–180]

(*Contd.*)

Material	Application	Ref.
PAMAM dendrimer composite membrane with a dendrimer and chitosan	CO_2 separation on porous substrates from a CO_2 and N_2 mixture	[181]
Amine modified PDLLA-PEG	VOCs	[182, 183]
Amphiphilic polyurethane NPs	Polynuclear aromatic hydrocarbons in soil	[184]
Silica nanoparticles made by mixing poly (propylene imine) and salicylic acid	Removal of PAHs	[185]
Polymer modified nanocomposites	Dyes and microorganisms	[186, 187]
Polyamine modified cellulose	VOCs	[188]

3.5 CONCLUSION

In the past few decades, there has been significant research in nanomaterial-based carbon sensing. An urgent need exists for the construction of reliable sensing platforms capable of accurately detecting and quantifying these pollutants. In order to monitor environmental pollutants, novel nanomaterials have the potential to be useful as sensors because of their unique properties, particularly their capacity to support required functionalities. A robust gas sensor should be selective, sensitive, and multiparametric so that it can be successfully commercialised, and in order to achieve this, the device should be comfortable and simple so that the user can manage and control it directly, and it should be integrated seamlessly into daily routines without requiring a lot of effort. To make a device that meets the needs of software engineers, it is necessary to select the right materials that are suitable for the device, determine manufacturing temperatures, and optimise the dimensions of the final device. In order to choose the right materials and their combination, careful thought needs to be given to the application at hand. Nanosensors can also impact employment and the economy. As a result, workers can ensure the safety of their work environment by detecting gas leaks successfully using this smart sensing technology, which minimises human error. This prevents catastrophic incidents in maintenance of factory equipment and ensures that goods and products are stored efficiently without being lost through storage. More research is however required for accelerating the application and development nanosensors in the coming decades.

REFERENCES

[1] Masciangoli T., Zhang W. 2003. Environmental Technologies. Environment Science and Technology. 37: 102–108.

[2] Adeleye, A.S., Conway, J.R., Garner, K., Huang, Y., Su, Y. and Keller, A.A. 2016. Engineered nanomaterials for water treatment and remediation: Costs, benefits, and applicability. Chemical Engineering Journal. 286: 640–662.

[3] Khan, F.I. and Ghoshal, A.K. 2000. Removal of volatile organic compounds from polluted air. Journal of Loss Prevention in the Process Industries. 13(6): 527–545.

[4] Vaseashta, A., Vaclavikova, M., Vaseashta, S., Gallios, G., Roy, P. and Pummakarnchana, O. 2007. Nanostructures in environmental pollution detection, monitoring, and remediation. Science and Technology of Advanced Materials. 8(1–2): 47.

[5] Kaur, S. and Roy, A. 2021. Bioremediation of heavy metals from wastewater using nanomaterials. Environment, Development and Sustainability. 23: 9617–9640.

[6] Guerra, F.D., Attia, M.F., Whitehead, D.C. and Alexis, F. 2018. Nanotechnology for environmental remediation: materials and applications. Molecules. 23(7): 1760.

[7] Aguilar-Pérez, K.M., Avilés-Castrillo, J.I. and Ruiz-Pulido, G. 2020. Nano-sorbent materials for pharmaceutical-based wastewater effluents-An overview. Case Studies in Chemical and Environmental Engineering. 2: 100028.

[8] Esmaeili, A. and Beni, A.A. 2019. Characterization of PVA/chitosan nano fiber membrane and increasing mechanical properties with cross-linking by heating. International Journal of Theoretical and Applied Mechanics. 4.

[9] Guo, Y., Zhang, L., Zhao, F., Li, G. and Zhang, G. 2021. Tribological behaviors of novel epoxy nanocomposites filled with solvent-free ionic SiO_2 nanofluids. Composites Part B: Engineering. 215: 108751.

[10] Dong, J., Xu, W., Liu, S., Du, L., Chen, Q., Yang, T., et al. 2021. Recent advances in applications of nonradical oxidation in water treatment: Mechanisms, catalysts and environmental effects. Journal of Cleaner Production. 321: 128781.

[11] Rasheed, T., Hassan, A.A., Kausar, F., Sher, F., Bilal, M. and Iqbal, H.M. 2020. Carbon nanotubes assisted analytical detection–Sensing/delivery cues for environmental and biomedical monitoring. TrAC Trends in Analytical Chemistry. 132: 116066.

[12] Sha, R. and Bhattacharyya, T.K. 2020. MoS_2-based nanosensors in biomedical and environmental monitoring applications. Electrochimica Acta. 349: 136370.

[13] Nath, S., Dey, A., Pachal, P., Sing, J.K. and Sarkar, S.K. 2021. Performance analysis of gas sensing device and corresponding IoT framework in mines. Microsystem Technologies. 27: 3977–3985.

[14] Brahmkhatri, V., Pandit, P., Rananaware, P., D'Souza, A. and Kurkuri, M.D. 2021. Recent progress in detection of chemical and biological toxins in water using plasmonic nanosensors. Trends in Environmental Analytical Chemistry. 30: e00117.

[15] Starecki, F., Charpentier, F., Doualan, J.L., Quetel, L., Michel, K., Chahal, R., et al. 2015. Mid-IR optical sensor for CO_2 detection based on fluorescence absorbance of Dy^{3+}: $Ga_5Ge_{20}Sb_{10}S_{65}$ fibers. Sensors and Actuators B: Chemical. 207: 518–525.

[16] Liu, L., Morgan, S.P., Correia, R., Lee, S.W. and Korposh, S. 2020. Multi-parameter optical fiber sensing of gaseous ammonia and carbon dioxide. Journal of Lightwave Technology. 38(7): 2037–2045.

[17] Starecki, F., Braud, A., Doualan, J.L., Ari, J., Boussard-Plédel, C., Michel, K. et al. 2019. All-optical carbon dioxide remote sensing using rare earth doped chalcogenide fibers. Optics and Lasers in Engineering. 122: 328–334.

[18] Xiong, Y., Xue, Q., Ling, C., Lu, W., Ding, D., Zhu, L., et al. 2017. Effective CO_2 detection based on LaOCl-doped SnO_2 nanofibers: Insight into the role of oxygen in carrier gas. Sensors and Actuators B: Chemical. 241: 725–734.

[19] George, R., Kumar, L.A. and Alagappan, M.J.D.J.N.B. 2019. Synthesis of nanotubular NiO-CNT composite and its application in temperature independent CO_2 gas sensors

fabricated using interdigitated silver electrode. Digest Journal of Nanomaterials and Biostructures. 14: 213–224.

[20] Abou Hammad, A.B., Elnahrawy, A.M. and Youssef, A.M. 2019. Sol gel synthesis of hybrid chitosan/calcium aluminosilicate nanocomposite membranes and its application as support for CO_2 sensor. International Journal of Biological Macromolecules. 125: 503–509.

[21] Stassen, I., Dou, J.H., Hendon, C. and Dincă, M. 2019. Chemiresistive sensing of ambient CO_2 by an autogenously hydrated Cu_3 (hexaiminobenzene) 2 framework. ACS Central Science. 5(8): 1425–1431.

[22] Ersoez, B., Schmitt, K. and Wöllenstein, J. 2020. Electrolyte-gated transistor for CO_2 gas detection at room temperature. Sensors and Actuators B: Chemical. 317: 128201.

[23] Santonico, M., Zompanti, A., Sabatini, A., Vollero, L., Grasso, S., Di Mezza, C. and Pennazza, G. 2020. CO_2 and O_2 detection by electric field sensors. Sensors. 20(3): 668.

[24] Lv, A., Pan, Y. and Chi, L. 2017. Gas sensors based on polymer field-effect transistors. Sensors. 17(1): 213.

[25] Huang, X.J., Aldous, L., O'Mahony, A.M., del Campo, F.J. and Compton, R.G. 2010. Toward membrane-free amperometric gas sensors: A microelectrode array approach. Analytical Chemistry. 82(12): 5238–5245.

[26] Bałaga, D., Siegmund, M., Kalita, M., Williamson, B.J., Walentek, A. and Małachowski, M. 2021. Selection of operational parameters for a smart spraying system to control airborne PM_{10} and $PM_{2.5}$ dusts in underground coal mines. Process Safety and Environmental Protection. 148: 482–494.

[27] Larionov, K.B., Zenkov, A.V., Mishakov, I.V. and Gromov, A.A. 2020. Decrease in carbon monoxide release and incomplete burning of fuel during activated combustion of coal. *In*: Journal of Physics: Conference Series. 1675(1): 012061. IOP Publishing.

[28] Manisalidis, I., Stavropoulou, E., Stavropoulos, A. and Bezirtzoglou, E. 2020. Environmental and health impacts of air pollution: A review. Frontiers in Public Health. 8: 14.

[29] Dhall, S., Mehta, B.R., Tyagi, A.K. and Sood, K. 2021. A review on environmental gas sensors: Materials and technologies. Sensors International. 2: 100116.

[30] Utari, L., Septiani, N.L.W., Nur, L.O., Wasisto, H.S. and Yuliarto, B. 2020. Wearable carbon monoxide sensors based on hybrid graphene/ZnO nanocomposites. IEEE Access. 8: 49169–49179.

[31] Kinoshita, H., Türkan, H., Vucinic, S., Naqvi, S., Bedair, R., Rezaee, R., et al. 2020. Carbon monoxide poisoning. Toxicology Reports. 7: 169–173.

[32] Ohata, J., Bruemmer, K.J. and Chang, C.J. 2019. Activity-based sensing methods for monitoring the reactive carbon species carbon monoxide and formaldehyde in living systems. Accounts of Chemical Research. 52(10): 2841–2848.

[33] Bhowmick, T., Ghosh, A., Nag, S. and Majumder, S.B. 2022. Sensitive and selective CO_2 gas sensor based on CuO/ZnO bilayer thin-film architecture. Journal of Alloys and Compounds. 903: 163871.

[34] CO_2 and Ocean Acidification: Causes, Impacts, Solutions. [accessed on 9 July 2020]; Available online: https://www.ucsusa.org/resources/co2-and-ocean-acidification.

[35] Kang, S., Zhao, K., Yu, D.G., Zheng, X. and Huang, C. 2022. Advances in biosensing and environmental monitoring based on electrospun nanofibers. Advanced Fiber Materials. 4(3): 404–435.

[36] Chang, J.W., Chen, H.L., Su, H.J., Liao, P.C., Guo, H.R. and Lee, C.C. 2011. Simultaneous exposure of non-diabetics to high levels of dioxins and mercury increases their risk of insulin resistance. Journal of Hazardous Materials. 185(2–3): 749–755.

[37] Sadowska, A., Nynca, A., Ruszkowska, M., Paukszto, L., Myszczynski, K., Swigonska, S., et al. 2021. Transcriptional profiling of Chinese hamster ovary (CHO) cells exposed to 2, 3, 7, 8-tetrachlorodibenzo-p-dioxin (TCDD). Reproductive Toxicology. 104: 143–154.

[38] Fagan, S.B., Santos, E.J.G., Souza Filho, A.G., Mendes Filho, J. and Fazzio, A. 2007. Ab initio study of 2, 3, 7, 8-tetrachlorinated dibenzo-p-dioxin adsorption on single wall carbon nanotubes. Chemical Physics Letters. 437(1–3): 79–82.

[39] Su, F., Lu, C. and Hu, S. 2010. Adsorption of benzene, toluene, ethylbenzene and p-xylene by NaOCl-oxidized carbon nanotubes. Colloids and Surfaces A: Physicochemical and Engineering Aspects. 353(1): 83–91.

[40] David, S.A., Rajkumar, R., Karpagavinayagam, P., Fernando, J. and Vedhi, C. 2022. Sustainable carbon nanomaterial-based sensors: Future vision for the next 20 years. *In*: Carbon Nanomaterials-Based Sensors. 429–443. Elsevier.

[41] Joseph, N. and Manoj, B. 2022. Nanomaterials-based chemical sensing. Nanotechnology for Electronic Applications. 131–147.

[42] Esrafili, M.D. and Abolghasemzadeh, S. 2022. Sc functionalized boron-rich C_{60} fullerene for efficient storage and separation of CO_2 molecules: A DFT investigation. Computational and Theoretical Chemistry. 1208: 113557.

[43] Rathinavel, S., Priyadharshini, K. and Panda, D. 2021. A review on carbon nanotube: An overview of synthesis, properties, functionalization, characterization, and the application. Materials Science and Engineering: B. 268: 115095.

[44] Sacco, L., Forel, S., Florea, I. and Cojocaru, C.S. 2020. Ultra-sensitive NO_2 gas sensors based on single-wall carbon nanotube field effect transistors: Monitoring from ppm to ppb level. Carbon. 157: 631–639.

[45] Ahmad, Z., Manzoor, S., Talib, M., Islam, S.S. and Mishra, P. 2020. Self-standing MWCNTs based gas sensor for detection of environmental limit of CO_2. Materials Science and Engineering: B. 255: 114528.

[46] Demon, S.Z.N., Kamisan, A.I., Abdullah, N., Noor, S.A.M., Khim, O.K., Kasim, N.A.M., et al. 2020. Graphene-based Materials in Gas Sensor Applications: A Review. Sensors and Materials. 32.

[47] Kim, S., Kwak, D.H., Choi, I., Hwang, J., Kwon, B., Lee, E., et al. 2020. Enhanced gas sensing properties of graphene transistor by reduced doping with hydrophobic polymer brush as a surface modification layer. ACS Applied Materials and Interfaces. 12(49): 55493–55500.

[48] Guo, X. and Facchetti, A. 2020. The journey of conducting polymers from discovery to application. Nature Materials. 19(9): 922–928.

[49] Akande, G., Ajayi, S.A., Fajobi, M.A., Oluwole, O.O. and Fayomi, O.S.I. 2021. Advancement in the production and applications of conductive polymers (CPs). Key Engineering Materials. 886: 12–29.

[50] Pavel, I.A., Lakard, S. and Lakard, B. 2022. Flexible sensors based on conductive polymers. Chemosensors. 10(3): 97.

[51] Riyazi, S. and Azim Araghi, M.E. 2020. Performance of interdigitated capacitive-type CO_2 sensor based on polypyrrole/copper phthalocyanine nanocomposite. Journal of Materials Science: Materials in Electronics. 31: 3539–3548.

[52] Riyazi, S. and Azim Araghi, M.E. 2020. Performance of interdigitated capacitive-type CO_2 sensor based on polypyrrole/copper phthalocyanine nanocomposite. Journal of Materials Science: Materials in Electronics. 31: 3539–3548.

[53] Farea, M.A., Bhanuse, G.B., Mohammed, H.Y., Farea, M.O., Sallam, M., Shirsat, S.M., et al. 2022. Ultrahigh sensitive and selective room-temperature carbon monoxide gas sensor based on polypyrrole/titanium dioxide nanocomposite. Journal of Alloys and Compounds. 917: 165397.

[54] Gil, B., Anastasova, S. and Lo, B. 2022. Graphene field-effect transistors array for detection of liquid conductivities in the physiological range through novel time-multiplexed impedance measurements. Carbon. 193: 394–403.

[55] Yang, K., Zhu, K., Wang, Y., Qian, Z., Zhang, Y., Yang, Z., et al. 2021. $Ti_3C_2T_x$ MXene-Loaded 3D Substrate toward On-Chip Multi-Gas Sensing with Surface-Enhanced Raman Spectroscopy (SERS) Barcode Readout. ACS Nano. 15(8): 12996–13006.

[56] Niu, G. and Wang, F. 2022. A review of MEMS-based metal oxide semiconductors gas sensor in Mainland China. Journal of Micromechanics and Microengineering. 32(5): 054003.

[57] Fine, G.F., Cavanagh, L.M., Afonja, A. and Binions, R. 2010. Metal oxide semiconductor gas sensors in environmental monitoring. Sensors. 10(6): 5469–5502.

[58] Zhao, C., Gong, H., Niu, G. and Wang, F. 2020. Ultrasensitive SO_2 sensor for sub-ppm detection using Cu-doped SnO_2 nanosheet arrays directly grown on chip. Sensors and Actuators B: Chemical. 324: 128745.

[59] Ma, J., Fan, H., Zheng, X., Wang, H., Zhao, N., Zhang, M., et al. 2020. Facile metal-organic frameworks-templated fabrication of hollow indium oxide microstructures for chlorine detection at low temperature. Journal of Hazardous Materials. 387: 122017.

[60] Li, Q., Li, Z., Zhang, Q., Zheng, L., Yan, W., Liang, X., et al. 2021. Porous γ-Fe_2O_3 nanoparticle decorated with atomically dispersed platinum: Study on atomic site structural change and gas sensor activity evolution. Nano Research. 14: 1435–1442.

[61] Fu, H., Yang, X., Wu, Z., He, P., Xiong, S., Han, D., et al. 2022. Gas-sensing performance of In_2O_3@MoO_3 hollow core-shell nanospheres prepared by a two-step hydrothermal method. Sensors and Actuators B: Chemical. 352: 131007.

[62] Guo, X., Ding, Y., Yang, X., Du, B., Zhao, C., Liang, C., et al. 2022. 2D $SnSe_2$ nanoflakes decorated with 1D ZnO nanowires for ppb-level NO_2 detection at room temperature. Journal of Hazardous Materials. 426: 128061.

[63] Godse, P.R., Mane, A.T., Navale, Y.H., Navale, S.T., Mulik, R.N. and Patil, V.B. 2021. Hydrothermally grown 1D ZnO nanostructures for rapid detection of NO_2 gas. SN Applied Sciences. 3: 1–18.

[64] Zhang, W.H., Ding, S.J., Zhang, Q.S., Yi, H., Liu, Z.X., Shi, M.L., et al. 2021. Rare earth element-doped porous In_2O_3 nanosheets for enhanced gas-sensing performance. Rare Metals. 40: 1662–1668.

[65] Umar, A., Ammar, H.Y., Kumar, R., Ibrahim, A.A. and Al-Assiri, M.S. 2020. Square disks-based crossed architectures of SnO_2 for ethanol gas sensing applications—An experimental and theoretical investigation. Sensors and Actuators B: Chemical. 304: 127352.

[66] Nguyen, H.T., Truong, T.H., Nguyen, T.D., Dang, V.T., Vu, T.V., Nguyen, S.T., et al. 2020. Ni-doped WO 3 flakes-based sensor for fast and selective detection of H_2S. Journal of Materials Science: Materials in Electronics. 31: 12783–12795.

[67] Shinde, R.B., Padalkar, N.S., Sadavar, S.V., Kale, S.B., Magdum, V.V., Chitare, Y.M., et al. 2022. 2D–2D lattice engineering route for intimately coupled nanohybrids of layered double hydroxide and potassium hexaniobate: Chemiresistive SO_2 sensor. Journal of Hazardous Materials. 432: 128734.

[68] Shakeel, A., Rizwan, K., Farooq, U., Iqbal, S. and Altaf, A.A. 2022. Advanced polymeric/inorganic nanohybrids: An integrated platform for gas sensing applications. Chemosphere. 294: 133772.

[69] Nadeem, M., Yasin, G., Arif, M., Tabassum, H., Bhatti, M.H., Mehmood, M., et al. 2021. Highly active sites of Pt/Er dispersed N-doped hierarchical porous carbon for trifunctional electrocatalyst. Chemical Engineering Journal. 409: 128205.

[70] Sun, Y., Li, X. and Liu, G. 2022. Enhanced pollutants removal and high-value cell inclusions accumulation with Fe^{2+} in heavy oil refinery treatment system using Rhodopseudomonas and Pseudomonas. Chemosphere. 294: 133520.

[71] Shakeel, A., Rizwan, K., Farooq, U., Iqbal, S. and Altaf, A.A. 2022. Advanced polymeric/inorganic nanohybrids: An integrated platform for gas sensing applications. Chemosphere. 294: 133772.

[72] Teijido, R., Ruiz-Rubio, L., Echaide, A.G., Vilas-Vilela, J.L., Lanceros-Mendez, S. and Zhang, Q. 2022. State of the art and current trends on layered inorganic-polymer nanocomposite coatings for anticorrosion and multi-functional applications. Progress in Organic Coatings. 163: 106684.

[73] Rahman, M.A., Habib, M.L., Chisty, A.H., Mallik, A.K., Khan, M.N., Haque, P., et al. 2022. Organic–inorganic polymer hybrids for water and wastewater treatment. Inorganic-Organic Composites for Water and Wastewater Treatment. 1: 29–54.

[74] Luo, M., Shaitan, K., Qu, X., Bonartsev, A.P. and Lei, B. 2022. Bioactive rare earth-based inorganic-organic hybrid biomaterials for wound healing and repair. Applied Materials Today. 26: 101304.

[75] Nadeem, M., Yasin, G., Arif, M., Tabassum, H., Bhatti, M.H., Mehmood, M., et al. 2021. Highly active sites of Pt/Er dispersed N-doped hierarchical porous carbon for trifunctional electrocatalyst. Chemical Engineering Journal. 409: 128205.

[76] Ghoorchian, A., Amouzegar, Z., Moradi, M., Khalili, S., Afkhami, A., Madrakian, T., et al. 2022. Use of Conductive Polymers in Detection Stage of Analysis/ Miniaturization Devices. *In*: Conductive Polymers in Analytical Chemistry. 165–184. American Chemical Society.

[77] Manjakkal, L., Szwagierczak, D. and Dahiya, R. 2020. Metal oxides based electrochemical pH sensors: Current progress and future perspectives. Progress in Materials Science. 109: 100635.

[78] MacDougall, S., Bayansal, F. and Ahmadi, A. 2022. Emerging methods of monitoring volatile organic compounds for detection of plant pests and disease. Biosensors. 12(4): 239.

[79] Tomić, M., Šetka, M., Vojkůvka, L. and Vallejos, S. 2021. VOCs sensing by metal oxides, conductive polymers, and carbon-based materials. Nanomaterials. 11(2): 552.

[80] Qin, X., Wu, T., Zhu, Y., Shan, X., Liu, C. and Tao, N. 2020. A paper based milli-cantilever sensor for detecting hydrocarbon gases via smartphone camera. Analytical Chemistry. 92(12): 8480–8486.

[81] Sun, M., Lv, J., Xu, H., Zhang, L., Zhong, Y., Chen, Z., et al. 2020. Smart cotton fabric screen-printed with viologen polymer: photochromic, thermochromic and ammonia sensing. Cellulose. 27: 2939–2952.

[82] Jung, W.T., Jeon, J.W., Jang, H.S., Lee, H.K. and Kim, B.H. 2020. Commercial silk-based electronic textiles for NO_2 sensing. Sensors and Actuators B: Chemical. 307: 127596.

[83] Libanori, A., Chen, G., Zhao, X., Zhou, Y. and Chen, J. 2022. Smart textiles for personalized healthcare. Nature Electronics. 5(3): 142–156.

[84] Davidson, C.E., Dixon, M.M., Williams, B.R., Kilper, G.K., Lim, S.H., Martino, R.A., et al. 2020. Detection of chemical warfare agents by colorimetric sensor arrays. ACS Sensors. 5(4): 1102–1109.

[85] Balogh, K., Jesus, J.M., Gouveia, C., Domingues, J.O., Markovics, A., Baptista, J.M., et al. 2013. Characterization of a novel dissolved CO_2 sensor for utilization in environmental monitoring and aquaculture industry. *In*: 8th Iberoamerican Optics Meeting and 11th Latin American Meeting on Optics, Lasers, and Applications. 8785: 1663–1669. SPIE.

[86] Jana, J., Ganguly, M. and Pal, T. 2016. Enlightening surface plasmon resonance effect of metal nanoparticles for practical spectroscopic application. RSC Advances. 6(89): 86174–86211.

[87] Njegovec, M. and Donlagic, D. 2020. A fiber-optic gas sensor and method for the measurement of refractive index dispersion in NIR. Sensors. 20(13): 3717.

[88] Shivananju, B.N., Yamdagni, S., Fazuldeen, R., Sarin Kumar, A.K., Hegde, G.M., Varma, M.M., et al. 2013. CO_2 sensing at room temperature using carbon nanotubes coated core fiber Bragg grating. Review of Scientific Instruments. 84(6).

[89] Chong, X., Kim, K.J., Li, E., Zhang, Y., Ohodnicki, P.R., Chang, C.H., et al. 2016. Near-infrared absorption gas sensing with metal-organic framework on optical fibers. Sensors and Actuators B: Chemical. 232: 43–51.

[90] Yi, S., Park, Y., Han, S., Min, N., Kim, E. and Ahn, T. 2005, June. Novel NDIR CO_2 sensor for indoor air quality monitoring. *In*: The 13th International Conference on Solid-State Sensors, Actuators and Microsystems, 2005. Digest of Technical Papers. TRANSDUCERS'05. 2: 1211–1214. IEEE.

[91] Frodl, R. and Tille, T. 2006. A High-Precision NDIR CO_2 gas sensor for automotive applications. IEEE Sensors Journal. 6(6): 1697–1705.

[92] Jha, R.K. 2021. Non-dispersive infrared gas sensing technology: A review. IEEE Sensors Journal. 22(1): 6–15.

[93] Boudaden, J., Klumpp, A., Hecker, C., Eisele, I. and Joseph, Y. 2018. Functionalized nanoparticles for CO_2 sensor. Funct. Nanostruct. Proc. 64–66.

[94] Frodl, R. and Tille, T. 2006. A High-Precision NDIR CO_2 gas sensor for automotive applications. IEEE Sensors Journal. 6(6): 1697–1705.

[95] Gibson, D. and MacGregor, C. 2013. A novel solid state non-dispersive infrared CO_2 gas sensor compatible with wireless and portable deployment. Sensors. 13(6): 7079–7103.

[96] Müller, L., Käpplinger, I., Biermann, S., Brode, W. and Hoffmann, M. 2014. Infrared emitting nanostructures for highly efficient microhotplates. Journal of Micromechanics and Microengineering. 24(3): 035014.

[97] Koppens, F.H.L., Mueller, T., Avouris, P., Ferrari, A.C., Vitiello, M.S. and Polini, M. 2014. Photodetectors based on graphene, other two-dimensional materials and hybrid systems. Nature Nanotechnology. 9(10): 780–793.

[98] Pusch, A., De Luca, A., Oh, S.S., Wuestner, S., Roschuk, T., Chen, Y., et al. 2015. A highly efficient CMOS nanoplasmonic crystal enhanced slow-wave thermal emitter improves infrared gas-sensing devices. Scientific Reports. 5(1): 17451.

[99] Alaloul, M., Al-Ani, A.M., As'Ham, K., Khurgin, J.B., Hattori, H.T. and Miroshnichenko, A.E. 2023. Mid-Wave Infrared Graphene Photodetectors with High Responsivity for On-Chip Gas Sensors. IEEE Sensors Journal. 23(3): 2040–2046.

[100] Comini, E., Faglia, G. and Sberveglieri, G. (Eds). 2008. Solid state gas sensing. Vol. 20. Springer Science & Business Media.

[101] Korotcenkov, G. 2013. Handbook of gas sensor materials. Conventional approaches. 1.

[102] Guo, Z., Song, N.R., Moon, J.H., Kim, M., Jun, E.J., Choi, J., et al. 2012. A benzobisimidazolium-based fluorescent and colorimetric chemosensor for CO_2. Journal of the American Chemical Society. 134(43): 17846–17849.

[103] Ko, K., Lee, J.Y. and Chung, H. 2020. Highly efficient colorimetric CO_2 sensors for monitoring CO_2 leakage from carbon capture and storage sites. Science of The Total Environment. 729: 138786.

[104] Kim, D.H., Cha, J.H., Lim, J.Y., Bae, J., Lee, W., Yoon, K.R., et al. 2020. Colorimetric dye-loaded nanofiber yarn: eye-readable and weavable gas sensing platform. ACS Nano. 14(12): 16907–16918.

[105] Cho, S.H., Suh, J.M., Eom, T.H., Kim, T. and Jang, H.W. 2021. Colorimetric sensors for toxic and hazardous gas detection: A review. Electronic Materials Letters. 17: 1–17.

[106] Davidson, C.E., Dixon, M.M., Williams, B.R., Kilper, G.K., Lim, S.H., Martino, R.A., et al. 2020. Detection of chemical warfare agents by colorimetric sensor arrays. ACS Sensors. 5(4): 1102–1109.

[107] Milam-Guerrero, J., Yang, B., To, D.T. and Myung, N.V. 2022. Nitrous oxide is no laughing matter: A historical review of nitrous oxide gas-sensing capabilities highlighting the need for further exploration. ACS Sensors. 7(12): 3598–3610.

[108] Yi, H., Laurent, O., Schilt, S., Ramonet, M., Gao, X., Dong, L., et al. 2022. Simultaneous Monitoring of Atmospheric CH_4, N_2O, and H_2O Using a Single Gas Sensor Based on Mid-IR Quartz-Enhanced Photoacoustic Spectroscopy. Analytical Chemistry. 94(50): 17522–17532.

[109] Vahora, S., Chhalotiya, U.K., Kachhiya, H., Tandel, J. and Shah, D. 2021. Simultaneous quantification of brexpiprazole and sertraline HCl in synthetic mixture by thin-layer chromatography method. JPC–Journal of Planar Chromatography–Modern TLC. 34(6): 549–557.

[110] Zhao, H., Wang, Z., Li, Y. and Yang, M. 2022. Single-sided and integrated polyaniline/poly (vinylidene fluoride) flexible membrane with micro/nanostructures as breathable, nontoxic and fast response wearable humidity sensor. Journal of Colloid and Interface Science. 607: 367–377.

[111] Mousa, S.A., Noby, S.Z. and Shalan, A.E. 2022. Graphene and its nanocomposites derivatives: Synthesis, properties, and their applications in water treatment, gas sensor, and solar cell fields. In: Advances in Nanocomposite Materials for Environmental and Energy Harvesting Applications. 95–128. Cham: Springer International Publishing.

[112] He, T., Li, X., Wang, Q., Zhou, Y., Wang, X., Wang, Z., et al. 2022. Poly (vinylidene fluoride)(PVDF) membrane fabrication with an ionic liquid via non-solvent thermally induced phase separation (N-TIPs). Applied Water Science. 12(3): 42.

[113] Xu, Y., Aljasar, S.A. and Qasaimeh, M.R. 2022, March. One step formation of laser reduced graphene oxide fabrication and characterization on novel flexible PET based

sensitive sensor. *In*: 2022 4th International Youth Conference on Radio Electronics, Electrical and Power Engineering (REEPE). 1–4. IEEE.

[114] Tan, P.H., Hsu, C.H., Shen, Y.C., Wang, C.P., Liou, K.L., Shih, J.R., et al. 2022. Complementary Metal–Oxide–Semiconductor Compatible 2D Layered Film-Based Gas Sensors by Floating-Gate Coupling Effect. Advanced Functional Materials. 32(13): 2108878.

[115] Wang, J., Ren, Y., Liu, H., Li, Z., Liu, X., Deng, Y., et al. 2022. Ultrathin 2D $NbWO_6$ perovskite semiconductor based gas sensors with ultrahigh selectivity under low working temperature. Advanced Materials. 34(2): 2104958.

[116] Kwon, E.H., Kim, M., Lee, C.Y., Kim, M. and Park, Y.D. 2022. Metal–Organic-Framework-Decorated Carbon Nanofibers with Enhanced Gas Sensitivity When Incorporated into an Organic Semiconductor-Based Gas Sensor. ACS Applied Materials and Interfaces. 14(8): 10637–10647.

[117] Wei, S., Li, Z., John, A., Karawdeniya, B.I., Li, Z., Zhang, F., et al. 2022. Semiconductor Nanowire Arrays for High-Performance Miniaturized Chemical Sensing. Advanced Functional Materials. 32(5): 2107596.

[118] Kwon, B., Bae, H., Lee, H., Kim, S., Hwang, J., Lim, H., et al. 2022. Ultrasensitive N-channel graphene gas sensors by nondestructive molecular doping. ACS Nano. 16(2): 2176–2187.

[119] Zhang, Y., He, G., Wang, L., Wang, W., Xu, X. and Liu, W. 2022. Ultraviolet-assisted low-thermal-budget-driven α-InGaZnO thin films for high-performance transistors and logic circuits. ACS Nano. 16(3): 4961–4971.

[120] Shin, E.S., Go, J.Y., Ryu, G.S., Liu, A. and Noh, Y.Y. 2021. Highly reliable organic field-effect transistors with molecular additives for a high-performance printed gas sensor. ACS Applied Materials and Interfaces. 13(3): 4278–4283.

[121] Liu, C., Hu, J., Wu, G., Cao, J., Zhang, Z. and Zhang, Y. 2021. Carbon nanotube-based field-effect transistor-type sensor with a sensing gate for ppb-level formaldehyde detection. ACS Applied Materials and Interfaces. 13(47): 56309–56319.

[122] Hong, S., Wu, M., Hong, Y., Jeong, Y., Jung, G., Shin, W., et al. 2021. FET-type gas sensors: A review. Sensors and Actuators B: Chemical. 330: 129240.

[123] Lu, H., He, B. and Gao, B. 2021. Emerging electrochemical sensors for life healthcare. Engineered Regeneration. 2, 175–181.

[124] Gil, B., Ip, H., Kassanos, P., Lo, B., Yang, G.Z. and Anastasova, S. 2022. Smart implanted access port catheter for therapy intervention with pH and lactate biosensors. Materials Today Bio. 15: 100298.

[125] Jiang, G., Cumberland, T., Fu, X., Li, M., Zhang, J., Xu, P., et al. 2020. Highly Stable Low-Cost Electrochemical Gas Sensor with an Alcohol-Tolerant N, S-Codoped Non-Precious Metal Catalyst Air Cathode. ACS Sensors. 6(3): 752–763.

[126] Gorbova, E., Tzorbatzoglou, F., Molochas, C., Chloros, D., Demin, A. and Tsiakaras, P. 2021. Fundamentals and principles of solid-state electrochemical sensors for high temperature gas detection. Catalysts. 12(1): 1.

[127] Bag, A. and Lee, N.E. 2021. Recent advancements in development of wearable gas sensors. Advanced Materials Technologies. 6(3): 2000883.

[128] Domènech-Gil, G., Samà, J., Fàbrega, C., Gràcia, I., Cané, C., Barth, S. and Romano-Rodríguez, A. 2023. Highly sensitive SnO_2 nanowire network gas sensors. Sensors and Actuators B: Chemical. 383: 133545.

[129] Freddi, S., Vergari, M., Pagliara, S. and Sangaletti, L. 2023. A Chemiresistor sensor array based on graphene nanostructures: From the detection of ammonia and possible interfering VOCs to chemometric analysis. Sensors. 23(2): 882.

[130] Chang, A., Uy, C., Xiao, X. and Chen, J. 2022. Self-powered environmental monitoring via a triboelectric nanogenerator. Nano Energy. 98: 107282.

[131] Huo, Z., Wei, Y., Wang, Y., Wang, Z.L. and Sun, Q. 2022. Integrated Self-Powered Sensors Based on 2D Material Devices. Advanced Functional Materials. 32(41): 2206900.

[132] Yang, Y., Guo, X., Zhu, M., Sun, Z., Zhang, Z., He, T., et al. 2023. Triboelectric nanogenerator enabled wearable sensors and electronics for sustainable internet of things integrated green earth. Advanced Energy Materials. 13(1): 2203040.

[133] Liu, B., Wang, S., Yuan, Z., Duan, Z., Zhao, Q., Zhang, Y., et al. 2020. Novel chitosan/ZnO bilayer film with enhanced humidity-tolerant property: Endowing triboelectric nanogenerator with acetone analysis capability. Nano Energy. 78: 105256.

[134] Chang, C.Y., Cheng, Y.H. and Ho, C.Y. 2022. Surface engineering of a triboelectric nanogenerator for room temperature high-performance self-powered formaldehyde sensors. Journal of Materials Chemistry A. 10(42): 22373–22389.

[135] Gao, B., Liu, L., Liu, J. and Yang, F. 2013. A photo-catalysis and rotating nano-$CaCO_3$ dynamic membrane system with Fe-$ZnIn_2S_4$ efficiently removes halogenated compounds in water. Applied Catalysis B: Environmental. 138: 62–69.

[136] Hosseini, M., Haghighi, M., Kahforoushan, D. and Zarrabi, M. 2017. Sono-dispersion of ceria and palladium in preparation and characterization of Pd/Al_2O_3-clinoptilolite-CeO_2 nanocatalyst for treatment of polluted air via low temperature VOC oxidation. Process Safety and Environmental Protection. 106: 284–293.

[137] Kempasiddaiah, M., Kandathil, V., Dateer, R.B., Baidya, M., Patil, S.A. and Patil, S.A. 2021. Efficient and recyclable palladium enriched magnetic nanocatalyst for reduction of toxic environmental pollutants. Journal of Environmental Sciences. 101: 189–204.

[138] Qu, M., Cheng, Z., Sun, Z., Chen, D., Yu, J. and Chen, J. 2021. Non-thermal plasma coupled with catalysis for VOCs abatement: A review. Process Safety and Environmental Protection. 153: 139–158.

[139] Beni, A.A. and Jabbari, H. 2022. Nanomaterials for environmental applications. Results in Engineering. 15: 100467.

[140] Goswami, A.D., Trivedi, D.H., Jadhav, N.L. and Pinjari, D.V. 2021. Sustainable and green synthesis of carbon nanomaterials: A review. Journal of Environmental Chemical Engineering. 9(5): 106118.

[141] Mishra, M., Sharma, M., Dubey, R., Kumari, P., Ranjan, V. and Pandey, J. 2021. Green synthesis interventions of pharmaceutical industries for sustainable development. Current Research in Green and Sustainable Chemistry. 4: 100174.

[142] Ahmed, S.F., Mofijur, M., Rafa, N., Chowdhury, A.T., Chowdhury, S., Nahrin, M., et al. 2022. Green approaches in synthesising nanomaterials for environmental nano-bioremediation: Technological advancements, applications, benefits and challenges. Environmental Research. 204: 111967.

[143] Wang, H., Sun, J., Wang, J., Jiang, L. and Liu, H. 2021. Green synthesis of nitrogen and fluorine co-doped porous carbons from sustainable coconut shells as an advanced synergistic electrocatalyst for oxygen reduction. Journal of Materials Research and Technology. 13: 962–970.

[144] Phan, S. and Luscombe, C.K. 2019. Recent advances in the green, sustainable synthesis of semiconducting polymers. Trends in Chemistry. 1(7): 670–681.

[145] Sedghi, S.M., Mortazavi, Y. and Khodadadi, A. 2010. Low temperature CO and CH_4 dual selective gas sensor using SnO_2 quantum dots prepared by sonochemical method. Sensors and Actuators B: Chemical. 145(1): 7–12.

[146] Hjiri, M., Bahanan, F., Aida, M.S., El Mir, L. and Neri, G. 2020. High performance CO gas sensor based on ZnO nanoparticles. Journal of Inorganic and Organometallic Polymers and Materials. 30: 4063–4071.

[147] Álvarez-Ramos, M.E., Necochea-Chamorro, J.I., Carrillo-Torres, R.C. and Sánchez-Zeferino, R. 2020. Room temperature CO_2 sensing using Au-decorated ZnO nanorods deposited on an optical fiber. Materials Science and Engineering: B. 262: 114720.

[148] Michel, C.R., Martínez-Preciado, A.H. and Rivera-Tello, C.D. 2015. CO_2 gas sensing response of YPO_4 nanobelts produced by a colloidal method. Sensors and Actuators B: Chemical. 221: 499–506.

[149] Korotcenkov, G., Cho, B.K., Gulina, L.B. and Tolstoy, V.P. 2012. Gas sensor application of Ag nanoclusters synthesized by SILD method. Sensors and Actuators B: Chemical. 166: 402–410.

[150] von Bültzingslöwen, C., McEvoy, A.K., McDonagh, C., MacCraith, B.D., Klimant, I., Krause, C., et al. 2002. Sol–gel based optical carbon dioxide sensor employing dual luminophore referencing for application in food packaging technology. Analyst. 127(11): 1478–1483.

[151] Willa, C., Schmid, A., Briand, D., Yuan, J. and Koziej, D. 2017. Lightweight, room-temperature CO_2 Gas sensor based on rare-earth metal-free composites—An impedance study. ACS Applied Materials and Interfaces. 9(30): 25553–25558.

[152] Choi, J.K., Hwang, I.S., Kim, S.J., Park, J.S., Park, S.S., Jeong, U., et al. 2010. Design of selective gas sensors using electrospun Pd-doped SnO_2 hollow nanofibers. Sensors and Actuators B: Chemical. 150(1): 191–199.

[153] Tanvir, N.B., Yurchenko, O., Wilbertz, C. and Urban, G. 2016. Investigation of CO_2 reaction with copper oxide nanoparticles for room temperature gas sensing. Journal of Materials Chemistry A. 4(14): 5294–5302.

[154] Kole, K., Halder, A., Singh, S., Samanta, A., Das, S., Kundu, A.K., et al. 2020. Chromogenic-Functionalized Silica Nanoflower Composites for the Detection of Carbon Dioxide. ACS Applied Nano Materials. 3(5): 4321–4328.

[155] Wang, C.T., Chen, H.Y. and Chen, Y.C. 2013. Gold/vanadium–tin oxide nanocomposites prepared by co-precipitation method for carbon monoxide gas sensors. Sensors and Actuators B: Chemical. 176: 945–951.

[156] Park, J.Y. and Lee, I.H. 2014. Photocatalytic degradation of 2-chlorophenol using Ag-doped TiO_2 nanofibers and a near-UV light-emitting diode system. Journal of Nanomaterials. 2014: 1–1.

[157] Bao, Q., Zhang, D. and Qi, P. 2011. Synthesis and characterization of silver nanoparticle and graphene oxide nanosheet composites as a bactericidal agent for water disinfection. Journal of Colloid and Interface Science. 360(2): 463–470.

[158] Sreeja, S. and Shetty, V. 2016. Microbial disinfection of water with endotoxin degradation by photocatalysis using Ag@ TiO_2 core shell nanoparticles. Environmental Science and Pollution Research International. 23(18): 18154.

[159] Nomura, A. and Jones, C.W. 2013. Amine-functionalized porous silicas as adsorbents for aldehyde abatement. ACS Applied Materials and Interfaces. 5(12): 5569–5577.

[160] Bollini, P., Didas, S.A. and Jones, C.W. 2011. Amine-oxide hybrid materials for acid gas separations. Journal of Materials Chemistry. 21(39): 15100–15120.

[161] Hicks, J.C., Drese, J.H., Fauth, D.J., Gray, M.L., Qi, G. and Jones, C.W. 2008. Designing adsorbents for CO_2 capture from flue gas—hyperbranched aminosilicas capable of capturing CO_2 reversibly. Journal of the American Chemical Society. 130(10): 2902–2903.

[162] Choi, S., Drese, J.H., Eisenberger, P.M. and Jones, C.W. 2011. Application of amine-tethered solid sorbents for direct CO_2 capture from the ambient air. Environmental Science and Technology. 45(6): 2420–2427.

[163] Hicks, J.C., Drese, J.H., Fauth, D.J., Gray, M.L., Qi, G. and Jones, C.W. 2008. Designing adsorbents for CO_2 capture from flue gas-hyperbranched aminosilicas capable of capturing CO_2 reversibly. Journal of the American Chemical Society. 130(10): 2902–2903.

[164] Nomura, A. and Jones, C.W. 2014. Enhanced formaldehyde-vapor adsorption capacity of polymeric amine-incorporated aminosilicas. Chemistry–A European Journal. 20(21): 6381–6390.

[165] Qi, G., Wang, Y., Estevez, L., Duan, X., Anako, N., Park, A.H.A., et al. 2011. High efficiency nanocomposite sorbents for CO_2 capture based on amine-functionalized mesoporous capsules. Energy and Environmental Science. 4(2): 444–452.

[166] Kuwahara, Y., Kang, D.Y., Copeland, J.R., Brunelli, N.A., Didas, S.A., Bollini, P., et al. 2012. Dramatic enhancement of CO_2 uptake by poly (ethyleneimine) using zirconosilicate supports. Journal of the American Chemical Society. 134(26): 10757–10760.

[167] Kuwahara, Y., Kang, D.Y., Copeland, J.R., Bollini, P., Sievers, C., Kamegawa, T., et al. 2012. Enhanced CO_2 adsorption over polymeric amines supported on heteroatom-incorporated SBA-15 silica: impact of heteroatom type and loading on sorbent structure and adsorption performance. Chemistry–A European Journal. 18(52): 16649–16664.

[168] Zhang, Y., Tang, Z.R., Fu, X. and Xu, Y.J. 2010. TiO_2–graphene nanocomposites for gas-phase photocatalytic degradation of volatile aromatic pollutant: is TiO_2–graphene truly different from other TiO_2–carbon composite materials? ACS Nano. 4(12): 7303–7314.

[169] Yang, M.Q., Zhang, N. and Xu, Y.J. 2013. Synthesis of fullerene–, carbon nanotube–, and graphene–TiO_2 nanocomposite photocatalysts for selective oxidation: A comparative study. ACS Applied Materials and Interfaces. 5(3). 1156–1164.

[170] Huang, H.Y., Yang, R.T., Chinn, D. and Munson, C.L. 2003. Amine-grafted MCM-48 and silica xerogel as superior sorbents for acidic gas removal from natural gas. Industrial and Engineering Chemistry Research. 42(12): 2427–2433.

[171] Chou, K.S., Lu, Y.C. and Lee, H.H. 2005. Effect of alkaline ion on the mechanism and kinetics of chemical reduction of silver. Materials Chemistry and Physics. 94(2–3): 429–433.

[172] Gupta, A. and Silver, S. 1998. Molecular genetics: silver as a biocide: will resistance become a problem? Nature Biotechnology. 16(10): 888–888.

[173] Bosetti, M., Massè, A., Tobin, E. and Cannas, M. 2002. Silver coated materials for external fixation devices: in vitro biocompatibility and genotoxicity. Biomaterials. 23(3): 887–892.

[174] Xiu, Z.M., Zhang, Q.B., Puppala, H.L., Colvin, V.L. and Alvarez, P.J. 2012. Negligible particle-specific antibacterial activity of silver nanoparticles. Nano Letters. 12(8): 4271–4275.

[175] Pal, S., Tak, Y.K. and Song, J.M. 2007. Does the antibacterial activity of silver nanoparticles depend on the shape of the nanoparticle? A study of the gram-negative bacterium *Escherichia coli*. Applied and Environmental Microbiology. 73(6): 1712–1720.

[176] Guo, M., Weng, X., Wang, T. and Chen, Z. 2017. Biosynthesized iron-based nanoparticles used as a heterogeneous catalyst for the removal of 2, 4-dichlorophenol. Separation and Purification Technology. 175: 222–228.

[177] Han, Y. and Yan, W. 2016. Reductive dechlorination of trichloroethene by zero-valent iron nanoparticles: reactivity enhancement through sulfidation treatment. Environmental Science and Technology. 50(23): 12992–13001.

[178] Cho, M., Chung, H., Choi, W. and Yoon, J. 2005. Different inactivation behaviors of MS-2 phage and Escherichia coli in TiO_2 photocatalytic disinfection. Applied and Environmental Microbiology. 71(1): 270–275.

[179] Alizadeh Fard, M., Aminzadeh, B. and Vahidi, H. 2013. Degradation of petroleum aromatic hydrocarbons using TiO_2 nanopowder film. Environmental Technology. 34(9): 1183–1190.

[180] Gu, J., Dong, D., Kong, L., Zheng, Y. and Li, X. 2012. Photocatalytic degradation of phenanthrene on soil surfaces in the presence of nanometer anatase TiO_2 under UV-light. Journal of Environmental Sciences. 24(12): 2122–2126.

[181] Kouketsu, T., Duan, S., Kai, T., Kazama, S. and Yamada, K. 2007. PAMAM dendrimer composite membrane for CO_2 separation: Formation of a chitosan gutter layer. Journal of Membrane Science. 287(1): 51–59.

[182] Guerra, F.D., Campbell, M.L., Whitehead, D.C. and Alexis, F. 2017. Tunable properties of functional nanoparticles for efficient capture of VOCs. ChemistrySelect. 2(31): 9889–9894.

[183] Campbell, M.L., Guerra, F.D., Dhulekar, J., Alexis, F. and Whitehead, D.C. 2015. Target-specific capture of environmentally relevant gaseous aldehydes and carboxylic acids with functional nanoparticles. Chemistry–A European Journal. 21(42): 14834–14842.

[184] Tungittiplakorn, W., Lion, L.W., Cohen, C. and Kim, J.Y. 2004. Engineered polymeric nanoparticles for soil remediation. Environmental Science and Technology. 38(5): 1605–1610.

[185] Diallo, M.S., Falconer, K., Johnson, J.H. and Goddard III, W.A. 2007. Dendritic anion hosts: Perchlorate uptake by G5-NH_2 poly (propyleneimine) dendrimer in water and model electrolyte solutions. Environmental Science and Technology. 41(18): 6521–6527.

[186] Mittal, H., Maity, A. and Ray, S.S. 2015. Synthesis of co-polymer-grafted gum karaya and silica hybrid organic–inorganic hydrogel nanocomposite for the highly effective removal of methylene blue. Chemical Engineering Journal. 279: 166–179.

[187] Sun, X.F., Liu, B., Jing, Z. and Wang, H. 2015. Preparation and adsorption property of xylan/poly (acrylic acid) magnetic nanocomposite hydrogel adsorbent. Carbohydrate Polymers. 118: 16–23.

[188] Guerra, F.D., Campbell, M.L., Attia, M.F., Whitehead, D.C. and Alexis, F. 2018. Capture of aldehyde VOCs using a series of amine-functionalized cellulose nanocrystals. Chemistry Select. 3(20): 5495–5501.

[189] Rezk, M.Y., Sharma, J. and Gartia, M.R. 2020. Nanomaterial-based CO_2 sensors. Nanomaterials. 10(11): 2251.

[190] Klar, T., Perner, M., Grosse, S., Von Plessen, G., Spirkl, W. and Feldmann, J. 1998. Surface-plasmon resonances in single metallic nanoparticles. Physical Review Letters. 80(19): 4249.

[191] Chan, G.H., Zhao, J., Schatz, G.C. and Van Duyne, R.P. 2008. Localized surface plasmon resonance spectroscopy of triangular aluminum nanoparticles. The Journal of Physical Chemistry C. 112(36): 13958–13963.

[192] Leong, H.S., Guo, J., Lindquist, R.G. and Liu, Q.H. 2009. Surface plasmon resonance in nanostructured metal films under the Kretschmann configuration. Journal of Applied Physics. 106(12).

[193] Jiang, N., Zhuo, X. and Wang, J. 2017. Active plasmonics: Principles, structures, and applications. Chemical Reviews. 118(6): 3054–3099.

[194] Homola, J. and Piliarik, M. 2006. Surface plasmon Resonance (SPR) Sensors. 45–67. Springer Berlin Heidelberg.

[195] Olaru, A., Bala, C., Jaffrezic-Renault, N. and Aboul-Enein, H.Y. 2015. Surface plasmon resonance (SPR) biosensors in pharmaceutical analysis. Critical Reviews in Analytical Chemistry. 45(2): 97–105.

[196] Hooshmand, S., Kassanos, P., Keshavarz, M., Duru, P., Kayalan, C.I., Kale, İ., et al. 2023. Wearable nano-based gas sensors for environmental monitoring and encountered challenges in optimization. Sensors. 23(20): 8648.

[197] Drese, J.H., Talley, A.D. and Jones, C.W. 2011. Amine-tethered solid adsorbents coupling high adsorption capacity and regenerability for CO_2 capture from ambient air. ChemSusChem. 4(3): 379–385.

Chapter 4

Carbon Storage Using Nanomaterials

Harwinder Singh*[1] and Arunima Verma[2]
[1]Sant Longowal Institute of Engineering and Technology, Longowal-148106, Punjab, India
[2]Department of Nanoscience and Materials, Central University of Jammu, Jammu-181143, Jammu and Kashmir India

4.1 INTRODUCTION

4.1.1 Overview of Carbon Storage Challenges

Carbon storage, or carbon sequestration, is a key tool the world will need to avoid total climate breakdown. Thus, the process of evolution means, storing these greenhouse gases i.e. carbon dioxide (CO_2) being captured from the atmosphere or reversely taken directly cumulatively captured between industry and other sources to decrease their quantity prevailing on the earth's surface. However, carbon storage has great potential but also faces multiple difficulties ranging from technological development to economic reasonability and necessary efficiency over decades [1]. Chief among these: creating technologies that are scalable and affordable. The currently proposed strategies, such as geological storage in which carbon dioxide is injected into rock formations deep underground are still maturing. However, sequestration may lead to problems like CO_2 leakage as well and is labour-intensive for surveillance purposes making it difficult to be a generalised approach. Advanced materials and techniques that improve the capture efficiency and lower costs are urgently needed [2]. Economic feasibility

*For Correspondence: harwinder4470@gmail.com

DOI: 10.1201/9781003492184-4

presents another significant barrier to the widespread implementation of carbon storage. High initial costs, combined with uncertain long-term returns, discourage investments. Additionally, the lack of fully developed regulatory frameworks creates ambiguities in the implementation and enforcement of carbon storage initiatives. The absence of clear policies and incentives further complicates the integration of carbon storage into existing industrial practices [3]. Beyond the technical and economic hurdles, carbon storage also raises environmental and social concerns. Environmental risks, such as groundwater contamination and induced seismicity, necessitate thorough assessment and management. Moreover, public perception and acceptance of carbon storage projects can vary, with communities often voicing concerns about safety and long-term impacts. Effective communication and stakeholder engagement are essential to addressing these challenges [4]. Ensuring the long term viability of stored carbon is another critical issue. Once CO_2 is sequestered, it must remain securely stored for centuries or even millennia to effectively contribute to climate change mitigation. However, predicting the behaviour of stored CO_2 over such extended periods is fraught with uncertainties, posing a significant challenge. Developing reliable monitoring systems that can detect leaks or other issues early is vital to maintaining the integrity of storage sites [5]. In conclusion, while carbon storage holds promise as a solution for reducing atmospheric CO_2 levels, a multitude of challenges must be overcome to make it a viable large scale option. Addressing technological limitations, economic and regulatory barriers, environmental risks, and ensuring long term storage security are crucial to realising the full potential of carbon storage in combatting climate change. Future research, development, and supportive policies will be instrumental in overcoming these challenges [6].

4.1.2 Importance of Nanomaterials in Carbon Storage

Nanomaterials play a pivotal role in advancing carbon storage technologies due to their exceptional surface area, porosity, and tunable chemical properties. These unique attributes enhance the adsorption and sequestration of carbon dioxide (CO_2), making them indispensable in efforts to mitigate climate change [7]. The high surface area of nanomaterials allows for greater CO_2 capture efficiency, while their adjustable pore structures enable selective adsorption, optimising storage capacity [8]. Furthermore, the ability to functionalise nanomaterials with various chemical groups offers tailored solutions for CO_2 binding and long term stability. Consequently, the integration of nanomaterials into carbon capture and storage systems represents a transformative approach to reducing atmospheric CO_2 levels, thus contributing significantly to global carbon management strategies [9]. Utilising nanomaterials for carbon storage encompasses a broad range of applications aimed at mitigating climate change by capturing and storing carbon dioxide (CO_2) effectively. Nanomaterials, with their enhanced surface area, tunable porosity, and unique chemical properties, offer significant potential to improve the efficiency of carbon capture processes. This includes their application in various industrial processes, power generation, and transportation sectors where CO_2 emissions are prevalent [10, 11]. The development and optimisation of nanomaterials for carbon storage also extend to their integration into existing Carbon Capture and Sequestration (CCS) systems, potentially revolutionizing current approaches by making them more cost effective and scalable. This chapter focuses on the use of nanomaterials for carbon storage in industrial settings, including the development of advanced materials for carbon storage in carbon as following:

1. **Enhancing Carbon Capture Efficiency:** The primary focus is to harness the superior adsorption capacities of nanomaterials to significantly enhance the efficiency of CO_2 capture. This involves designing nanostructures that maximise surface interactions with CO_2 molecules, thereby increasing the amount of CO_2 that can be stored per unit mass of the material [11].
2. **Developing Sustainable and Scalable Solutions:** Another critical focus is to create sustainable carbon storage solutions that can be easily scaled to meet the demands of global CO_2 emissions. Nanomaterials, due to their versatile synthesis methods and abundant raw materials, present an opportunity to develop cost effective and scalable carbon storage technologies [12].
3. **Improving the Durability and Stability of Storage Materials:** The long term stability and durability of nanomaterials in storing CO_2 are essential to ensure that captured carbon remains sequestered without leakage over extended periods. Research focuses on enhancing the chemical and thermal stability of these materials under various environmental conditions to ensure reliable long term storage.
4. **Integrating with Renewable Energy Technologies:** The integration of nanomaterial-based carbon storage systems with renewable energy technologies, such as solar and wind, aims to create synergistic solutions that reduce overall carbon footprints. This includes using renewable energy to power the synthesis and regeneration of nanomaterials, thus making the entire process more environmentally friendly [13].
5. **Exploring Functionalised Nanomaterials for Selective CO_2 Capture:** Functionalising nanomaterials with specific chemical groups or coatings can enhance their selectivity for CO_2 over other gases. The focus is to design materials that preferentially capture CO_2 from mixed gas streams, such as flue gases, thus improving the purity of captured CO_2 and reducing the energy required for its separation [13, 14].

4.2 TYPES OF NANOMATERIALS FOR CARBON STORAGE

4.2.1 Carbon-Based Nanomaterials

Carbon-based nanomaterials, such as carbon nanotubes (CNTs) and carbon nanofibres (CNFs), have garnered significant attention in scientific research due to their exceptional mechanical, electrical, and thermal properties. These materials, composed entirely of carbon atoms arranged in various configurations, hold immense potential for applications in fields ranging from electronics and energy storage to biomedicine and materials science [15].

- **Carbon Nanotubes (CNTs):** Carbon nanotubes are cylindrical nanostructures composed of graphene sheets rolled into a seamless tube [15]. They can be classified into Single-Walled (SWCNTs) and Multi-Walled carbon nanotubes (MWCNTs), depending on the number of concentric graphene layers. CNTs exhibit remarkable properties, such as high tensile strength (up to 63 GPa), exceptional electrical conductivity, and thermal conductivity (as high as 3500 W/m·K). These properties arise from the strong sp^2 carbon bonds and the unique quantum confinement effects in the nanotube's one-dimensional structure [16].
- **Carbon Nanofibres (CNFs):** Carbon nanofibres are cylindrical structures with diameters ranging from tens to hundreds of nanometres and lengths extending to several micrometres [17]. CNFs can be produced through various methods, including electrospinning and Chemical Vapour Deposition (CVD). Like CNTs, CNFs exhibit excellent mechanical strength, electrical conductivity, and thermal stability. However, CNFs differ from CNTs in their structural morphology, as they can have a more disordered or graphitic structure.

4.2.2 Metal-Organic Frameworks (MOFs)

Metal-Organic Frameworks (MOFs) represent a class of porous materials that have garnered immense attention due to their remarkable structure and versatile applications. Comprising metal ions or clusters coordinated to organic ligands, MOFs are known for their exceptional surface areas, tunable pore sizes, and high porosity. These unique characteristics make MOFs particularly attractive for a number of applications, most notably in the field of carbon capture [18].

4.2.2.1 Structural Characteristics of MOFs

The architecture of MOFs is built upon a framework of metal ions (often referred to as Secondary Building Units or SBUs) and organic linkers as shown in Figure 4.1. These components assemble periodically to form a crystalline, porous network. The metal ions act as nodes, while organic molecules, typically carboxylates, phosphonates, or azolates, serve as linkers that connect these nodes [18]. This modular assembly allows for an extensive variety of MOFs, each with distinct structural attributes. The structural flexibility of MOFs is one of their defining features. By altering the metal ions or organic linkers, it is possible to design MOFs with specific pore sizes, shapes, and functionalities. This adaptability is critical for tailoring MOFs to target specific gases, such as CO_2, in carbon capture processes [19, 20].

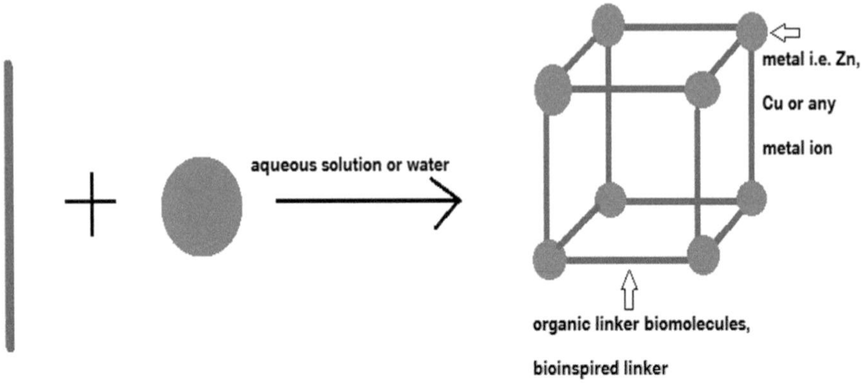

Figure 4.1 Metal-organic framework [21].

4.2.2.2 Properties of MOFs Relevant to Carbon Capture

1. **High Surface Area:** MOFs possess some of the highest surface areas of any known materials, often exceeding 7,000 m^2/g. This expansive surface area provides ample space for gas molecules to adsorb.
2. **Tunable Pore Size:** The ability to customise pore dimensions allows for selective adsorption of CO_2 over other gases, enhancing the efficiency of carbon capture.
3. **Chemical Functionality:** MOFs can be functionalised with various chemical groups to enhance their affinity for CO_2 molecules, further improving selectivity and capacity.
4. **Thermal and Chemical Stability:** MOFs designed for carbon capture are often engineered to withstand the harsh conditions of industrial processes, maintaining their integrity and performance over extended periods [18, 19].

4.2.2.3 Applications of MOFs in Carbon Capture

The unique properties of MOFs make them highly effective in carbon capture applications. These materials can be employed in various stages of the carbon capture process, including:

1. **Post Combustion Capture:** MOFs can be integrated into systems that capture CO_2 from flue gases of power plants. Due to their high selectivity and adsorption capacity, MOFs can effectively isolate CO_2 from other gases, allowing for its subsequent storage or utilisation.
2. **Pre Combustion Capture:** In processes where fossil fuels are converted to hydrogen, MOFs can be used to capture CO_2 before combustion occurs, thus reducing overall emissions.
3. **Direct Air Capture:** MOFs are also being explored for Direct Air Capture (DAC), a process that involves extracting CO_2 directly from the

atmosphere. The tunable nature of MOFs allows them to be optimised for this application, potentially offering a scalable solution to reducing atmospheric CO_2 levels [20].

4.3 MECHANISMS OF CARBON CAPTURE USING NANOMATERIALS

Carbon capture using nanomaterials involves several intricate mechanisms, with adsorption, surface chemistry, and functionalisation playing pivotal roles. Adsorption mechanics is the primary process where nanomaterials such as Metal-Organic Frameworks (MOFs), carbon nanotubes (CNTs), and graphene-based materials physically or chemically trap CO_2 molecules on their surfaces. These materials offer high surface area-to-volume ratios, enhancing their adsorption capacities [21, 22]. Surface chemistry significantly influences the efficiency of CO_2 capture, as it dictates how CO_2 molecules interact with the nanomaterial surface. The surface can be tailored to increase selectivity for CO_2 over other gases like nitrogen, by introducing specific functional groups [23]. Functionalisation nonmaterials involves modifying their surface properties by attaching chemical groups that can react with CO_2, such as amine groups, which form stable carbamate bonds with CO_2. This functionalisation improves the interaction between CO_2 and the nanomaterial, enhancing both the adsorption capacity and selectivity [24]. Additionally, the interaction of nanomaterials with other carbon compounds, such as methane, can be modulated through similar surface modifications, allowing for a broad range of applications in Carbon Capture and Storage (CCS) technologies. These mechanisms collectively make nanomaterials promising candidates for efficient and selective carbon capture solutions.

4.4 NANOMATERIALS SYNTHESIS AND FUNCTIONALISATION TECHNIQUES

1. **Chemical Vapour Deposition (CVD):** Carbon Vapour Deposition is a widely used technique for creating thin films, coatings, and nanostructures on various substrates. The process involves the chemical reaction of gaseous precursors at elevated temperatures to produce a solid material that adheres to the surface of a substrate. In CVD, a substrate is exposed to one or more volatile precursors, which react or decompose on the substrate surface, forming a thin film or coating. This method is popular in the semiconductor industry for producing high purity, high performance materials like silicon dioxide or silicon nitride. CVD is also utilised in the production of coatings for tools and in the creation of advanced materials for various technological applications [25].
2. **Sol-Gel Process:** The sol-gel process is a versatile chemical technique used to create materials, particularly ceramics and glasses, through a

series of transformations from a liquid sol to a solid gel. The process begins with the preparation of a sol, which is a colloidal suspension of nanoparticles in a liquid, often composed of metal alkoxides or metal chlorides. This sol is then processed to form a gel through hydrolysis and condensation reactions. During hydrolysis, the metal alkoxides react with water to form hydroxyl groups, which then condense to form a network structure. The gel is then dried and typically undergoes further heat treatment or sintering to improve its mechanical properties and achieve the desired final material structure [26]. The sol-gel process is favoured for its ability to produce highly pure, homogeneous, and well-controlled materials with a wide range of compositions and structures, making it valuable in fields like materials science, optics, and catalysis.

3. **Electrospinning:** Electrospinning is a widely used technique for producing fine fibres with diameters ranging from nanometres to micrometres. This process involves applying a high voltage to a polymer solution or melt, which creates an electric field that forces the polymer to eject from a charged needle or spinneret. As the polymer jet travels toward a grounded collector, it undergoes stretching and thinning due to the electrostatic forces, resulting in the formation of ultra fine fibres. These fibres are collected on a substrate to form a nonwoven mat or other desired structures [27]. Electrospinning is prized for its versatility and ability to produce fibres with high surface area-to-volume ratios, making it valuable in various applications, including tissue engineering, filtration, and drug delivery.
4. **Functionalisation Methods for Enhanced Storage** focus on modifying materials to improve their performance in storing energy or other substances. One effective approach involves surface modification, where functional groups are introduced to increase surface area and enhance interactions with the stored material [26]. For instance, the introduction of hydrophobic or hydrophilic groups can improve the interaction between storage materials and the substances they hold, leading to better efficiency and stability. Another method is the incorporation of nanoparticles or nanostructures, which can increase the surface area and provide more active sites for storage, resulting in improved capacity and performance [27]. Additionally, chemical functionalisation, such as grafting specific molecules into storage materials, can tailor their properties for specific applications, such as increasing electrical conductivity or chemical stability. These methods have been extensively studied and applied to various storage systems, including batteries, supercapacitors, and gas storage materials.

4.5 CHALLENGES AND LIMITATIONS

1. **Scalability of Nanomaterial Synthesis:** The scalability of nanomaterial synthesis is a critical consideration for their application in various

fields, including electronics, medicine, and energy [28]. The scalability of these processes refers to the ability to produce nanomaterials in larger quantities while maintaining consistency and quality. Traditional synthesis methods, such as chemical vapour deposition (CVD) and sol-gel processes, often face challenges when scaled up due to issues related to uniformity, cost, and reproducibility [29]. Advances in scalable synthesis techniques, such as high-throughput synthesis, continuous flow reactors, and improved surface modification methods, are addressing these challenges. For instance, high-throughput synthesis allows for the rapid production and screening of nanomaterials, while continuous flow reactors offer better control over reaction conditions and scale-up feasibility. Additionally, new methods like microfluidics are enabling precise control over nanoparticle size and distribution, making them suitable for large scale production [30].

2. **Environmental Impact of Nanomaterial Release:** The use of nanomaterials in carbon storage, while promising, raises concerns about their potential release into the environment. Nanoparticles can enter soil, water, and air, potentially causing unintended ecological effects. These materials are often engineered to be highly reactive or possess unique properties that could disrupt local ecosystems if not properly contained or managed [31]. For instance, the release of carbon nanotubes or other nanomaterials could have adverse effects on soil health, water quality, and the health of plant and animal life.
3. **Health Risks to Humans:** Nanomaterials used in carbon storage may pose health risks if they become airborne or enter the water supply. Inhalation or ingestion of nanoparticles can lead to respiratory issues, cardiovascular problems, and other health effects. These particles' small size and high surface area increase their potential for toxicity [32]. For example, carbon nanotubes have been shown to cause lung inflammation and damage in animal models, raising concerns about their long term health effects on humans.

4.6 FUTURE PERSPECTIVES AND DEVELOPMENT

High CO_2 adsorption capability is achieved through the combination of a microporous structure and a large surface area, as demonstrated in previous studies. Activated carbon has garnered significant attention from researchers as a promising alternative adsorbent for CO_2 capture on an industrial scale due to its low disposal challenges, cost effectiveness in recycling, and ease of regeneration. However, I have not found any notable research references detailing the specific industries that utilise activated porous carbons. Scientists and researchers in the field of CO_2 adsorption, whether in academia or industry, should urgently apply these materials in large-scale applications and evaluate their CO_2 capture efficiency. Several factors need to be considered when employing carbon nanomaterials in large scale industrial applications:

1. Developing cost effective sorbents would promote a more equitable approach.
2. The adsorbent should exhibit long term stability in flue gases.
3. It must be largely recyclable with low energy requirements for regeneration.
4. Understanding the regeneration mechanism is crucial.
5. It is essential to investigate the influence of additional factors, including various gas mixtures, environmental conditions, external ions, and water vapour.

The use of Direct Air Capture (DAC) technology for CO_2 removal is currently a highly debated topic in the context of climate change mitigation. Combining DAC technology with low cost, reusable carbonaceous porous adsorbents presents an excellent opportunity to create economically viable DAC systems. With its CO_2 adsorption potential in low concentration environments, this approach could significantly contribute to climate change mitigation efforts.

4.7 CONCLUSION

Innovations in science and technology are continuously unfolding, driving progress that benefits both humanity and the planet. With its capacity to manipulate the properties of bulk materials at the microscopic level, nanotechnology has captivated the interest of researchers across various disciplines. Nanomaterials have found valuable applications in numerous fields, and 0D carbon nanomaterials are no exception. These materials are being utilised in biomedicine, sensors, environmental monitoring, robotics, agriculture, and beyond. Due to their exceptionally small size, 0D carbon nanomaterials possess unique and adjustable properties, opening the door to a multitude of potential applications. Their tunable microporosities, hierarchical porosities, expansive surface area, and other distinctive features make these carbon nanomaterials highly effective agents for CO_2 capture. Enhancing CO_2 capture efficiency can be achieved by modifying and functionalising the surface of carbon nanomaterials. For instance, carbon nanoparticles have undergone various treatments such as imine and amine functionalisation, acid treatment, the formation of composites or hybrids, chemical or physical activation, metal oxide impregnation, and mono- or co-doping with different atoms. In this chapter, we explored the challenges of reducing carbon emissions, the significance of nanomaterials in achieving this goal, and the practical and theoretical approaches to utilising them for zero-carbon emission capture and storage.

REFERENCES

[1] Wilcox, J., Psarras, P. and Liguori, S. 2020. Carbon sequestration: A critical strategy for climate mitigation. Annual Review of Environment and Resources. 45: 231–259. https://doi.org/10.1146/annurev-environ-012420-050113.

[2] Advances and challenges in carbon capture and storage technologies. Journal of Environmental Science and Technology. 55(3): 1234–1250.

[3] Economic and Regulatory Frameworks for Carbon Capture and Storage. Environmental Policy and Governance. 29(4): 280–295. Public perception and social implications of carbon storage projects. Nature Climate Change. 12(2): 110–117.

[4] Ensuring long-term carbon sequestration: challenges and monitoring strategies. Energy and Environmental Science. 13(8): 2456–2470.

[5] Global CCS Institute 2021. The Global Status of CCS 2021: Challenges and Opportunities.

[6] Samanta, A., Zhao, A., Shimizu, G.K.H., Sarkar, P. and Gupta, R. 2015. A review on solid adsorbents for carbon dioxide capture. Journal of Industrial and Engineering Chemistry. 23: 1–11. https://doi.org/10.1016/j.jiec.2014.09.001

[7] Zhu, Q.-L. and Xu, Q. 2017. Nanomaterials in carbon dioxide capture and storage: Progress and challenges. Nano Energy. 34: 269–280. https://doi.org/10.1016/j.nanoen.2017.03.023

[8] Wang, Q., Luo, J., Zhong, Z. and Borgna, A. 2020. Nanomaterials for the efficient capture and utilization of carbon dioxide. Chemical Reviews. 120(14): 7441–7515. https://doi.org/10.1021/acs.chemrev.9b00730.

[9] Advancements in Nanomaterials for Carbon Capture. 2022. A review. Journal of Environmental Science and Technology.

[10] Scalable Nanomaterials for Carbon Sequestration, Carbon Capture and Storage Journal. 2021.

[11] Martinez, E. and Brown F. 2020. Durability of nanomaterials in CO_2 storage applications. Materials Chemistry and Physics.

[12] Martinez, E. and Brown, F. 2020. Durability of nanomaterials in CO_2 storage applications. Materials Chemistry and Physics.

[13] Ahmed, I. and Zhao, J. 2022. Selective CO_2 capture using functionalized nanomaterials. Journal of Nanoscience.

[14] Iijima, S. 1991. Helical microtubules of graphitic carbon. Nature. 354(6348): 56–58. https://doi.org/10.1038/354056a0

[15] Dresselhaus, M.S., Dresselhaus, G. and Avouris, P. 2000. Carbon nanotubes: Synthesis, structure, properties, and applications. Springer Science and Business Media. https://doi.org/10.1007/978-3-662-04286-1.

[16] Meyyappan, M. 2004. Carbon nanotubes: Science and applications. CRC Press. https://doi:10.1201/9781420027471(https://www.crcpress.com/Carbon-Nanotubes-Science-and-Applications/Meyyappan/p/book/9780849321115).

[17] Li, H., Wang, K., Sun, Y., Lollar, C.T., Li, J. and Zhou, H.C. 2018. Recent advances in gas storage and separation using metal–organic frameworks. Materials Today. 21(2): 108–121. https://doi.org/10.1016/j.mattod.2017.07.006

[18] Sumida, K., Rogow, D.L., Mason, J.A., McDonald, T.M., Bloch, E.D., Herm, Z.R., et al. 2012. Carbon dioxide capture in metal–organic frameworks. Chemical Reviews. 112(2): 724–781.

[19] Zaworotko, M.J. 2008. Chemistry. Organic chemistry: Crystal engineering of diamondoid networks. Science. 321(5888): 911–912. https://doi.org/10.1126/science.1162224

[20] Bui, M., Adjiman, C.S., Bardow, A., Anthony, E.J., Boston, A., Brown, S., et al. 2018. Carbon capture and storage (CCS): The way forward. Energy and Environmental Science. 11(5): 1062–1176. https://doi.org/10.1039/C7EE02342A

[21] Li, J.R., Sculley, J. and Zhou, H.C. 2012. Metal-organic frameworks for separations. Chemical Reviews. 112(2): 869–932.

[22] Samanta, A., Zhao, A., Shimizu, G.K., Sarkar, P. and Gupta, R. 2012. Post-combustion CO_2 capture using solid sorbents: A review. Industrial and Engineering Chemistry Research. 51(4): 1438–1463.

[23] Vossen, J.L. and Kern, W. (Eds). 2001. Thin Film Processes (2nd ed.). Academic Press.

[24] Brinker, C.J. and Scherer, G.W. 1990. Sol-gel Science: The Physics and Chemistry of Sol-gel Processing. Academic Press.

[25] Reneker, D.H. and Chun, I. 1996. Electrospinning of nanofibers from polymer solutions. Nanotechnology. 7(6): 216–223. https://doi.org/10.1088/0957-4484/7/6/003

[26] Shvedova, A.A., Kisin, E.R., Porter, D., Schulte, P., Kagan, V.E., Fadeel, B., et al. 2009. Mechanisms of pulmonary toxicity and medical applications of carbon nanotubes: Two faces of Janus?. Pharmacology & Therapeutics. 121(2): 192–204.

[27] Moradi, S., Kundu, S. and Saidaminov, M.I. 2022. High-throughput synthesis of thin films for the discovery of energy materials: a perspective. ACS Materials Au. 2(5): 516–524. https://doi.org/10.1021/acsmaterialsau.2c00028

[28] Nguyen, T.D. 2013. From formation mechanisms to synthetic methods toward shape-controlled oxide nanoparticles. Nanoscale. 5(20): 9455–9482. http://dx.doi.org/10.1039/C3NR01810E

[29] Piccolo, A. and Mazzarino, M.J. 2021. Environmental implications of engineered nanomaterials in carbon storage applications. Science of the Total Environment. 768: 144543. https://doi.org/10.1016/j.scitotenv.2020.144543.

[30] Donaldson, K. and Stone, V. 2022. The health effects of nanomaterials: A review of recent research and implications for future safety assessments. Environmental Health Perspectives. 130(4): 46001. https://doi.org/10.1289/EHP8320.

[31] Handy, R.D. and Shaw, B.J. 2007. Toxic effects of nanoparticles and nanomaterials: implications for public health, risk assessment and the public perception of nanotechnology. Health, Risk & Society. 9(2): 125–144.

[32] Morimoto, Y., Izumi, H., Tomonaga, T., Nishida, C. and Higashi, H. 2025. Adverse effects of nanoparticles on humans. Journal of Occupational Health. 67(1): uiaf002.

Chapter 5

Carbon Sequestration Using Nanomaterials

Pooja Dhiman and Ajay Kumar*
Department of Chemistry, School of Basic and Applied Sciences,
Maharaja Agrasen University, Solan, (HP), India-174103

5.1 INTRODUCTION

In today's age, not only climate change but also global warming are one of the two most concerning global issues to be solved at the global environmental platform for the sake of both biotic and abiotic components of the biosphere. Extreme alterations in climate is a long-impacting and irreversible change in the weather profile owing to the alterations in the quality of atmospheric air, degrees of temperature, speed of wind, extent of precipitation and other factors directly or indirectly affecting weather conditions [1]. Not only this but, rapid urbanisation has also proceeded to an alarming rise in greenhouse effect gases emissions including O_3 (Ozone), perfluorocarbon groups of gases, water vapours, sulphur hexafluoride (SF_6), methane (CH_4), carbon dioxide (CO_2), and Nitrous Oxide (N_2O) and also include hydrofluorocarbon, etc. [2]. In order to meet the requirements of elevated requirement and supply of the global energy, direct burning of fossil fuel (natural gas, coal and oil) has resulted in an increase in the level of carbon dioxide in the atmosphere [3]. This boosted level of greenhouse effect gas CO_2 concentration in the environment is a prime cause for global level climatic changes owing to unproportionate ratio between atmospheric oxygen and the gas accountable for greenhouse effect i.e. carbon dioxide ratio [4].

*For Correspondence: ajaydogra972@gmail.com

DOI: 10.1201/9781003492184-5

The disturbance of carbon cycle, indirectly has an impact on the biogeochemical cycles consisting of several components that directly impacts the formation of Ozone (O_3) in the oxygen cycle, subsequently adversely affect the ozone layer in the stratosphere lying 15–30 km above the Earth's surface. The combustion of fossil fuels had been documented as one of the major and accountable causes of boosted emission of carbon dioxide (CO_2) into the environment. Carbon dioxide (CO_2) is primarily accountable for ~81% of greenhouse gases (GHG) [5]. Scientists of this domain had estimated and forecasted that in the upcoming 25 years, there will be an elevation in global surface temperature from 1.5 to 5.9°C owing to elevated concentration of atmospheric CO_2 in the environment [6]. Previously published reports across the globe have documented that the world's average temperature is increasing and will continue at the same steady rate if the generation and emissions of Greenhouse Effect Gases (GHGs) remain ignored and unresolved. Because of the ongoing reasons climate change has resulted in several extreme weather occurrences throughout the world, which are notable because these anthropogenic activities lead to global warming [7]. Numerous unseasonal and unwelcomed weather conditions including melting of ice cover, natural heating of the planetary surfaces, a rise in the sea water level, hot and cold waves, droughts, floods, tropical cyclones and violent storms [8]. The climatic conditions of the Earth are seasonal and vary according to all available time zones, naturally variable. Due to the industrial revolution, has led to dramatic alterations in industrial material manufacturing, infrastructures, transportation facilities and global agriculture production, economics and ultimately revenue generation [9]. In order to meet demands of the most populated nations including India and China regarding employment and provision of services has led to increase in production on industrial scale. To achieve targets in the production of required commodities, upto 70% requirement of energy is supplied by fossil fuels [10]. In these conditions, releases, and emission of carbon dioxide (CO_2) into the atmosphere can no longer be avoided. Because of the latest advanced technology named as Carbon Capture and Sequestration (CCS) technology has high infrastructural installation cost and hence is an expensive technique to resolve the ongoing concern. One of the better and practical ways to omit release of CO_2 emissions is the capture and sequestration of carbon dioxide by deploying nanomaterials including graphene (GO and RGO), biochar, g-C_3N_4, BN based nanocomposite as carbon capture and sequestration technology for the removal carbon dioxide from the atmosphere [11]. However, in order to mitigate the release of CO_2 from the atmosphere, numerous constraints acting as hurdles need to be overcome before carbon sequestration by nanomaterials is able to gain its ground as a CO_2 mitigation technology. For more explorations in the area of carbon sequestration by nanomaterials based composites, the first and foremost task is the identification of the legal and regulatory concerns pertaining to the current issue which really need to be resolved [12]. In order to optimise the capturing of ~6Gt of carbon dioxide (CO_2) annually, contribution and support from government, research institutions, private bodies and the energy industry is important [13]. The implementation of strict laws pertaining to the release of CO_2 by power production plants enabled nations including UK to cut short its CO_2 emission by ~80% matching the CO_2 emission level as it was in 1990 [14]. To meet the set agenda, the level of excessive CO_2 prevailing in the atmosphere should be captured, stowed, redeployed to fabricate several carbon-based commercially valuable products and additionally releasing carbon dioxide (CO_2) need to be minimised by replacing the use of fossil fuel with more eco-friendly available biofuels. One of the most feasible and plausible way to redeploy the seized and stowed CO_2 is its deployment as a starting feedstock material or reactant to either fabricate or produce biofuels and other commercially valuable products.

5.2 CARBON SEQUESTRATION

The removal of excess carbon dioxide present in the atmosphere by storing and utilising it as a starting feedstock in the formation of the valuable commercial products at a later point of time, is called carbon sequestration. There are two methods of sequestration of CO_2 from the atmosphere; natural and artificial. Natural carbon can be sequestered by the involvement of plants and microorganisms, specifically by the engagement of algae, bacteria, yeast and fungi through two diversified pathways, primarily by the photosynthesis or by non-photosynthetic route [15]. Both autotrophs and heterotrophs assimilate CO_2 into several organic natural or other carbonaceous valuable products including lignin, lignocellulose, chitin, cellulose, hemicellulose etc. [16]. Attempts are being made by researchers of this domain to develop more eco-friendly CO_2 sequestration pathways by proper involvement of these microorganisms for carbon capture and their utilisation at a later stage. These micro-organisms follow diversified pathways for their conversion mechanisms and varying potential to generate biomass/bioenergy [17]. But these methods are associated with their own set of drawbacks. Contrary to these, carbon sequestration by the utilisation of metallic and metal free based nanomaterials composites have their own set of advantages over the natural methods of carbon sequestration such as their better efficiency, cost effectiveness and speed [18].

5.3 BIOLOGICAL METHODS OF CO_2 CAPTURE AND SEQUESTRATION

Biomass is the outcome when, inside the photosynthetic organisms, atmospheric CO_2 has undergone through a sequence of biochemical processes [19]. At present, biomass is providing 12% of global energy to run the economy of its nations by fulfilling its energy demand to run all its industrial production processes smoothly, continuously, ultimately boosting the economic status of a nation. The mature plants function as an endless natural resource to harness energy by renewing it with new saplings and self-forestation process. Biomass is present in several sources including agricultural residues, wood, natural gas, and biofuel etc.

Biomass is yet another alternate pathway to capture and stow CO_2 from the atmosphere [20]. A number of species generally remain involved in the transformation of CO_2 into biomass by engaging both plants and microorganisms, e.g. fungi, bacteria, algae and yeast [21]. A greater quantity of resources are generally required to generate biomass and energy from plants to meet the current demand of energy in order for the economic cycle of a nation to run smoothly by continuously regulating the industrial production processes steadily by supplying energy as per their requirement. The scope of algal cultivation in ponds, small tanks and lakes, and ultimately in the sea, etc. has not only broadened the horizon for CO_2 capture but has also unlocked a major pathway for generating bioenergy [4]. Algae have reasonably high capacity and potential as microsized factories to manufacture and produce biomass as well as energy with optimum yield [22].

5.3.1 Carbon Sequestration by Plants

The CO_2 of the atmosphere is assimilated inside the plant's body during the photosynthesis process. During photosynthesis, chlorophyll contributes the central role in the transformation of atmospheric CO_2 into the biobased molecule by absorption and the utilisation of the Sun's radiant energy, to carry out both dark and light reactions [23]. Energy is produced in the form of biomolecules including NADPH and ATP from $NADP^+$ and ADP by the involvement of light energy based reactions by utilisation of solar radiations [24]. The occurrence of the dark reaction takes place by involving the Calvin–Benson cycle in addition to bioenergy molecules from the light reaction [25].

5.3.2 Carbon Sequestration by Microorganisms

Most microbes are able to fix atmospheric carbon dioxide (CO_2) from the air by involving several related and inter-connected mechanisms, routes, and pathways. Microorganisms are capable to utilize both photosynthetic and non-photosynthetic metabolic paths to remove, and convert atmospheric CO_2 into biomass and energy [26]. High potential to bioremediate atmospheric CO_2, biofixation at optimum rate, augmentation at its ease with generation of several additives are the adventitious features associated with the CO_2 capture by microbes. To keep the growth of the culture in continuous in the bioreactors, micro-organisms are generally employed without any competition and scarcity of nutrition [27]. It includes fungi, algae, and yeast etc. Bacteria are small, unicellular organisms that have a wide classification area, out of these the majority of these fall under categories: Actinomycetes, Mycoplasma, Cyanobacteria, Archaebacteria, Eubacteria, and Rickettsia, etc. [28]. From these categories, the Eubacteria, Cyanobacteria and Archaebacteria are autotrophs which are able to fix carbon dioxide (CO_2) for the production of organic carbon-based valuable products [4].

5.3.3 Carbon Sequestration by Anaerobic Gram-negative Bacteria

Anaerobic gram-negative bacteria has been playing a substantial contribution to the central role in the degradation of organic carbon materials, carbon cycle and acid synthesis [29]. Several species have been mainly engaged in the biofixation of CO_2 including *Clostridium autoethanogenum, C. formicoaceticum, Clostridium pasteurianum, Clostridium thermoaceticum, Acetobacterium woodi* which have been reported to be incorporated by atmospheric carbon dioxide into acetyl-CoA by the involvement of Wood–Ljungdahl pathway/Reductive acetyl CoA pathway which has been documented to be direct and the most effective and efficient thermodynamic biochemical pathways for the fixation of atmospheric carbon dioxide (CO_2) in valuable organic products [30].

5.3.4 Carbon Sequestration by Cyanobacteria

Cyanobacteria are prokaryotes; bluish-greenish bacteria and algae conjugate between microscopic bacteria and photosynthetic green plants. They are photoautotrophs gram negative bacteria available, and occurring and prevailing in both micro- and macroscopic forms [31]. For CO_2 fixation, these organisms such as photoautotrophs and chemoautotrophs have carboxysomes in their cytoplasm. All of these are thought to be involved in the Calvin Benson cycle due to an environmental and biochemical adaptation known as the Carbon dioxide Concentration Mechanism (CCM) [32].

5.3.5 Carbon Sequestration by Algae

On the earth, algae are considered as natural and prominent bio-factories to convert CO_2 into biomass and energy [33]. Macroalgae generate high quantities of lipid; which is why these are directly used to extract and produce of biodiesel. The use of microsized algae had been documented in the previously published records including diatoms, cyanobacteria, euglenoids, greenish, bluish, reddish, brownish, goldish, and yellow coloured algal species with great potential to fix atmospheric CO_2 enzymatically by deploying RuBisCO in the Calvin–Benson cycle. For instance ~1 kg of micro-sized algae have been estimated to fix 1.84 kg of atmospheric carbon dioxide (CO_2) [4]. The atmospheric carbon dioxide (CO_2) fixed by anabaena has been reported to be 1.46 g/L/d as well as 6.24 g/L/d by chlorella vulgaris. These have the potential to fix atmospheric carbon dioxide (CO_2) to generate biomass and bioenergy [34].

5.3.6 Carbon Sequestration by Fungi

Fungi are eukaryotes, multicellular organisms protected by a solid rigid cell wall. They have been documented to be heterotrophs and act as chief decomposers and primary organisms to capture and sequester atmospheric carbon from the terrestrial ecosystem [4]. Keeping their decomposing organic matter in mind, they are bifurcated into two broad categories saprophytes and mycorrhizal fungi [35]. Saprophytic fungi have been reported to produce proteinaceous enzymes to degrade matters including hemicellulose, pectin, lignin, and cellulose, etc. which for years have been accountable for degradation, mineralisation as well as the carbon cycle. As per documented records, the decomposition by mycorrhizal fungi is considerably poor owing to the complete absence of required enzymes [36].

5.3.7 Carbon Sequestration by Genetically Engineered Microorganism

In order to fix, atmospheric carbon dioxide (CO_2), the genetically engineered microorganism for effective and efficient fixation of CO_2 a few strains of yeast including E. coli bacteria, and other micro and macro-algae have been recorded [37]. Biological carbon sequestration is considerably significant to be engaged for the reasonable improvement and upgradation of climate alterations. From the several methods of atmospheric carbon sequestration by involving diversified microalgal fixation pathways, CO_2 had been recorded to possess several advantages in comparison to other available competing techniques by which biomass can be biotransformed to generate energy, keeping economic gains in mind. This method is also considered promising as the long- term sequestration strategy. Researchers and the authors of this manuscript have reported that biological carbon sequestration from point sources by the involvement of microalgae have been effective, efficient, economical and eco-friendly [38].

5.3.8 Metal Organic Frameworks

Another frame of materials known as metal-organic frameworks (MOF) has garnered significant interest for their exploration to absorb CO_2 owing to their remarkable high exposed surface. These are much flexible in altering the surface characteristic and pore architectures [39]. MOF are solid reinforced networks made possible by either metal ion or cluster connected to organic spacers. The choice of organic motif as linkers component in metal networks enhance the chance for carbon capturing ability, CO_2 selectivity, and related kinetics [40]. While the involved metal-organic frameworks (MOF) have expressed higher loading potential of pure CO_2, in the availability of several other interfering gases, the adsorption of CO_2 has been observed to be diminished dramatically. Significantly, considerable in depth research had been conducted and still going on for up-gradating the stability, recycling and selectivity of metal-organic frameworks (MOF) for several multiple cycles [41]. The physical adsorbents have small capacities and poor relative selectivity at lesser pressures. Contrary to physical adsorption, chemisorption had been reported to be involved in the reaction of CO_2 with the several other reactive groups on the adsorbent's surface, and the heat of adsorption has also been reported to be in the range of −60 to −90 kJ/mol [42]. Chemisorbents have been classified into two classes, chiefly amine- supported adsorbents as well as alkali metal-supported adsorbents [43]. Amine-supported adsorbents have been synthesised with the assistance of grafting as well as impregnation techniques. In impregnation supported adsorbents, generally weak interactions have been noticed among the different sites between the base support and amine group, on the other hand grafted amine have been observed to demonstrate relatively strong covalent bonding as the binding force. Keeping their mutual interactions in mind, the impregnated adsorbents are supposed to demonstrate relatively poor thermal stability if compared with grafted amine-supported adsorbents. A far more improved performance toward CO_2 capture, has been demonstrated by the several alkali metal-supported adsorbents, including the carbonates of sodium, lithium, potassium, and calcium [44]. Although, the commercial utilisation of lithium-supported adsorbents has been reported to be hampered because of elevated adsorption potential and their relatively greater diffusional resistance [45]. Potassium and sodium supported carbonated chemisorbents have been documented to express greater adsorption capabilities for CO_2 as 7.23 and 9.43 mmol/g, respectively [44]. In post combustion processes, for carbon capture in the presence of flue gases, the average adsorption- desorption cycle temperatures ranging from 60–200°C enables them as potential adsorbents. Owing to their relatively greater exposed surface area, microporous materials have also proved effective with other available adsorbents. The outcome of the research conducted on polyphenylene material (PPN-4) has e revealed its adsorption potential in close proximity to saturation owing to its relatively greater exposed surface area of as high as 6460 m^2/g [46]. It is extremely difficult to regenerate the used composite after CO_2 is captured in order to overcome the energy-related constraints that apply to the overall process cost-related implementation. Nowadays, different regeneration processes available in the market and the practice includes vacuum swing, temperature swing, pressure swing, hybrid (temperature and pressure) and electric swing adsorption [47]. Taking their practical operations into consideration, it has been noticed that pressure swing-based adsorption works at higher pressures contrary to this desorption-based models operating at atmospheric pressures. As the name suggests, the vacuum swing adsorption has been seen to operate under vacuum and normal air pressure, in contrast to pressure swing adsorption [48]. Furthermore, in a number of situations, gas should typically be prepared to remove any interfering gases that could contaminate the adsorbent or reduce selectivity through undesired attachment to the accessible active sites. Temperature Swing (TS) and Electric Swing (ES) are adsorption-supported regeneration procedures where changes in the working temperature have an impact on the adsorption potential [49]. Either steam or hot air is deployed as the reusability channel. In ES adsorption, the operating temperature of the adsorbent has been reported to increase with the assistance of the Joule effect of electricity to detach the adsorbate molecules from the adsorbent [50]. In the electric swing, because of *in situ* heating of adsorbents, adsorption consequently results in relatively fast kinetics, dynamics without any constraints of heat transfer or flow rate and relatively lower energy demand, unlike the case of vacuum and temperature swing based adsorption [51]. PS based adsorption is comparatively more cost-effective alternate because it requires a few seconds longer regeneration time than TS based adsorption, which requires longer hours. As an hybrid, the combination of PS and TS based adsorption processes has also been explored for CO_2 capture [52]. Adsorption also finds potential utilisation in pre combustion processes particularly in water-gas shift based reaction as well as steam reforming of hydrocarbons [53]. Enhancement in the productivity of molecular Hydrogen (H_2) has been observed as soon as the capture of CO_2 takes place, because of the observed shift of equilibrium toward the right (Product) forward direction [54].

5.4 NANOMATERIAL FOR CARBON SEQUESTRATION

Researchers in various domains of science have reported the contribution of nanomaterial based composites in the carbon sequestration including BN, g-C_3N_4, biomass derived biochar and graphene based composites, predominantly working on the principle of adsorption and adsorption led photo degradation [55].

5.4.1 Metal-free Nanomaterial

Investigating metallic co-catalysts to enhance and assist photocatalytic CO_2 reduction is currently the main focus of study. Nevertheless, these are typically more costly, scarce, and environmentally harmful. However, the metal-free co-catalysts based on carbon (such as graphene, RGO, GO and various composites based on biochar) not only facilitated various extraordinary properties including outstanding electric conductivity, enormous exposed surface area, and a unique set of morphological features, but they are also non-toxic, robust, inexpensive, and available in abundance. Specifically, graphene-supported co-catalysts have been vastly deployed as comparative low cost co-catalysts in conjunction with several other available Z scheme based semi-conductor as photocatalyst including Cu_2O, TiO_2, CdS and graphitic-C_3N_4 for photocatalytic reduction of CO_2.

5.4.1.1 Biochar

The amendment of soil with biochar has been observed to elevate the carbon content of soil which has been considered as major approach among others to eliminate carbon for a longer span of time as more than 80% of the soil is the preserved organic carbon [56]. Biochar has been observed to be one of the promising and potential bio resource enriched recalcitrant carbon, enabling it a long standing pool of carbon. Biochar is produced at high temperature (400–700°) treatment or gasification of biomass under inert surrounding environment. The various by-products are obtained such as (i) crude bio-oil, (ii) Tar, (iii) gas components including CH_4, H_2, CO_2 and CO etc. [57]. It have been noticed that biochar mainly consists of the recalcitrant aromaticforms of carbon which is difficult to get oxidised and discharged into the atmosphere in the form of CO_2 [58]. The documented presence of biochar in Amazonian terra preta soil provides archaeological evidence that biochar has the capacity to retain carbon in the soil for millennia. This has made it possible to use biochar in a carbon-negative manner to sequester CO_2 from the atmosphere. Biochar is able to capture and sequester atmospheric CO_2 by excluding other interfering oxides of gases such as N and S from exhaust gas by the air purification method. Biochar spread in soil has helped to decline` the emissions of N_2O and CH_4 gases which had been reported to possess potencies of multiple times higher than CO_2 respectively [58]. The sequestration of carbon by the deployment of biochar is totally different from that of afforestation, without tillage agriculture and their conversion into trees. As soon as agricultural areas had been transformed into no tillage farming area can be the reason to lose carbon capturing potential in two decades as forest lands had initiated maturing over the years and discharged CO_2 in the atmosphere [59]. These strategies have made biochar a long term carbon capture to lower emissions of atmospheric CO_2. The previously published plethora stated that unpyrolyzed carbon-rich materials, including decomposed vegetative or faunal residues, had decomposed in the soil years ago to discharge carbon at a slow rate over a few decades. Nevertheless, the total depletion of carbon from the composition of biochar is extremely slow and is challenging to judge and predict the storage of carbon over several centuries to numerous millennia. Integration of carbon sequestration with pyrolysis have assisted biorefineries to collect more carbon credits. Not only this but, pyrolysis of biomass has been reported to convert ~50% of the associated starting feedstock raw material as biomass carbon to bio oil and gases, whereas the rest 50% of carbon has been preserved in the form of biochar [60]. As per reports, the burning of combustible gases including hydrogen (H_2) and methane (CH_4) and bio oil for the sake of energy can discharge CO_2, as it is also assimilated by plants to carry out their photosynthesis process thereby making the whole process carbon neutral. As an outcome, amendment of biochar in the soil is capable to sequester the stable biomass carbon for relatively longer durations enabling the whole refinery process carbon negative [61]. The results of which have assisted in the abatement of CO_2 by earning more carbon credits. According to reports, in Brazil, proliferation of biochar by carbonisation from sugarcane bagasse has been reported to demonstrate the optimisation of the pyrolytic process for its deployment in household and related industrial aspects [62]. Carbon credits have been documented in the ratio of 1:1 in conjunction to the tons of CO_2 stowed or sequestered by carbon sequestration strategies. Approximately ~8 Gt of CO_2 is being captured and sequestered on the annual scale by burning of fossil fuels [63]. This statistics has expressed that ~8–10 billion carbon credits are still being generated annually. Currently, the trading price for CO_2 according to Chicago Climate Exchange have been U.S. $4/ton [64]. Under such conditions, the carbon credit economy ought to be comparable in size to the current economy of fossil fuels. This is why; the Earth has demonstrated preservation capacity for biochar as limitless. The evaluation of the budget of carbon and its credit have been an integrated bioenergy system, keeping the following parameters into consideration for decision making into mind such as: (i) Emissions of Green House Effect Gases (GHG) in association with biomass production; (ii) missions of CO_2 and input of energy during transportation, harvesting of biomass, pre-treatment, and their conversion; (iii) up gradation of energy input in the processing of product, purification of flue gases, and handling of biochar. All the above mentioned factors have the potential to resolve the amount of emissions because of Green House Effect Gases (GHG) and balances carbon in each sequential during overall biomass chain conversion with potential carbon sequestration of biochar [65]. Biochar has also demonstrated diversified advantages to the soil in which it has been amended. Biochar had been observed to boost the potential water-holding capacity, population of soil faunal development of the plants' root [66]. The overall bulk density of the soil has been reported to decrease because the microporous structure of biochar permitting the diffusion of water and air through the pores. Biochar, on its amendment with the soil, has raised the internal surface area for more improved and better adsorption of humus and soil nutrients. Biochar had been reported to retain considerable amount of soil mineral matter to facilitate their availability to plants penetration through their roots. Researchers [67] have suggested two main pathways by which biochar has the potential to decline pollution of either soil or groundwater (i) by holding and proper management and circulation of nutrients and minerals throughout the soil, thereby decreasing their possibilities of leaching into the groundwater; and (ii) by considerable improvement in the availability of nutrients in the soil thereby declining the spread of chemicals in the form of fertilisers [67].

5.4.1.2 Boron Nitride (BN)

Researchers have focused on porous Boron Nitride (BN) as potential adsorbents. Porous BN had demonstrated its primary features of being an idealistic photocatalyst. Shankar et al. recently investigated porous BN for CO_2 photoreduction [68]. The compelling metal-free compositional behaviour in addition to its tailored engineered chemistry and structural composition of BN allowed tunable band structural features. Porous BN has been reported to be an amorphous analogue of h-Boron Nitride (h-BN), with a wide band gap of (~5.5 eV) acting as an insulator. The concept of doping of boron nitride (BN) for its utilisation in nanomaterial based composites has been used in order to decline its band gap to facilitates the formation of Z scheme based resultant hetero- junctions [69]. At present the reports of BN supported materials for CO_2 photoreduction are on carbon-doped BN, which have expressed its instability in the presence of moisture because of the B–C bond breakage [70]. We have reported the utilisation of perforated BN dual application via for CO_2 adsorption as well as photoreduction catalyst under the light energy in beweeten 200–800 nm [71]. Furthermore, researchers have gone in depth into the photochemistry of porous BN by elucidating and predicting the energy potential of bands. The prediction of the energy potential of the conduction band (CB) and valence band (VB), along with the shift in the Fermi level on the absolute energy scale, is regarded as an important step in order to explore the reaction pathway of CO_2 photoreduction [72].

Porous BN materials had been documented to facilitate combined photoreduction and CO_2 capture in an ordinary gas/solid phase assembly under ambient conditions, without using any co-catalysts [73]. Unlike its crystalline counterpart, porous BN have demonstrating its semiconducting characteristic features supporting the behaviour and potential to photoreduce carbon dioxide (CO_2) to carbon monoxide oxide (CO) under irradiation of both single visible light as well as UV-Visible with considerably greater evolution rate of carbon monooxide (CO). Porous BN have expressed a steady evolution rate of CO in a single cycle and has maintained ~85% of its initial efficiency for multiple photocatalytic reactions [74]. The whole band structure and profile of porous BN illustrate it as an n-type semiconductor, which possesses CB and VB of the straddling type, enabling it to offer enough redox edge potentials for the conversion of CO_2/CO and H^+/H_2 [73]. Considering tailorable chemical and structural properties BN, a significant scope to expand its photocatalytic potential and performance in the region of visible light has revealed the rise in its rate of photocatalytic activity. The greater exposed surface area of porous BN because of its porous structure is the prime factor for interfacial adsorption and photocatalytic performance, enhancing the approach to active CO_2 absorption for efficient reduction. This strategy is significantly limited, particularly in the gaseous phase including the reduction of atmospheric CO_2 reduction because the ability of the gaseous adsorbate molecules to get adsorbed on the catalyst surface, in close proximity to the available active photocatalytic sites [70]. The porous structure of the h-BN has more advantages in case of more upgraded and improved adsorption led reduction of CO_2.

5.4.1.3 g-C_3N_4

Graphitic carbon nitride (g-C_3N_4) has been reported to be a two-dimensional (2-D) material with a unique set of characteristic features, having promising potential and deployment in CO_2 capture and sequestration [75]. Although, the poor alkalinity (basicity) of nitrogen species in g-C_3N_4 its interaction with CO_2 is very poor [76]. With the objective, to support and upgrade the deployment of graphitic-C_3N_4 for atmospheric CO_2 capture, a novel class of amine-functionalised graphitic-C_3N_4 as adsorbents had been and are still being fabricated by prominent researchers working in this area. The amine- group functionalised g-C_3N_4 adsorbent are fabricated by physical saturation of polyethylenimine (PEI), which are facilitated with numerous amine based functional moieties and had previously revealed extremely poor volatility, into the stacked available pores of graphitic-C_3N_4 [77]. It has been observed that, after the amine functionalisation, the efficiency of graphitic-C_3N_4 for atmospheric CO_2 capture with inclusion of all its plausible capacities and selectivity has improved. According to Balakrishnan et al. the highest reported atmospheric CO_2 capacity of PEI– graphitic-C_3N_4 composites had been 3.77 mmol/g examined at 100°C and ambient conditions of pressure, had been exhibited as superior to most of the other graphitic-C_3N_4 supported photocomposites and solid amines documented in the previously published plethora of outcomes research conducted in this area [78]. Additionally, the CO_2 adsorbed by PEI–graphitic-C_3N_4 composites can facilely be denuded by the concentrated swing and the reported loss in CO_2 capacity has been negligible after consecutive 10 cycles of adsorption and desorption [79].

According to Zhao et al. graphitic-C_3N_4 is a robust metal-free photocatalyst, adaptable because of its tunable and engineered conjugated structure, thermal resistance, reasonably strong chemical stability and visible light driven tunable band gap of ~2.7 eV [80]. These unique and novel characteristic features of graphitic-C_3N_4 enable it as highly adaptable to eliminate hazardous pollutants including reduction of atmospheric CO_2. However, its practical uses have been restricted by several constraints, including its, relatively smaller surface area, limited potential for visible light harvesting, limited mobility of charge and crystalline structure and accordingly behaviour [81]. Xu et al. made a number of attempts to overcome these restrictions with the assistance of numerous design strategies including heterojunction construction, elemental doping, band gap engineering and co doping in order to facilitate it as adsorbent to capture and sequester atmospheric CO_2 [82]. Huang et al. also made several attempts to a straight forward environmental-friendly, inexpensive method to adjust the characteristics features of graphitic-C_3N_4 by chemical functionalisation [83].

According to Majdoub et al. chemical functionalisation is the incorporation of different active functional groups into the graphitic-C_3N_4 framework to tune characteristic features of graphitic-C_3N_4. Chemical functionalisation is usually performed through the interaction of either covalent or non-covalent bonds and is the most effective and efficient way to adjust the characteristic features of graphitic-C_3N_4 [84]. Covalent functionalisation is one of the most effective strategies to incorporate the different active functional moieties in graphitic-C_3N_4 surface by involvement of covalent bonding between the chemical agent and graphitic-C_3N_4. Different chemicals including acetic acid (CH_3COOH), sulphuric acid (H_2SO_4), sodium hydroxide (NaOH) and nitric acid (HNO_3) have been used to tune the graphitic-C_3N_4 features. The covalent functionalisation assures greater photocatalytic performance of the graphitic-C_3N_4 from absorptions under visible light, for improved and better separation efficiency with greater exposed surface area [85].

5.4.1.4 Graphene and Reduced Graphene Oxide

Pure, original, and unoxidized graphene is referred to as "pristine graphene" because of its special qualities, which include (i) high thermal conductivity of about 5000 $Wm^{-1}K^{-1}$ [86], (ii) a charge transfer potential at normal room temperature of about 200000 $cm^2 V^{-1}s^{-1}$ [87], (iii) high transparency (97.7%), (iv) strong ambipolar, (v) unconventional quantum Hall effect, electric field effect, (vi) relativistic Dirac fermions, (vii) high adsorptive capacity for both organic and inorganic molecules, and cost-effective production etc. are as shown in Figure 5.1 [88].

Due to these characteristics, graphene is being explored in wide area of applications such as adsorbent, electrochemical sensors, energy storage, electroanalysis, and the removal of pollutants from the biosphere through CO_2 reduction [89]. GO is changed into RGO following further reductive exfoliation treatment [90]. RGO is developing into a very good intermediary between graphene and graphene oxide and has properties similar to graphene [91]. Additionally, by employing various microwave photoassisted, thermal, and electrochemical treatment techniques to be used later in the reduction of CO_2, it is simpler to synthesize RGO in the required quantity from less expensive graphene oxide [92].

Table 5.1 summarize some of the latest studies on metal-free photocatalysis for CO_2 photoreduction.

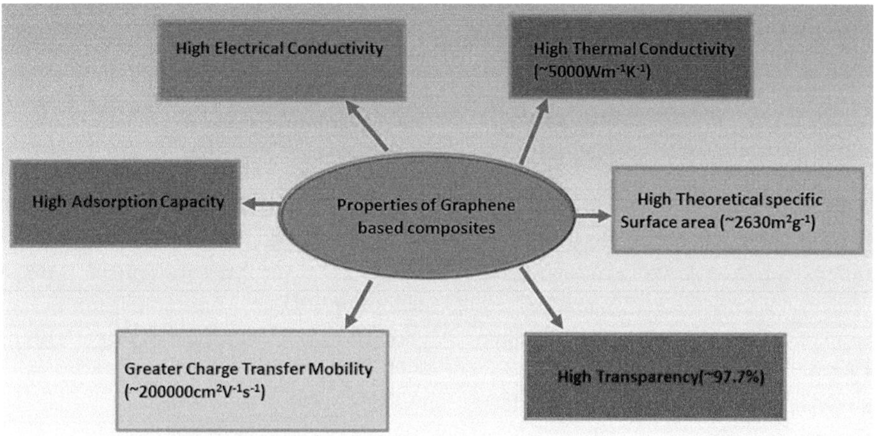

Figure 5.1 Properties of graphene and graphene-based composites.

5.4.2 Metal Based Nano Material

The loading of redox co-catalysts at the outer exposed surface of the target photocatalyst enables the availability of the photo-induced e^- and h^+ to the surface. Therefore, the synergistic effect can be facilitated through co-loading techniques. Eminent researchers reported the co-loading of MnOx and Au on BiOI surface and observed the greatest photocatalytic performance based efficiency of carbon dioxide (CO_2) transformation to carbon monooxide under simulated UV radiation, greater than that of pristine materials [117]. The resulting Au/BiOI/MnOx photo catalyst expressed the high performance for CO release under simulated visible light radiation approximately ~19 folds higher than that of pristine BiOI, revealing that the e^- extracting Au cocatalyst contributed significantly in comparison to the h^+ from MnOx co-catalyst in photo-induced CO_2 reduction. The more improved and upgraded photocatalytic CO release performance on BiOI/Au/MnOx derived from decline in the rate of charge recombination. co-loaded RuO_2 and Pt onto nanoplate-structured octahedron like Zn_2SnO_4 for elevated CO_2 photo-reduction to CH_4 under aqueous vapour as a reaction medium [118]. Different photocatalysts for photocatalytic CO_2 reduction have been shown in Table 5.2. Researchers found that co-depositing RuO_2 and Pt on Zn_2GeO_4 nano-ribbons greatly enhanced photoconversion of CO_2 to CH_4 (about 25 μmol g^{-1} h^{-1}. For initial 1 h) in aqueous vapor as reaction medium [119]. In order to accomplish the co-deposition, RuO_2 was first loaded using an impregnation-thermal treatment strategy that was guided by an in situ photo-reduction of Pt. Researchers revealed that the co loading of RuO_2 and Pd on TiO_2 surface led to the synergistic beneficially favoured improved photocatalytic performance because palladium (Pd) as a cocatalyst is able to catalyse the reduction of CO_2 to RuO_2 and HCOOH

Table 5.1 Metal free photocatalysts for photo-catalytic reduction of CO_2

Photocatalyst/ Co-Catalyst as Composite	**Reaction Medium**	**Loading Method (LM)**	**CO_2 Reduction Products**	**Ref.**
Graphitic/C_3N_3/Pt	Water (H_2O) vapour	Chemical Reduction (CR)	Methane (CH_4)	[93]
graphitic-C_3N_4/Pt	Water (H_2O) vapour	Impregnation and Chemical Reduction (CR)	Methane (CH_4)	[94]
Red phosphorous/ graphitic-C_3N_4/Pt	Water (H_2O) vapour	Photo Deposition (PD)	Methane (CH_4)	[95]
In_2O_3/graphitic-C_3N_4/Pt	Water (H_2O) medium	Photo deposition (PD)	Methane (CH_4)	[96]
Graphitic-C_3N_4/$NaNbO_3$/ Pt	H_2O medium	Photo deposition (PD)	Methane (CH_4)	[97]
Graphitic-C_3N_4/Pd	Solution-phase	in situ growth	CO, CH_4, C_2H_5OH	[98]
Layered double hydroxide/ graphitic-C_3N_4/Pd	Water (H_2O)	Photo deposition (PD)	Carbon monoxide (CO)	[99]
Graphitic-C_3N_4/Ru complex	TEOA and $CH_3CON(CH_3)_2$	Adsorption	Formic acid (HCOOH)	[100]
g-C_3N_4/Ru complex	TEOA/CH_3CN	Adsorption	CO, Formic acid (HCOOH)	[101]
Graphitic-C_3N_4 /Ru complex	TEOA/CH_3CN	Adsorption	Formic acid (HCOOH)	[102]
Graphene oxide (GO)/Cu	Water (H_2O) vapour	Microwave-assisted one-pot method	Methanol (CH_3OH)	[103]
Graphitic-C_3N_4/ Co-porphyrin	Bubbled CO_2 TEOA and CH_3CN	Stirring in Ethanol solution	Carbon monooxide (CO)	[104]
Graphitic-C_3N_4/Co/ZIF-9	Bipyridine, H_2O CH_3CN and TEOA	Mixing in solution	Carbon monooxide (CO)	[105]
Graphitic-C_3N_4/Graphene	H_2O vapour	One-pot impregnation–thermal reduction (TR)	Methane (CH_4)	[106]
Graphene/Ti0.91O_2 hollow spheres (0.01 g)	H_2O vapour and CO_2	–	Carbon monooxide (CO), Methane (CH_4)	[107]
Sandwich-like graphene-TiO_2 hybrid sheets (0.1 g)	H_2O vapour and CO_2	–	Ethane (C_2H_6), Methane (CH_4)	[108]
Graphene-WO_3 nanobelt (0.1 g)	H_2O vapour and CO_2	–	Methane (CH_4), Oxygen (O_2)	[109]

Photocatalyst/ Co-Catalyst as Composite	Reaction Medium	Loading Method (LM)	CO_2 Reduction Products	Ref.
Cu_2O/reduced graphene oxide (RGO) (0.5 g)	H_2O vapour and CO_2	–	Carbon monooxide (CO)	[110]
Graphene oxide (GO)/ CdS nanorods (0.01 g)	H_2O vapour and CO_2	–	Methane (CH_4)	[111]
Graphene–graphitic-C_3N_4 hybrid	H_2O vapour and CO_2	–	Methane (CH_4)	[106]
Metal Organic Frame Workv (MOF)-525/ Single atom Co	TEOA and CH_3CN	Post–metalation	Carbon monooxide (CO), Methane (CH_4)	[112]
Graphene oxides (GO)	Water (H_2O) medium	–	Methanol (CH_3OH)	[97]
Graphitic-C_3N_4/$NaNbO_3$ nanowires	Water (H_2O) medium	–	Methane (CH_4)	[97]
In_2O_3/graphitic-C_3N_4	Water (H_2O) phase	–	Methane (CH_4)	[113]
SnO_2–x/graphitic-C_3N_4 composite	Water (H_2O) phase	–	Carbon monooxide (CO), Methanol (CH_3OH), and methane (CH_4)	[113]
AgX/graphitic-C_3N_4 (X = Br and Cl) nano-composite	Water (H_2O) medium	–	Methane (CH_4)	[114, 115]
Ag supported on Ag_2SO_3 (Ag/Ag_2SO_3)	Water (H_2O) medium	–	Methane (CH_4) and Carbon monooxide (CO)	[116]

also carry out oxidation of SO_3^{2-} to SO_4^{2-} [120]. As an example for improved and better performance supported activity has been reported to be consistent with the assistance of the surface photovoltage spectrum.

5.5 ADSORPTION BASED CO_2 SEQUESTRATION

The drawbacks including area, low loading of CO_2, loss of adsorbent during absorption processes and poor gas-liquid contact had suggested adsorption would have been a better alternative for capture and sequestration of CO_2 [134]. Adsorption is a CO_2 capture technique that uses a fluid–solid interface to give outcomes more efficiently before and after combustion. CO_2 molecules from the gaseous form get adsorbed as soon as they come in contact with the active sites on the solid adsorbent surface [57]. The metallic or metal-free nanocomposite that adsorbs the CO_2 is commonly referred to as the adsorbent, whilst the adsorbed species itself is called the adsorbate. The molecules of the fluid get stuck to the adsorbent surface either by action of physical forces or by a chemical reaction called physiosorption

Table 5.2 Photo-catalysts for photocatalytic CO_2 reduction.

Photocatalyst/Co-Catalyst as Composite	Reaction Medium	Loading Method	CO_2 Reduction Products	Ref.
$TiO_2/RuO_2/Pd$	CO_2 bubbled NaOH and Na_2SO_3 aqueous solution	–	Formic acid (HCOOH)	[121]
TiO_2/Cu/Pt	H_2O vapour	Direct current (DC) sputtering	CH_4 and Olefin branched paraffin	[122]
$Zn_2GeO_4RuO_2$/Pt/)	H_2O vapour	Photo-deposition (Pt), calcination (RuO_2) and Impregnation	Methane (CH_4)	[123]
TiO_2/Cu/Ce	CO_2 saturated– sodium hydroxide (NaOH) aqueous solution	Impregnation–Reduction (IR)	Methanol (CH_3OH)	[124]
Zn_2SnO_4/RuO_2/Pt/	H_2O vapour	–	CH_4	[125]
Ti^{3+} self-doped TiO_2/Cu(I)/Pd	H_2O vapour	–	CH_4	[126]
Au/sMnO_x/BiOI	H_2O medium	Photodeposition	Carbon monooxide (CO)	[117]
$CuCo_2O_4$/CdS	TEOA/CH_3CN	Wet-chemical method	Carbon monooxide (CO)	[127]
TiO_2-pillared $Cu_2O/K_2Ti_4O_9$	H_2O vapour	Chemical reduction	Methanol (CH_3OH)	[128]
Fe_3O_4/coated optical fiber CuO/ TiO_2	H_2O vapour	Impregnation–calcination	Methane (CH_4)	[127]
PbS/Cu/TiO_2	H_2O vapour	Impregnation–calcination–reduction	Carbon monooxide (CO)	[128]
N-doped/Ni@NiO/$InTaO_4$	Bubbled H_2O	Impregnation–reduction–oxidation CO_2	Methanol (CH_3OH)	[129]
NiO/$InTaO_4$	H_2O vapour	Sol–gel	Methanol (CH_3OH), and CH_3CHO	[130]
NiO/$InNbO_4$	CO_2 saturated $KHCO_3$	Incipient wetness impregnation	CH_3OH	[131]
NiO/$InVO_4$/	CO_2 bubbled $KHCO_3$	Incipient wetness impregnation	Methanol (CH_3OH)	[132]
Carbon nanotube/Ni/TiO_2	H_2O vapour	Precipitation-calcination	Methane (CH_4)	[131]
Co–ZIF-9/CdS	Stirring in ethanol solution	CO_2 bubbled bipyridine, H_2O, CH_3CN and TEOA	Carbon monooxide (CO)	[132]
Co–ZIF-9/ TiO_2	H_2O Medium	*In situ* growth	Carbon monooxide (CO)	[133]

and chemical adsorption [135]. Adsorption being a surface phenomenon, the properties of solid adsorbents such as porosity along with surface-reactive species and active functional groups species, exposed surface area, polarity and play significantly in deciding the idealistic adsorbent to capture and sequester carbon dioxide (CO_2) from released gases [136]. The characteristic requirements of effective adsorbents are: (i) fast adsorption-desorption kinetics; (ii) high CO_2 loading and selectivity; (iii) abundance; (iv) low heat capacity; (v) low cost; and (vi) thermal and chemical stability both while the process of adsorption as well as its substantial regeneration [137]. The adsorption process favours low temperatures as it is an exothermic process, contrary to this, desorption that is a regeneration process is favoured by elevated temperatures. Physical adsorption had been supported on the mutual interactions among the molecules of adsorbate carbon dioxide (CO_2) and the solid surface of adsorbent including biochar, C_3N_4, BN and graphene based nanocomposites etc [138, 139]. The solute molecules of gaseous carbon dioxide (CO_2) preferentially mutually interact as well as stick to the surface of the deployed adsorbent keeping the intermolecular forces among the adsorbent and adsorbate gaseous molecules predominate the ongoing interactive forces among the gases [140]. The recorded and reported heat of adsorption for physical adsorption process should be in the range of −20 to −50 kJ/mol [141]. As in previously documented records the operating conditions were maintained accordingly with the CO_2 partial pressures which is approx 0.1 MPa at ambient temperature (25°C), by considering the exothermic nature of the adsorption process. [142]. Carbonaceous materials, including C_3N_4, biochar, chemically modified polymeric surface materials graphene, reduced graphene oxide, mesoporous and microporous zeolites, in addition to metal-organic frameworks (MOF) had been observed as the potential adsorbents [143]. Carbonaceous composite materials including carbon molecular sieves, carbon nanotubes and activated carbon, can not only capture but retain CO_2 by the physical adsorption process by working on default principle and mechanism. Activated carbons with greater thermal stability have demonstrated substantially poor adsorption potential. In order to have improved adsorption capacities, the two fundamental conditions are believed to be met by the working carbonaceous material based deployed adsorbent. These conditions are the (i) Chemical modification of the surface of the carbonaceous adsorbent with basic groups (ii) Greater carbonaceous adsorbents exposed surface area in order to support the acidic nature CO_2. From literature, activated carbons have been reported to demonstrate speeded kinetics in a restricted time for both adsorption as well as desorption cycles [144]. However, activated carbon has illustrated poor selectivity with respect to carbon dioxide (CO_2). According researchers in this area, micro-porous carbon molecular sieves facilitates a narrow pore size distribution as well as greater pore volume believed to support and raise both loading capacity as well as selectivity. The demand for both single walled - as well as multiwalled carbon nanotubes has been high for years because of their selectivity toward CO_2 in comparison to other available adsorbents. Researchers of this field have already conducted their investigation on both natural as well as synthetic zeolites capture and thereof sequester CO_2 from flue gas streams [145]. Ferrierite, mordenite, chabazite, and clinoptilolite, the investigated usual zeolites, have already shown significantly better and enhanced results in the segregation and separation of carbon dioxide from nitrogen gas. According to reports, zeolites' ordered framework falls within the restricted dimensional range of ~0.5–1.2 nm pore size distribution, which functions as size exclusion materials to separate gases from their mixtures [146]. Greater adsorption potential with significantly improved adsorption kinetics has been demonstrated by the adsorption cum segregation performance of reported natural sodium zeolites with higher sodium ion concentrations and their extended exposed surface [147]. In order to create traditional zeolites, the ratio of silica to aluminum is altered, and either alkali or alkaline earth metals are substituted to assist the pores with charges [148]. The structural modifications of adsorbents can also be done by the insertion of the reactive basic functional groups including either amine or alkaline earth metals/ alkali metals to boost loading capacities. Keeping the structure and adsorption properties in mind the possible adsorption mechanism of various adsorbents for the remediation of CO_2 from the atmosphere is illustrated in the Figure 5.2.

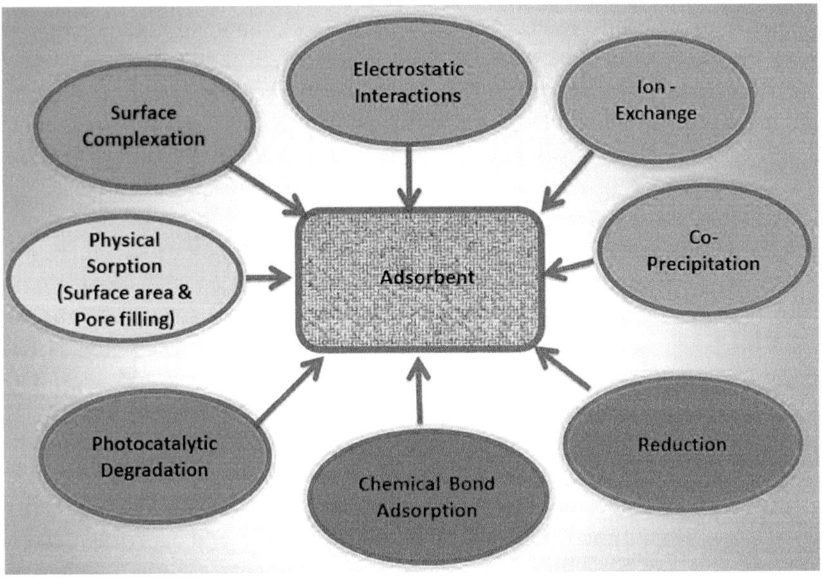

Figure 5.2 Schematic diagram for possible mechanisms for the sequestration of CO_2 by photocatalysts as adsorbents.

The existence of strengthened mutual interaction among the different basic functional groups and acidic CO_2 not only substantially upgrades the adsorption process but also additionally boosts the selectivity of acidic CO_2 gas over other prevailing gases of the mixture. The adsorption potential of both either natural or synthetic zeolites is governed by the size of the pore as well as specific sodium ions (Na^+) inserted in those pores for better adsorption. The preferential adsorption of oppositely charged molecules, such as acidic CO_2, is enhanced by the charge created by sodium ions in zeolite [146]. Researchers had observed that kinetics of adsorption process on zeolites beds had been recorded as quickly achieving the state of equilibrium after few minutes of loadings. Weak selectivity of natural/synthetic zeolites toward acidic CO_2 has challenged researchers to explore the insertion of several different cations in the adsorbent channels [149]. The adsorptive performance of anthropogenic/simulated/synthetic/manmade adsorbents is governed by the working parameters including pore size, and pore volume, high partial pressures and low temperature of the gas, basicity such as nature of cations inserted into the pore structure of zeolites [150].

5.6 BASICS OF CO_2 PHOTOREDUCTION SYSTEMS

The mechanism of CO_2 photoreduction generally takes place through a set of sequential steps in a system. By keeping in mind semi-conductor supported photocatalysts as an example, to reduce CO_2, the semi-conductor must facilitate the incidence of solar/simulated light energy (Ehv) equals to (=) or elevated than (>) its band gap (Eg) (for an instance, Ehv $\geq$ 3.2 eV in case of anatase TiO_2). The photo-excitation propagates via the generation of electrons in CB and holes in VB, which acts as the specific charge carriers for photo-reduction as well as photo-oxidation to takes place, respectively. [151]. The photocatalyst is also expected to have its electrons (e^-) with greater energy in comparison to the reduction potential of CO_2. As an ideal system, it has been recorded that the conduction band (CB band edge) potential of the photocatalyst must be at a higher position means (more negative in comparison to) the reduction (edge) potential for a specific photocatalyst at a predetermined experimental pH value. Simultaneously, the holes should oxidise water (H_2O) to molecular (O_2) in order to generate protons (H^+), meaning that the Valence Band (VB) edge potential must be substantially below (more +ive in comparison to) the oxidation level of water for effective and efficient photolysis. The plausible reactions for CO_2 reduction can takes place in aqueous (H_2O) media. The e^-/h^+ pair formation is the initial stage of the sequence in the photo-reduction of CO_2, as shown in Figure 5.3. Following these photo-excitation, the e^-/h^+ pairs should be appropriately spaced apart to enable their transit to redox active species across the junction and to slow down the electron-hole pair recombination rate. Figure 5.3 explains the plausible transfer of charge pathways of the photocatalyst, which have been dealt with numerous critical charge transfer sequential steps [152]. Even though the electron–hole pair's lifespan is only a few nanoseconds, it is sufficient to promote redox processes. Nevertheless, compared to other electron migration processes, the time scale of charge recombination is faster by a magnitude of two to three times. In photo-assisted reaction processes, this recombination phenomenon is considered as one of the major constraints. Consequently, any method that prevents charge recombination would significantly enhance CO_2 photoreduction efficiency and rates. Water is typically oxidized by holes following photoexcitation to produce molecular oxygen and protons, which are necessary for the photoreduction of carbon dioxide [153]. The sequential steps of photoreduction of CO_2 are mentioned below:

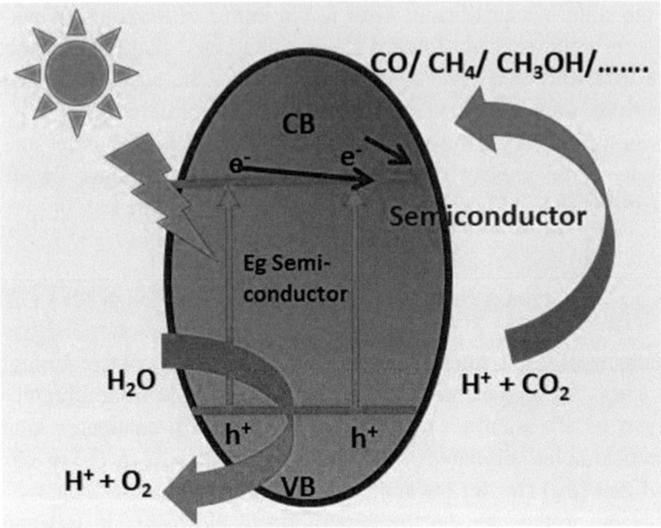

Figure 5.3 Possible mechanism of photoreduction of carbon dioxide (CO_2).

In the initial step (i) the energy of incident photons of solar energy equal to or greater than the band gap energy (Eg) has been absorbed by the photocatalyst for electron–hole (e^--h^+) pairs generation. (ii) The newly generated electrons (e^-) and holes (h^+) are then translocated independently to the particles interface of the photocatalysts or the cathode and the anode during second step. In the third step (iii), the newly photogenerated electrons (e^-) carry out CO_2 reduction into fuels, although the holes (h^+) are accountable to carry out oxidation of water (H_2O) into molecular oxygen (O_2). During steps (ii) and (iii) a greater proportion of electrons (e^-) and holes (h^+) can be deployed through recombination (iv) From the above mentioned three steps, the initial two steps are similar to those in solar energy powered splitting of water (H_2O). The chief difference among splitting of water and CO_2 reduction are the surface reaction of the newly photogenerated electrons (e^-). The proton (h^+) reduction often takes place in competition with reduction of carbon dioxide (CO_2) particularly in the availability of water (H_2O) [153].

5.7 FUTURE PROSPECTS

Exploration of novel cocatalyst with better selectivity, reasonable stability, high activity, and economic in cost to carry out semiconductor-based photocatalytic carbon dioxide (CO_2) reduction needs to be carried out. Previously conducted research had already illustrated that the extraordinary reduction of acidic carbon dioxide (CO_2), electrocatalysts can be deployed as an effective and efficient co-catalyst to carry out reduction of carbon dioxide (CO_2). Several electrocatalysts, including Pt/Cu alloy, and Cu/Au alloy and Cu, had recorded of being largely active co-catalysts in order to perform photo-assisted reduction of CO_2. This is why, it has been urgently required to explore novel co-catalysts to carry out reduction of carbon dioxide (CO_2) from the reported organo- and in-organo-metallic electrocatalysts. Inorganic electrocatalysts, including several ternary and binary metal alloys, Metal Nitrides (MN), Metal Oxides (MO) hybrids, Metal Carbides (MC), Metal Phosphide (MP), Metal Selenides (MSe), spinels, perovskites, metal free and metallic CO_2 photocatalyst/reductant as well as adsorbents which can also be used as co-catalysts. The synergetic upgraded effects observed in the multi-candidate inorganic electro-catalysts might also result in their outstanding activity-based performance and reasonable selectivity as co-catalysts in photocatalytic reduction of carbon dioxide (CO_2). However, apart from electrocatalysts, co-catalysts have generally had a substantial contribution in upgrading the separation and transfer efficiency of newly photoexcited carriers for charge transfer in semi-conductor based photocatalysts. For example, Pt demonstrated low selectivity for electro-catalytic CO_2 reduction. However, Platinum (Pt) co-catalyst can reasonably improve the photocatalytic reduction of carbon dioxide (CO_2) activity-based performance because of its outstanding electronextracting capability. The multi-functional inorganic electrocatalysts exceptional activity-based performance and respectable specificity as co-catalysts in the photocatalytic reduction of carbon dioxide may also be the consequence of their collaboratively enhanced effects. Co-catalysts, in contrast to electrocatalysts, have typically made a significant contribution to improving the charge separation and transfer efficiency of recently photoexcited carriers. Pt, for instance, showed poor selectivity in electrocatalytic CO_2 reduction. However, due to its exceptional electron-extracting capacity, platinum co-catalyst can significantly enhance the photo-assisted reduction of CO_2 activity-based performance. The significant distinction thus needs to be considered while choosing efficient and successful co-catalysts from a group of electrocatalysts. Additionally, in order to improve the photo-assisted CO_2 reduction efficiency, a number of newly created 2D nanomaterials should be studied as co-catalysts. The utilization of biological agents, such as bacteria or enzymes, as co-catalysts demonstrates that additional enhancements in CO_2 reduction can be achieved. Furthermore, maintaining co-catalysts long-term performance on semiconductor-based photocatalysts has shown to be extremely difficult. Highly active, specific, and stable photocatalytic reduction of acidic CO_2 has been shown to be possible with single-atom co-catalysts supported on semiconductor-based photocatalysts. As a result, creating co-catalysts based on single atoms has proven difficult but also quite promising. Additionally, metal organic frameworks (MOF), organometallic molecules are yet another category of potential photo cocatalyst because of being ultra-small in size and facilitated with countless number of reactive sites. Specifically, their characteristic features can be engineered by regulating their final compositional structures.

5.8 CONCLUSION

Either capture of CO_2 or its conversion into any other suitable and reusable form of energy has been an attractive and challenging task. However, it remains at an early stage of technological development and presently suffers from several drawbacks, limitations and loopholes which makes it more expensive and limits its practical use t on an industrial scale. The utilisation of biological assemblies generally result in poor productivity owing to ongoing demand for cofactors and the substantial sensitivity of involved microorganisms to a greater concentrations of carbon dioxide (CO_2). This is why, robust enzymatic assemblies can tolerate extremely challenging operating parameters pertaining to pollutants and elevated temperatures etc. which have a greater potential to capture and sequester atmospheric CO_2 from the environment. On the other hand, anthropogenic nanomaterials based adsorbents such as graphene, RGO, g-C_3N_4, BN and biochar based heterojunction composites keeping their unique structural features, porosity, surface area and adsorption capacity with respect to time into consideration, also have considerable potential in the CO_2 capture and its sequestration from the atmosphere.

REFERENCES

[1] Harde, H. 2022. How Much CO_2 and the Sun Contribute to Global Warming: Comparison of Simulated Temperature Trends with Last Century Observations. Science of Climate Change. 2(1): N1–1.

[2] Mikhaylov, A., Moiseev, N., Aleshin, K. and Burkhardt, T. 2020. Global climate change and greenhouse effect. Entrepreneurship and Sustainability Issues. 7(4): 2897.

[3] Wang, J. and Azam, W. 2021. Natural resource scarcity, fossil fuel energy consumption, and total greenhouse gas emissions in top emitting countries. Geoscience Frontiers. 15(2): 101757.

[4] Gayathri, R., Mahboob, S., Govindarajan, M., Al-Ghanim, K.A., Ahmed, Z., Al-Mulhm, N., et al. 2021. A review on biological carbon sequestration: A sustainable solution for a cleaner air environment, less pollution and lower health risks. Journal of King Saud University-Science. 33(2): 101282.

[5] Qian, H., Zhu, X., Huang, S., Linquist, B., Kuzyakov, Y., Wassmann R., et al. 2023. Greenhouse gas emissions and mitigation in rice agriculture. Nature Reviews Earth and Environment. 4(10): 716–732.

[6] Scafetta, N. 2021. Testing the CMIP6 GCM Simulations versus surface temperature records from 1980–1990 to 2011–2021: High ECS is not supported. Climate. 9(11): 161.

[7] Clarke, B., Otto, F., Stuart-Smith, R. and Harrington, L. 2022 Extreme weather impacts of climate change: An attribution perspective. Environmental Research: Climate. 1(1): 12001.

[8] Yadav, M., Gosai, H.G., Singh, G., Singh, A., Singh, A.K., Singh, R.P., et al. 2023. Major Impact of global climate change in atmospheric, hydrospheric and lithospheric context. *In*: Global Climate Change and Environmental Refugees: Nature, Framework and Legality. Springer. 35–55.

[9] Alavipanah, S.K., Mansourmoghaddam, M., Gomeh, Z., Galehban, E. and Hamzeh, S. 2022. The reciprocal effect of global warming and climatic change (new perspective): A review. Desert. 27(2): 291–305.

[10] Ahmed, M., Shuai, C. and Ahmed, M. 2023. Analysis of energy consumption and greenhouse gas emissions trend in China, India, the USA, and Russia. International Journal of Environmental Science and Technology. 20(3): 2683–2698.

[11] Salvi, B.L. and Jindal, S. 2019. Recent developments and challenges ahead in carbon capture and sequestration technologies. SN Applied Sciences. 1: 1–20.

[12] Wang, X. and Cui Y. 2023. Impact of nanomaterials on microbial communities in the carbon sequestration processes of urban river ecosystems. Materials Express. 13(8): 1400–1406.

[13] Chakrabartty, I. and Hakeem, K.R. 2024. Green Nanomaterials in Energy Conversion and Storage Applications. CRC Press.

[14] Lees, E. and Eyre, N. 2021. Thirty years of climate mitigation: Lessons from the 1989 options appraisal for the UK. Energy Efficiency. 14(4): 37.

[15] Zhu, X., Lei, C., Qi, J., Zhen, G., Lu, X., Xu, S., et al. 2022. The role of microbiome in carbon sequestration and environment security during wastewater treatment. Science of the Total Environment. 837: 155793.

[16] Ali, M., Javeed, H.M.R., Tariq, M., Khan, A.A., Qamar, R., Nawaz, F., et al. 2023. Use of biochar for biological carbon sequestration. *In*: Climate Change Impacts on Agriculture: Concepts, Issues and Policies for Developing Countries. Springer. 393–411.

[17] Liang, J., Nabi, M., Zhang, P., Zhang, G., Cai, Y., Wang, Q., et al. 2020. Promising biological conversion of lignocellulosic biomass to renewable energy with rumen microorganisms: A comprehensive review. Renewable and Sustainable Energy Reviews. 134: 110335.

[18] Dutta, V., Verma, R., Gopalkrishnan, C., Yuan, M.H., Batoo, K.M., Jayavel, R., et al. 2022. Bio-inspired synthesis of carbon-based nanomaterials and their potential environmental applications: A state-of-the-art review. Inorganics. 10(10): 169.

[19] Daneshvar, E., Wicker, R.J., Show, P.L. and Bhatnagar, A. 2022. Biologically-mediated carbon capture and utilization by microalgae towards sustainable CO_2 biofixation and biomass valorization–A review. Chemical Engineering Journal. 427: 130884.

[20] Aghel, B., Behaein, S. and Alobaid, F. 2022. CO_2 capture from biogas by biomass-based adsorbents: A review. Fuel. 328: 125276.

[21] Onyeaka, H. and Ekwebelem, O.C. 2023. A review of recent advances in engineering bacteria for enhanced CO_2 capture and utilization. International Journal of Environmental Science and Technology. 20(4):4635–4648.

[22] Lim, Y.A., Ilankoon, I., Chong, M.N. and Foo, S.C. 2023. Improving microalgae growth and carbon capture through micro-size bubbles generation in flat-panel photobioreactors: Impacts of different gas sparger designs on mixing performance. Renewable and Sustainable Energy Reviews. 171: 113001.

[23] Mandal, R. and Dutta, G. 2020. From photosynthesis to biosensing: Chlorophyll proves to be a versatile molecule. Sensors International. 1: 100058.

[24] Walter, J. and Kromdijk, J. 2022. Here comes the sun: How optimization of photosynthetic light reactions can boost crop yields. Journal of Integrative Plant Biology. 64(2): 564–591.

[25] Gurrieri, L., Sparla, F., Zaffagnini, M. and Trost, P. 2023. Dark complexes of the Calvin-Benson cycle in a physiological perspective. *In*: Seminars in Cell and Developmental Biology. Elsevier.

[26] Ferretti, M.V., Hussien, R.A., Ballicora, M.A., Iglesias, A.A., Figueroa, C.M. and Diez MDA. 2022. The ADP-glucose pyrophosphorylase from Melainabacteria: A comparative study between photosynthetic and non-photosynthetic bacterial sources. Biochimie. 192: 30–37.

[27] Garg, S. 2020. Bioremediation of agricultural, municipal, and industrial wastes. In: Waste Management: Concepts, Methodologies, Tools, and Applications. IGI Global. 948–970.

[28] Wang, Q., Cao, X., Jiang, H. and Guo, Z. 2021. Straw Application and Soil Microbial Biomass Carbon Change: A Meta-Analysis. CLEAN–Soil, Air, Water. 49(2): 2000386.

[29] Ranjan, M.R., Bhardwaj, P. and Tripathi, A. 2021. Microbial Sequestration of Atmospheric Carbon Dioxide. Climate Change and the Microbiome: Sustenance of the Ecosphere. 199–216.

[30] Pal, T.K., De, D. and Bharadwaj, P.K. 2020. Metal–organic frameworks for the chemical fixation of CO_2 into cyclic carbonates. Coordination Chemistry Reviews. 408: 213173.

[31] Blankenship, R.E. 2021. Molecular mechanisms of photosynthesis. John Wiley & Sons.

[32] Yang, B., Liu, J., Ma, X., Guo, B., Liu, B., Wu, T., et al. 2017. Genetic engineering of the Calvin cycle toward enhanced photosynthetic CO_2 fixation in microalgae. Biotechnology for biofuels. 10(1): 1–13.

[33] Cheng, L., Xia, G. and Thome, J.R. 2021. Flow boiling heat transfer and two-phase flow phenomena of CO_2 in macro-and micro-channel evaporators: Fundamentals, applications and engineering design. Applied Thermal Engineering. 195: 117070.

[34] Bhatia, S.K., Bhatia, R.K., Jeon, J.M., Kumar, G. and Yang, Y.H. 2019. Carbon dioxide capture and bioenergy production using biological system–A review. Renewable and Sustainable Energy Reviews. 110: 143–158.

[35] Procter, A.C., Ellis, J.C., Fay, P.A., Polley, H.W. and Jackson, R.B. 2014. Fungal community responses to past and future atmospheric CO_2 differ by soil type. Applied and Environmental Microbiology. 80(23): 7364–7377.

[36] Habeeb, T.H., Abdel-Mawgoud, M., Yehia, R.S., Khalil, A.M.A., Saleh, A.M. and AbdElgawad, H. 2020. Interactive impact of arbuscular mycorrhizal fungi and elevated CO_2 on growth and functional food value of Thymus vulgare. Journal of Fungi. 6(3): 168.

[37] Cho, K.H., Wolny, J., Kase, J.A., Unno, T. and Pachepsky Y. 2022. Interactions of *E. coli* with algae and aquatic vegetation in natural waters. Water Research. 209: 117952.

[38] Nowicki, M., DeVries, T. and Siegel, D.A. 2022. Quantifying the carbon export and sequestration pathways of the ocean's biological carbon pump. Global Biogeochemical Cycles. 36(3): e2021GB007083.

[39] Ibrahim, A.O., Adegoke, K.A., Adegoke, R.O., AbdulWahab, Y.A., Oyelami, V.B. and Adesina, M.O. 2021. Adsorptive removal of different pollutants using metal-organic framework adsorbents. Journal of Molecular Liquids. 333: 115593.

[40] Dutta, A., Pan, Y., Liu, J.Q. and Kumar, A. 2021. Multicomponent isoreticular metal-organic frameworks: Principles, current status and challenges. Coordination Chemistry Reviews. 445: 214074.

[41] Hashemi, L., Masoomi, M.Y. and Garcia, H. 2022. Regeneration and reconstruction of metal-organic frameworks: Opportunities for industrial usage. Coordination Chemistry Reviews. 472: 214776.

[42] Raganati, F., Miccio, F. and Ammendola, P. 2021. Adsorption of carbon dioxide for post-combustion capture: A review. Energy and Fuels. 35(16): 12845–12868.

[43] Soujanya, Y. 2020. CO_2 adsorption by functionalized sorbents. *In*: Advances in Carbon Capture. Elsevier. 229–240.

[44] Samanta, A., Zhao, A., Shimizu, G.K.H., Sarkar, P. and Gupta, R. 2012. Post-combustion CO_2 capture using solid sorbents: a review. Industrial and Engineering Chemistry Research. 51(4): 1438–1463.

[45] Hao, H., Hutter, T., Boyce, B.L., Watt, J., Liu, P. and Mitlin, D. 2022. Review of multifunctional separators: Stabilizing the cathode and the anode for alkali (Li, Na, and K) metal–sulfur and selenium batteries. Chemical Reviews. 122(9): 8053–8125.

[46] Huang, N., Day, G., Yang, X., Drake, H. and Zhou, H.C. 2017. Engineering porous organic polymers for carbon dioxide capture. Science China Chemistry. 60: 1007–1014.

[47] Plaza, M.G. and Rubiera, F. 2019. Evaluation of a novel multibed heat-integrated vacuum and temperature swing adsorption post-combustion CO_2 capture process. Applied Energy. 250: 916–925.

[48] Liu, W., Lin, Y.C., Jiang, L., Ji, Y., Yong, J.Y. and Zhang, X.J. 2022. Thermodynamic exploration of two-stage vacuum-pressure swing adsorption for carbon dioxide capture. Energy. 241: 122901.

[49] Das, A., Peu, S.D., Hossain, M.S., Nahid, M.M.A., Karim, F.R. Bin Chowdhury, H., et al. 2023. Advancements in adsorption based carbon dioxide capture technologies—A comprehensive review. Heliyon.

[50] Ding, J., Xia, Q., Li, L. and Gao, G. 2022. Joule heat enabled efficient *In-situ* regeneration of carbon nanotubes as model adsorbents. Chemical Engineering Journal. 444: 136600.

[51] Dhoke, C., Zaabout, A., Cloete, S. and Amini, S. 2021. Review on reactor configurations for adsorption-based CO_2 capture. Industrial and Engineering Chemistry Research. 60(10): 3779–3798.

[52] Zhao, R., Liu, L., Zhao, L., Deng, S., Li, S. and Zhang, Y. 2019. A comprehensive performance evaluation of temperature swing adsorption for post-combustion carbon dioxide capture. Renewable and Sustainable Energy Reviews. 114: 109285.

[53] Pruvost, F., Cloete, S., del Pozo, C.A. and Zaabout, A. 2022. Blue, green, and turquoise pathways for minimizing hydrogen production costs from steam methane reforming with CO_2 capture. Energy Conversion and Management. 274: 116458.

[54] Soltani, S.M., Lahiri, A., Bahzad, H., Clough, P., Gorbounov, M. and Yan, Y. 2021. Sorption-enhanced steam methane reforming for combined CO_2 capture and hydrogen production: A state-of-the-art review. Carbon Capture Science and Technology. 1: 100003.

[55] Li, Y., Liu, Z., Li, Z. and Wang, Q. 2022. Renewable biomass-derived carbon-supported g-C_3N_4 doped with Ag for enhanced photocatalytic reduction of CO_2. Journal of Colloid and Interface Science. 606: 1311–1321.

[56] Han, L., Sun, K., Yang, Y., Xia, X., Li, F., Yang, Z., et al. 2020. Biochar's stability and effect on the content, composition and turnover of soil organic carbon. Geoderma. 364: 114184.

[57] Gupta, S., Kua, H.W. and Dai Pang, S. 2020. Effect of biochar on mechanical and permeability properties of concrete exposed to elevated temperature. Construction and Building Materials. 234: 117338.

[58] Kimani, S.M., Bimantara, P.O., Hattori, S., Tawaraya, K., Sudo, S., Xu, X., et al. 2020. Co-application of poultry-litter biochar with Azolla has synergistic effects on CH_4 and N_2O emissions from rice paddy soils. Heliyon. 6(9).

[59] Dalal, R., Wathore, R. and Labhasetwar, N. 2022. Sustainable production of biochar, bio-gas and bio-oil from lignocellulosic biomass and biomass waste. *In*: Zero Waste Biorefinery. Springer. 177–205.

[60] Kharissova, A.B., Kharissova, O.V., Kharisov, B.I. and Méndez, Y.P. 2024. Carbon negative footprint materials: A review. Nano-Structures and Nano-Objects. 37: 101100.

[61] Zafeer, M., Menezes, R.A., Venkatachalam, H., Bhat, K.S. 2023. Sugarcane bagasse-based biochar and its potential applications: a review. Emergent Materials. 1–29.

[62] Nagabhushan, D., Russell, R.H., Waltzer, K., Thompson, J., Beck, L., Jaruzel, M. 2021. Carbon capture: Prospects and policy agenda for CO_2-neutral power generation. The Electricity Journal. 34(7): 106997.

[63] Balmford, A., Keshav, S., Venmans, F., Coomes, D., Groom, B., Madhavapeddy, A., et al. 2023. Realizing the social value of impermanent carbon credits. Nature Climate Change. 13(11): 1172–1178.

[64] Tan, K., Qin, Y. and Wang, J. 2022. Evaluation of the properties and carbon sequestration potential of biochar-modified pervious concrete. Construction and Building Materials. 314: 125648.

[65] Llovet, A., Mattana, S., Chin-Pampillo, J., Gascó, G., Sánchez, S., Mondini, C., et al. 2021. Long-term effects of gasification biochar application on soil functions in a Mediterranean agroecosystem: Higher addition rates sequester more carbon but pose a risk to soil faunal communities. Science of the Total Environment. 801: 149580.

[66] Fang, L., Huang, T., Lu, H., Wu, X.L. Chen, Z., Yang, H., et al. 2023. Biochar-based materials in environmental pollutant elimination, H_2 production and CO_2 capture applications. Biochar. 5(1): 42.

[67] Warnock, D.D., Lehmann, J., Kuyper, T.W. and Rillig, M.C. 2007. Mycorrhizal responses to biochar in soil – concepts and mechanisms. Plant and Soil [Internet]. 300(1): 9–20. Available from: https://doi.org/10.1007/s11104-007-9391-5

[68] Shankar, R., Sachs, M., Francàs, L., Lubert-Perquel, D., Kerherve, G., Regoutz, A., et al. 2019. Porous boron nitride for combined CO_2 capture and photoreduction. Journal of Materials Chemistry A. 7(41): 23931–23940.

[69] Laghaei, M., Ghasemian, M., Lei, W., Kong, L. and Chao, Q. 2023. A review of boron nitride-based photocatalysts for carbon dioxide reduction. Journal of Materials Chemistry A.

[70] Itskou, I., L'Hermitte, A., Marchesini, S., Tian, T. and Petit, C. 2023. How to Tailor Porous Boron Nitride Properties for Applications in Interfacial Processes. Accounts of Materials Research. 4(2): 143–155.

[71] Kumari, S., Gusain, R., Kumar, A., Manwar, N., Jain, S.L. and Khatri, O.P. 2020. Direct growth of nanostructural MoS_2 over the h-BN nanoplatelets: An efficient heterostructure for visible light photoreduction of CO_2 to methanol. Journal of CO_2 Utilization. 42: 101345.

[72] Wu, D., Guo, J., Wang, H., Zhang, X., Yang, Y., Yang, C., et al. 2021. Green synthesis of boron and nitrogen co-doped TiO_2 with rich BN motifs as Lewis acid-base couples for the effective artificial CO_2 photoreduction under simulated sunlight. Journal of Colloid and Interface Science. 585: 95–107.

[73] Shankar, R., Lubert-Perquel, D., Mistry, E., Nevjestić, I., Heutz, S. and Petit, C. 2020. Water-stable boron-doped boron nitride photocatalyst for visible light-driven H_2 evolution and CO_2 photoreduction.

[74] Chen, H., Jiang, D-en., Yang, Z. and Dai, S. 2022. Engineering nanostructured interfaces of hexagonal boron nitride-based materials for enhanced catalysis. Accounts of Chemical Research. 56(1): 52–65.

[75] Cheng, L., Song, Y., Chen, H., Liu, G., Liu, G. and Jin, W. 2020. g-C_3N_4 nanosheets with tunable affinity and sieving effect endowing polymeric membranes with enhanced CO_2 capture property. Separation and Purification Technology. 250: 117200.

[76] Pomilla, F.R., Cortes, M.A., Hamilton, J.W.J., Molinari, R., Barbieri, G., Marcì, G., et al. 2018. An Investigation into the Stability of Graphitic C_3N_4 as a Photocatalyst for CO_2 Reduction. The Journal of Physical Chemistry C. 122(50): 28727–28738.

[77] Yu, Y. and Huang, H. 2023. Coupled adsorption and photocatalysis of g-C_3N_4 based composites: Material synthesis, mechanism, and environmental applications. Chemical Engineering Journal. 453: 139755.

[78] Peng, H.L., Zhong, F.Y., Zhang, J.B., Zhang, J.Y., Wu, P.K., Huang, K., et al. 2018. Graphitic carbon nitride functionalized with polyethylenimine for highly effective capture of carbon dioxide. Industrial and Engineering Chemistry Research. 57(32): 11031– 11038.

[79] Sun, X., Zhu, L., Wang, P., Zhao, W. and Chen, X. 2021. CO_2 removal from natural gas by moisture swing adsorption. Chemical Engineering Research and Design. 176: 162–168.

[80] Wang, W., Xu, P., Chen, M., Zeng, G., Zhang, C., Zhou, C., et al. 2018. Alkali metal-assisted synthesis of graphite carbon nitride with tunable band-gap for enhanced visible-light-driven photocatalytic performance. ACS Sustainable Chemistry and Engineering. 6(11): 15503–15516.

[81] Ponnaiah, S.K. and Periakaruppan, P. 2024. Fabrication of tin oxide and carbon black nanocomposite for effectual electron-hole separation and visible light-harvesting: Enhanced photocatalytic degradation of an organophosphorus pesticide. Materials Chemistry and Physics. 313: 128806.

[82] Ezugwu, C.I., Liu, S., Li, C., Zhuiykov, S., Roy, S. and Verpoort, F. 2022. Engineering metal-organic frameworks for efficient photocatalytic conversion of CO_2 into solar fuels. Coordination Chemistry Reviews. 450: 214245.

[83] Majdoub, M., Anfar, Z. and Amedlous, A. 2020. Emerging chemical functionalization of g-C_3N_4: covalent/noncovalent modifications and applications. ACS Nano. 14(10): 12390–12469.

[84] Hayat, A., Sohail, M., Ajmal, Z., Abd El-Gawad, H.H., Ghernaout, D., Al-Hadeethi, Y., et al. 2024. Advances/Scope and prospects of g-C_3N_4 derived fascinating photocatalyst as a leading route towards solar energy adaption. Journal of Cleaner Production. 140568.

[85] Ji, X., Guo, Y., Hua, S., Li, H. and Zhang, S. 2020. Interaction-determined sensitization photodegradation of dye complexes by boron nitride under visible light irradiation: experimental and theoretical studies. New Journal of Chemistry. 44(22): 9238–9247.

[86] Anjum, Q., Nasir, N., Cheema, S.A., Imran, M., Rahman, A.R., Tanveer, Z., et al. 2022. Multiscale modeling investigation into the thermal conductivity dynamics of graphene-silver nano-composites: a molecular dynamic study. Digest Journal of Nanomaterials and Biostructures (DJNB). 17(2).

[87] Du, X., Skachko, I., Barker, A. and Andrei, E.Y. 2008. Approaching ballistic transport in suspended graphene. Nature Nanotechnology. 3(8): 491–495.

[88] Hina, M., Kamran, K., Bashir, S., Ahmed, J., Ameer, D., Jahanzaib, M., et al. 2023. Extra Ordinary Properties of Graphene. *In*: Graphene: Fabrication, Properties and Applications. Springer. 21–52.

[89] Kakade, P.M., Kachere, A.R., Rondiya, S.R. and Bhosale, S.V. 2021. Graphene oxide assisted synthesis of magnesium oxide nanorods. ES Materials and Manufacturing. 12(2): 63–71.

[90] Saleem, H., Haneef, M. and Abbasi, H.Y. 2018. Synthesis route of reduced graphene oxide via thermal reduction of chemically exfoliated graphene oxide. Materials Chemistry and Physics. 204: 1–7.

[91] Ma, T., Gao, H., Cong, H., Yao, H., Wu, L., Yu, Z., et al. 2018. A bioinspired interface design for improving the strength and electrical conductivity of graphene-based fibers. Advanced materials. 30(15): 1706435.

[92] Fawaz, M., Bahadur, R., Dharmarajan, N.P., Yang, J.H., Sathish, C.I., Sadanandan, A.M., et al. 2023. Emerging trends of carbon nitrides and their hybrids for photo-/electro-chemical energy applications. Carbon. 214: 118345.

[93] Ong, W.J., Tan, L.L., Chai, S.P. and Yong, S.T. 2015. Heterojunction engineering of graphitic carbon nitride (gC_3N_4) via Pt loading with improved daylight-induced photocatalytic reduction of carbon dioxide to methane. Dalton Transactions. 44(3): 1249–1257.

[94] Tahir, B., Tahir, M. and Amin, N.A.S. 2017. Photo-induced CO_2 reduction by CH_4/H_2O to fuels over Cu-modified g-C_3N_4 nanorods under simulated solar energy. Applied Surface Science. 419: 875–885.

[95] Yuan, Y.P., Cao, S.W., Liao, Y.S., Yin, L.S. and Xue, C. 2013. Red phosphor/g-C_3N_4 heterojunction with enhanced photocatalytic activities for solar fuels production. Applied Catalysis B: Environmental. 140: 164–168.

[96] Cao, S.W., Liu, X.F., Yuan, Y.P., Zhang, Z.Y., Liao, Y.S., Fang, J., et al. 2014. Solar-to-fuels conversion over In_2O_3/g-C_3N_4 hybrid photocatalysts. Applied Catalysis B: Environmental. 147: 940–946.

[97] Shi, H., Chen, G., Zhang, C. and Zou, Z. 2014. Polymeric g-C_3N_4 coupled with $NaNbO_3$ nanowires toward enhanced photocatalytic reduction of CO_2 into renewable fuel. ACS Catalysis. 4(10): 3637–3643.

[98] Bai, S., Wang, X., Hu, C., Xie, M., Jiang, J. and Xiong, Y. 2014. Two-dimensional gC_3N_4: an ideal platform for examining facet selectivity of metal co-catalysts in photocatalysis. Chemical Communications. 50(46): 6094–6097.

[99] Hong, J., Zhang, W., Wang, Y., Zhou, T. and Xu, R. 2014. Photocatalytic reduction of carbon dioxide over self-assembled carbon nitride and layered double hydroxide: the role of carbon dioxide enrichment. ChemCatChem. 6(8): 2315–2321.

[100] Kuriki, R., Sekizawa, K., Ishitani, O. and Maeda, K. 2015. Visible-light-driven CO_2 reduction with carbon nitride: enhancing the activity of ruthenium catalysts. Angewandte Chemie International Edition. 54(8): 2406–2409.

[101] Maeda, K., Sekizawa, K. and Ishitani, O. 2013. A polymeric-semiconductor–metal-complex hybrid photocatalyst for visible-light CO_2 reduction. Chemical Communications. 49(86): 10127–10129.

[102] Maeda, K., Kuriki, R., Zhang, M., Wang, X. and Ishitani, O. 2014. The effect of the pore-wall structure of carbon nitride on photocatalytic CO_2 reduction under visible light. Journal of Materials Chemistry A. 2(36): 15146–15151.

[103] Shown, I., Hsu, H.C., Chang, Y.C., Lin, C.H., Roy, P.K., Ganguly, A., et al. 2014. Highly efficient visible light photocatalytic reduction of CO_2 to hydrocarbon fuels by Cu-nanoparticle decorated graphene oxide. Nano Letters. 14(11): 6097–6103.

[104] Zhao, G., Pang, H., Liu, G., Li, P., Liu, H., Zhang, H., et al. 2017. Co-porphyrin/carbon nitride hybrids for improved photocatalytic CO_2 reduction under visible light. Applied Catalysis B: Environmental. 200: 141–149.

[105] Wang, S., Lin, J. and Wang, X. 2014. Semiconductor–redox catalysis promoted by metal–organic frameworks for CO_2 reduction. Physical Chemistry Chemical Physics. 16(28): 14656–14660.

[106] Ong, W.J., Tan, L.L., Chai, S.P. and Yong, S.T. 2015. Graphene oxide as a structure-directing agent for the two-dimensional interface engineering of sandwich-like graphene–gC_3N_4 hybrid nanostructures with enhanced visible-light photoreduction of CO_2 to methane. Chemical Communications. 51(5): 858–861.

[107] Tu, W., Zhou, Y., Liu, Q., Tian, Z., Gao, J., Chen, X., et al. 2012. Robust hollow spheres consisting of alternating titania nanosheets and graphene nanosheets with high photocatalytic activity for CO_2 conversion into renewable fuels. Advanced Functional Materials. 22(6): 1215–1221.

[108] Tu, W., Zhou, Y., Liu, Q., Yan, S., Bao, S., Wang, X., et al. 2013. An in situ simultaneous reduction-hydrolysis technique for fabrication of TiO_2-graphene 2D sandwich-like hybrid nanosheets: graphene-promoted selectivity of photocatalytic-driven hydrogenation and coupling of CO_2 into methane and ethane. Advanced Functional Materials. 23(14): 1743–1749.

[109] Wang, P.Q., Bai, Y., Luo, P.Y. and Liu, J.Y. 2013. Graphene–WO_3 nanobelt composite: Elevated conduction band toward photocatalytic reduction of CO_2 into hydrocarbon fuels. Catalysis Communications. 38: 82–85.

[110] An X., Li, K. and Tang, J. 2014. Cu_2O/reduced graphene oxide composites for the photocatalytic conversion of CO_2. ChemSusChem. 7(4): 1086–1093.

[111] Yu, J., Jin, J., Cheng, B. and Jaroniec, M. 2014. A noble metal-free reduced graphene oxide–CdS nanorod composite for the enhanced visible-light photocatalytic reduction of CO_2 to solar fuel. Journal of Materials Chemistry A. 2(10): 3407–3416.

[112] Zhang, H., Wei, J., Dong, J., Liu, G., Shi, L., An, P., et al. 2016. Efficient visible-light-driven carbon dioxide reduction by a single-atom implanted metal–organic framework. Angewandte Chemie. 128(46): 14522–14526.

[113] Xu, M., Zhao, X., Jiang, H., Chen, S. and Huo, P. 2021. MOFs-derived C-In_2O_3/g-C_3N_4 heterojunction for enhanced photoreduction CO_2. Journal of Environmental Chemical Engineering. 9(6): 106469.

[114] Murugesan, P., Narayanan, S. and Manickam, M. 2017. Experimental studies on photocatalytic reduction of CO_2 using AgBr decorated g-C_3N_4 composite in TEA mediated system. Journal of CO_2 Utilization. 22: 250–261.

[115] He, Y., Zhang, L., Fan, M., Wang, X., Walbridge M.L., Nong, Q., et al. 2015. Z-scheme SnO_2–x/g-C_3N_4 composite as an efficient photocatalyst for dye degradation and photocatalytic CO_2 reduction. Solar Energy Materials and Solar Cells. 137: 175–184.

[116] Nahar, S. 2020. Photocatalytic degradation of organic pollutants and reduction of carbon dioxide over silver-based photocatalysts. UKM, Bangi.

[117] Bai, Y., Ye, L., Wang, L., Shi, X., Wang, P. and Bai, W. 2016. A dual-cocatalyst-loaded Au/BiOI/MnO_x system for enhanced photocatalytic greenhouse gas conversion into solar fuels. Environmental Science: Nano. 3(4): 902–909.

[118] Özdokur, K.V., Koçak, S. and Ertaş, F.N. 2018. Nanostructured Metal–Metal Oxides and Their Electrocatalytic Applications. Advanced Coating Materials. 275.

[119] Gao, J., Yu, B. and He, L.N. 2011. Reduction of carbon dioxide to energy-rich products. *In*: Production and Purification of Ultraclean Transportation Fuels. ACS Publications. 143–174.

[120] Ismail, A.A., Bahnemann, D.W. and Al-Sayari, S.A. 2012. Synthesis and photocatalytic properties of nanocrystalline Au, Pd and Pt photodeposited onto mesoporous RuO_2-TiO_2 nanocomposites. Applied Catalysis A: General. 431: 62–68.

[121] Domínguez-Espíndola, R.B., Arias, D.M., Rodríguez-González, C. and Sebastian, P. 2022. Advances in TiO_2-Based Photocatalytic Systems for CO_2 Reduction Using Solar Energy. Available at SSRN 4057325.

[122] Pollap, A., Serafin, J., Serrano, I., Sreńscek-Nazzal, J. and Llorca, J. 2022. Promising Cheap Photocatalysts Based on Nanoshaped TiO_2-RGO Composite Doped with Metals (Pt and Cu) for Hydrogen Photoproduction. Available at SSRN 4156660.

[123] Cedeño Morales, E.M. 2022. Desarrollo de estructuras a base de carbono a partir de orgánometálico tipo MOFs para fines de adsorción y conversión de CO_2. Universidad Autónoma de Nuevo León.

[124] Rajalakshmi, K. 2011. Photocatalytic Reduction of Carbon dioxide in conjunction with decomposition of water on oxide semiconductor surfaces. Indian Institute of Technology, Madras, India.

[125] Jin, B., Yan, Q. and Dou, Y. 2012. Materials for energy storage and conversion based on metal oxides. Recent Patents on Materials Science. 5(3): 199–212.

[126] Xiu, Z., Guo, M., Zhao, T., Pan, K., Xing, Z., Li, Z., et al. 2020. Recent advances in Ti^{3+} self-doped nanostructured TiO_2 visible light photocatalysts for environmental and energy applications. Chemical Engineering Journal. 382: 123011.

[127] Miao, G., Zhou, M., Zhang, Y., Yang, Z. and Liu, Y. 2022. Nanoarchitectonic composites with direct z-scheme heterojunction of $CsPbBr_3$ quantum dot/$CuCo_2O_4$ nanosheet for efficient photocatalytic CO_2 reduction. Journal of Alloys and Compounds. 925: 166733.

[128] Júnior, M.A.M., Morais, A. and Nogueira, A.F. 2016. Boosting the solar-light-driven methanol production through CO_2 photoreduction by loading Cu_2O on TiO_2-pillared $K_2Ti_4O_9$. Microporous and Mesoporous Materials. 234: 1–11.

[129] Tsai, C.W., Chen, H.M., Liu, R.S., Asakura, K. and Chan, T.S. 2011. Ni@NiO core–shell structure-modified nitrogen-doped $InTaO_4$ for solar-driven highly efficient CO_2 reduction to methanol. The Journal of Physical Chemistry C. 115(20): 10180–10186.

[130] Singhal, N., Goyal, R. and Kumar, U. 2017. Visible-light-assisted photocatalytic CO_2 reduction over $InTaO_4$: selective methanol formation. Energy and Fuels. 31(11): 12434–12438.

[131] Lu, M.F., Zhou, C.P., Li, Q.Q., Zhang, C.L. and Shi H.F. 2018. Effects of B site doping on electronic structures of $InNbO_4$ based on hybrid density functional calculations. *In*: IOP Conference Series: Materials Science and Engineering. IOP Publishing. 12020.

[132] Chen, H., Yang, X., Lin, M., Li, D., Wang, B., Shen, J., et al. 2023. Dual Z-Scheme Heterojunction SiC-Sandwiched $InVO_4$ Nanoflowers for Boosting Visible-Light Photocatalytic CO_2 Reduction with H_2O. ACS Sustainable Chemistry and Engineering. 11(51): 18029–18040.

[133] Crake, A. 2017. Metal-organic frameworks based materials for photocatalytic CO_2 reduction. Materials Science and Technology. 33(15): 1737–1749.

[134] Karimi, M., Shirzad, M., Silva, J.A.C. and Rodrigues, A.E. 2022. Biomass/Biochar carbon materials for CO_2 capture and sequestration by cyclic adsorption processes: A review and prospects for future directions. Journal of CO_2 Utilization. 57: 101890.

[135] Abd, A.A., Naji, S.Z., Hashim, A.S. and Othman, M.R. 2020. Carbon dioxide removal through physical adsorption using carbonaceous and non-carbonaceous adsorbents: a review. Journal of Environmental Chemical Engineering. 8(5): 104142.

[136] Zeghioud, H., Fryda, L., Djelal, H., Assadi, A. and Kane, A. 2022. A comprehensive review of biochar in removal of organic pollutants from wastewater: Characterization, toxicity, activation/functionalization and influencing treatment factors. Journal of Water Process Engineering. 47: 102801.

[137] Mallesh, D., Anbarasan, J., Kumar, P.M., Upendar, K., Chandrashekar, P., Rao, B., et al. 2020. Synthesis, characterization of carbon adsorbents derived from waste biomass and its application to CO_2 capture. Applied Surface Science. 530: 147226.

[138] Yadav, G., Yadav, N., Sultana, M. and Ahmaruzzaman, M. 2023. A comprehensive review on low-cost waste-derived catalysts for environmental remediation. Materials Research Bulletin. 112261.

[139] Lu, L., Yang, Q., Xu, Q., Sun, Y., Tang, S. and Tang, X., et al. 2022. Two-dimensional materials beyond graphene for the detection and removal of antibiotics: A critical review. Critical Reviews in Environmental Science and Technology. 52(19): 3493–3524.

[140] Zhang, Y., Wang, S., Feng, D., Gao, J., Dong, L., Zhao, Y., et al. 2022. Functional biochar synergistic solid/liquid-phase CO_2 capture: A review. Energy and Fuels. 36(6): 2945–2970.

[141] Sevim, F., Lacin, O., Ediz, E.F. and Demir, F. 2021. Adsorption capacity, isotherm, kinetic, and thermodynamic studies on adsorption behavior of malachite green onto natural red clay. Environmental Progress and Sustainable Energy. 40(1): e13471.

[142] Boscherini, M., Miccio, F., Papa, E., Medri, V., Landi, E., Doghieri, F., et al. 2021. The relevance of thermal effects during CO_2 adsorption and regeneration in a geopolymer-zeolite composite: Experimental and modelling insights. Chemical Engineering Journal. 408: 127315.

[143] Maftouh, A., El Fatni, O., El Hajjaji, S., Jawish, M.W. and Sillanpää, M. 2023. Comparative review of different adsorption techniques used in heavy metals removal in water. Biointerface Research in Applied Chemistry. 13: 397.

[144] Kamran, U. and Park, S.J. 2021. Chemically modified carbonaceous adsorbents for enhanced CO_2 capture: A review. Journal of Cleaner Production. 290: 125776.

[145] Matito-Martos, I., Sepulveda, C., Gomez, C., Acien, G., Perez-Carbajo, J. and Delgado, J.A., et al. 2021. Potential of CO_2 capture from flue gases by physicochemical and biological methods: A comparative study. Chemical Engineering Journal. 417: 128020.

[146] Boer, D.G., Langerak, J. and Pescarmona, P.P. 2023. Zeolites as Selective Adsorbents for CO_2 Separation. ACS Applied Energy Materials. 6(5): 2634–2656.

[147] Rad, L.R. and Anbia, M. 2021. Zeolite-based composites for the adsorption of toxic matters from water: A review. Journal of Environmental Chemical Engineering. 9(5): 106088.

[148] Lin, Y.J. and Chen, J.C. 2021. Resourcization and valorization of waste incineration fly ash for the synthesis of zeolite and applications. Journal of Environmental Chemical Engineering. 9(6): 106549.

[149] Bahmanzadegan, F. and Ghaemi, A. 2023. Modification and functionalization of zeolites to improve the efficiency of CO_2 adsorption: A review. Case Studies in Chemical and Environmental Engineering. 100564.

[150] Zhang, Q., Yu, J. and Corma, A. 2020. Applications of zeolites to C1 chemistry: recent advances, challenges, and opportunities. Advanced Materials. 32(44): 2002927.

[151] Yin S., Zhou, Y., Liu, Z., Wang, H., Zhao, X., Zhu, Z., et al. 2024. Elucidating protonation pathways in CO_2 photoreduction using the kinetic isotope effect. Nature Communications. 15(1): 437.

[152] Wang, K., Cheng, Q., Hou, W., Guo, H., Wu, X., Wang, J., et al. 2024. Unlocking the Charge-Migration Mechanism in S-Scheme Junction for Photoreduction of Diluted CO_2 with High Selectivity. Advanced Functional Materials. 34(8): 2309603.

[153] Kumar, A., Singh, P., Khan, A.A.P., Van Le, Q., Nguyen, V.H., Thakur, S., et al. 2022. CO_2 photoreduction into solar fuels via vacancy engineered bismuth-based photocatalysts: Selectivity and mechanistic insights. Chemical Engineering Journal. 439: 135563.

Chapter **6**

CO_2 Conversion to Valuable Products Using Nanomaterials

Urvika[1], Indu Sharma[2], Rekha Gaba*[1] and Ramesh Kataria*[2]
[1]Department of Chemistry, DAV University, Jalander-144012, Punjab, India
[2]Department of Chemistry, Panjab University, Chandigarh-160014, India

6.1 INTRODUCTION

The primary cause of atmospheric pollution and the global enhancement of the greenhouse effect is the excessive release of greenhouse gases, specifically, the gases carbon dioxide (CO_2) or methane (CH_4) [1]. The burning of fossil fuels to generate power and energy is the main factor contributing to this occurrence. The primary source of energy in the modern age is the combustion of fossil fuels, which produces a large excretion of atmospheric greenhouse gases. The calculation of the Climatic Forcing Potential, which is sometimes referred to as the Global Warming Potential power source, relies on the dynamic interaction of several essential components. The US Environmental Protection Agency (2005) estimated that the country's total greenhouse gas emissions in 2004 were roughly 84.6% carbon dioxide, 7.9% methane, 5.5% nitrous oxide, and 2% from sulfur hexafluorides, hydrochlorofluorocarbons, and chlorofluorocarbons [1].

There has been a noticeable emphasis in recent years on the concern that CO_2 buildup in the atmosphere is causing global warming. Despite the existence of various proposed CO_2 capture technologies, there is increasing interest in harnessing captured CO_2 from power plants (Figure 6.1) [2]. This preference primarily arises from apprehensions regarding the safety implications associated with underground and oceanic storage of CO_2. The atmospheric concentration of

*For Correspondence: rekhagaba@gmail.com; rkataria@pu.ac.in

DOI: 10.1201/9781003492184-6

CO_2 is now nearing 400 parts per million (ppm), significantly exceeding the pre-industrial level of about 300 ppm [3]. The range of methodologies and a variety of materials are utilized in CO_2 conversion procedures. Additionally, the absorption of excess CO_2 by the oceans initiates ocean acidification, a phenomenon that carries extensive implications. This chemical process entails the dissolution of CO_2 in seawater, leading to the formation of carbonic acid and substantial disruption to marine ecosystems. Organisms harboring calcareous carbonate structures, including corals, mollusks, and some plankton species, are particularly susceptible as their shells or skeletons experience diminished strength, impairing growth, and reproductive capacities [4].

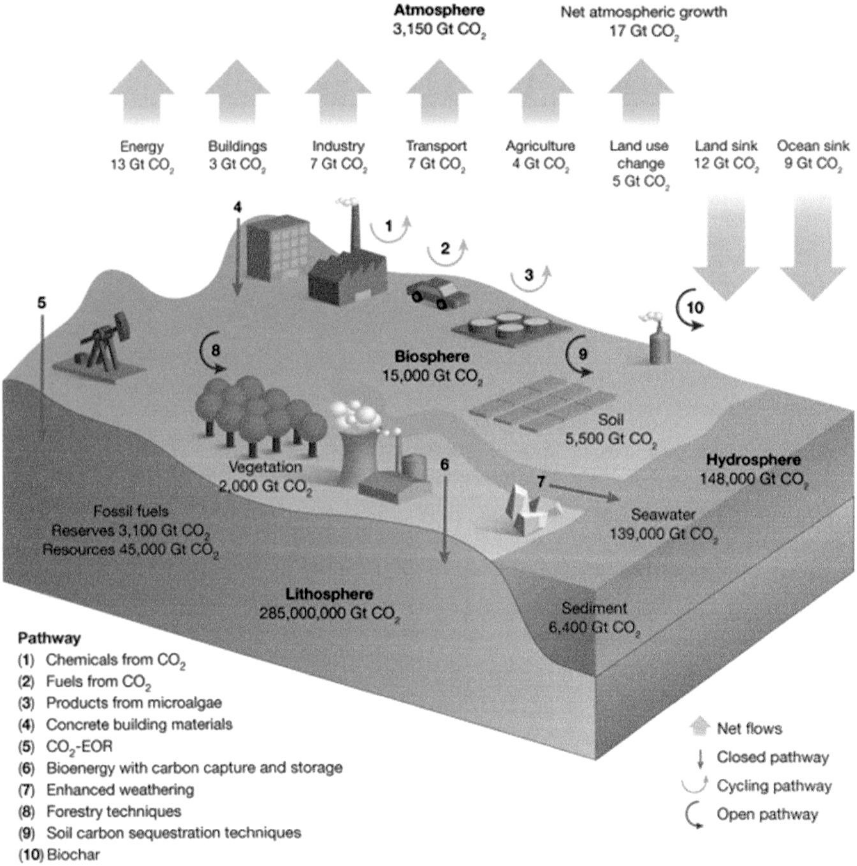

Figure 6.1 Representation of open utilization pathways, closed pathways and cycling utilization. Reproduced with permission from [3].

Furthermore, nanostructured catalysts demonstrate significant efficacy in the process of CO_2 photoreduction. Since the middle of 2010, scholars have consistently broadened their areas of interest to enhance CO_2 utilization's efficiency, stability, selectivity, and environmental friendliness while attempting to lower related expenses [5]. Nanomaterials function as catalysts in various

CO_2 conversion processes, including photocatalysis, electrocatalysis, and thermal catalysis showcasing adaptability and efficiency across a spectrum of energy inputs. Photocatalytic nanomaterials harness solar energy to trigger CO_2 conversion reactions, providing a sustainable and eco-friendly avenue for carbon utilization. Electrocatalytic nanomaterials facilitate the abatement of carbon dioxide electrochemically, resulting in the production of valuable fuels and chemicals through the application of renewable electricity sources [6].

This chapter illustrates the use of nanomaterials to transform CO_2 into goods with additional value including the production of fuels, chemicals, and building materials. Hence, it provides an outline to use nanomaterials for various thermal and electrocatalytic reactions that may help in the mitigation of CO_2 from the atmosphere.

6.2 EXERTION OF CO_2

Fossil fuel combustion emerges as a towering force with its immense smokestacks spewing forth vast clouds of carbon dioxide into the environment. From the bustling furnaces of power plants to the congested streets teeming with vehicles, the burning of coal, oil, and natural gas casts a profound shadow over the global carbon balance. Yet, human influence extends beyond industry and transportation [7]. Agriculture plays a pivotal role as fields are cultivated, forests cleared and livestock roam. The expansion of agricultural frontiers through deforestation liberates carbon stored in trees and soil further tilting the scales in favor of CO_2. Volcanic eruptions and primal forces of creation and destruction release significant amounts of CO_2 into the atmosphere [8]. There are various sources of exertion of CO_2 that are enlisted below:

6.2.1 Nature Calls

Although they contribute very little to global CO_2 levels in comparison to human activity, volcanic eruptions, and wildfires, respectively, release CO_2 held in the Earth's crust and vegetation.

6.2.2 Industrial Processes

Carbon dioxide is emitted due to the combustion and chemical reactions that occur in numerous industrial activities including manufacturing, cement production and chemical processing.

6.2.3 Using Fossil Fuels for Energy

When natural gas, oil, and coal are burned to provide energy for transportation, power plants, and industrial activities, significant volumes of CO_2 are released into the atmosphere.

6.2.4 Decomposition Process

The disintegration of organic matter, natural decay, and decomposition processes in ecosystems emit CO_2, which adds to ambient emissions.

6.2.5 Agricultural Practices

Raising livestock, clearing forests to make room for more agriculture, and applying fertilizer and tillage to the soil all contribute to emissions by releasing carbon dioxide into the sky.

6.2.6 Transportation

Within the cities with a high vehicle density, the combustion of petrol and diesel fuels in automobiles, trucks, ships, and aircraft results in a considerable contribution to CO_2 emissions.

6.2.7 Process of Deforestation

The carbon balance is thrown off when forests are destroyed for construction, farming, or other purposes, releasing CO_2 that was stored within plants and trees into the atmosphere.

6.3 CO_2: A CONCERN FOR HUMAN HEALTH AND THE ENVIRONMENT

This aims to provide a summary of the most current information regarding the risks associated with underground CO_2 storage, highlighting areas that require further investigation to enhance our understanding of these risks. The dangers associated with underground CO_2 storage such as CO_2 and CH_4 leakage into water sources, seismic activity, land shifting and displacement of brine are not fully comprehended. The primary goal of research and development is to determine the mechanisms that govern the release through or alongside wells, faults, and fractures in order to evaluate the rate of leakage and its effect on marine ecosystems. Carbon dioxide (CO_2) is undeniably crucial for sustaining life on our planet but when its presence exceeds natural levels, it poses significant hazards to the impact of surroundings on human well-being. The heightened concentrations of CO_2 in indoor environments such as in poorly ventilated spaces can trigger a range of adverse effects. Symptoms like headaches, dizziness, fatigue, and difficulty focusing often collectively termed "sick building syndrome" are becoming prevalent. Furthermore, prolonged exposure to elevated CO_2 levels can worsen respiratory conditions like chronic obstructive pulmonary disorder and asthma due to CO_2 acting as a respiratory irritant. Additionally, cognitive functions and decision making abilities may suffer impacting productivity and overall mental wellness [9].

The impact of CO_2 extends beyond human health, profoundly influencing global climate and ecosystems. As a potent greenhouse gas, carbon dioxide plays a role in the occurrence of climate change by retaining heat within the atmosphere. This leads to modifications and variations in the frequency and severity of extreme weather phenomena including storms, heat waves and floods along with shifts in weather patterns. The repercussions are felt across various sectors posing risks to human lives, infrastructure, and agricultural productivity. Furthermore, the acceleration of polar ice caps and glaciers melting, driven by rising temperatures raises the risk of sea level rise, endangering vulnerable ecosystems and coastal communities [10].

In addition, the oceans' digestion of excess CO_2 causes a condition called ocean acidification. This process, whereby CO_2 dissolves in seawater to form carbonic acid, disrupts marine ecosystems profoundly. Creatures having calcium carbonate structures, such as certain forms of plankton, mollusks, and corals, are particularly vulnerable, as their shells or skeletons weaken, impairing growth and reproductive capabilities. This disturbance ripples across marine food webs, affecting fish populations and endangering the means of subsistence for coastal people who depend on fishing [11].

Elevated CO_2 levels in terrestrial environments can alter plant physiology and distribution, which can have profound effects on ecosystem health and biodiversity. While some plant species may benefit from increased CO_2 concentrations through enhanced photosynthesis and growth, others may suffer from reduced nutrient uptake and water efficiency rendering them more susceptible to stressors like drought, pests, and diseases. Such changes potentially impact wildlife populations that depend on these plants for food in a cascade manner and habitat, leading to shifts in species distributions and loss of biodiversity.

In light of these profound impacts, urgent action is imperative to mitigate CO_2 emissions and transition to sustainable, low carbon energy sources. Addressing the root causes of CO_2 emissions and implementing strategies to adjust to the consequences of global warming are essential steps toward safeguarding human well-being and the resilience of the Earth's ecosystems for present and future generations [12].

The following list of variables that negatively affect the environment and people as a result of CO_2.

6.3.1 Pollution

Unwanted waste of human origin that is disposed of in the atmosphere, soil freshwater and marine environment garbage without thinking about the cost or the consequences poses an existential danger to the health of people and the planet and imperils the viability of contemporary society. Fine Particulate Matter (PM2.5) in the air, ozone, sulfur, and nitrogen oxides, freshwater pollution, mercury, phosphate, nitrogen, plastic, and oil and gas waste in the ocean, and lead, mercury, pesticides, chemical residues from industry, electronic waste, and radiation poisoning of the land are all considered forms of pollution.

6.3.2 Change in Weather

The increased CO_2 concentrations in the environment have a huge impact on global change by enhancing the greenhouse effect. Global temperatures rise as a result, putting lives and livelihoods at risk from increasingly frequent and severe heatwaves, storms, droughts, and other extreme weather events.

6.3.3 Ecosystem Disturbance

Elevated CO_2 levels have the potential to disrupt ecosystems and biodiversity by altering temperature and precipitation patterns. This includes altered habitats, rearranged species ranges, heightened extinction risk for vulnerable species, and shifted timing of biological processes like migration and flowering.

6.3.4 Effects on Health

Excessive indoor CO_2 concentrations, including those in crowded or inadequately ventilated buildings, can lead to fatigue, headaches, and cognitive decline. Asthma and Chronic Obstructive Pulmonary Disease, also called (COPD) are two respiratory conditions that might worsen with prolonged exposure to high CO_2 levels.

6.3.5 Impact on Water Resources

Changes in rainfall patterns carried on by warming temperatures caused by CO_2 may cause floods in some places and droughts in others. These modifications may influence water resources used for energy generation, agriculture, industry, and drinking, escalating rivalry and possible conflicts over water. Additionally, the oceans absorb a large amount of excess CO_2 released into the atmosphere, which causes ocean acidification. This process harms marine life by decreasing the pH of seawater. Species that form shells, such the corals, and mollusks, along with certain types of plankton, are particularly damaged and can hurt fisheries or marine ecosystems.

6.4 NEED TO CONVERT CO_2 INTO VALUABLE MATERIALS

Converting CO_2 into usable fuels or chemicals is a viable way to lower CO_2 emissions and investigate a possible energy storage technique. Formerly seen as a waste product with harmful environmental effects, CO_2 now presents untapped potential as a valuable resource for producing various commodities. This conversion process not only helps mitigate greenhouse gas emissions by capturing CO_2 from the atmosphere or industrial sources but also generates marketable goods. Innovative technologies such as carbon storage and capture

(CCS) and carbon utilization and capture (CCU) facilitate this transformation and allow CO_2 to be turned into synthetic fuels, building materials, chemicals, and even food products (Figure 6.2) [13]. Utilizing CO_2 in this manner not only aids in decarbonizing key sectors of the economy but also promotes resource circularity and strengthens resilience to climate change impacts. A summary and classification of the mechanisms and characteristics of several conversion methods are made, including mineralization, photochemical/photoelectrochemical, thermochemical, enzymatic, and electrochemical processes.

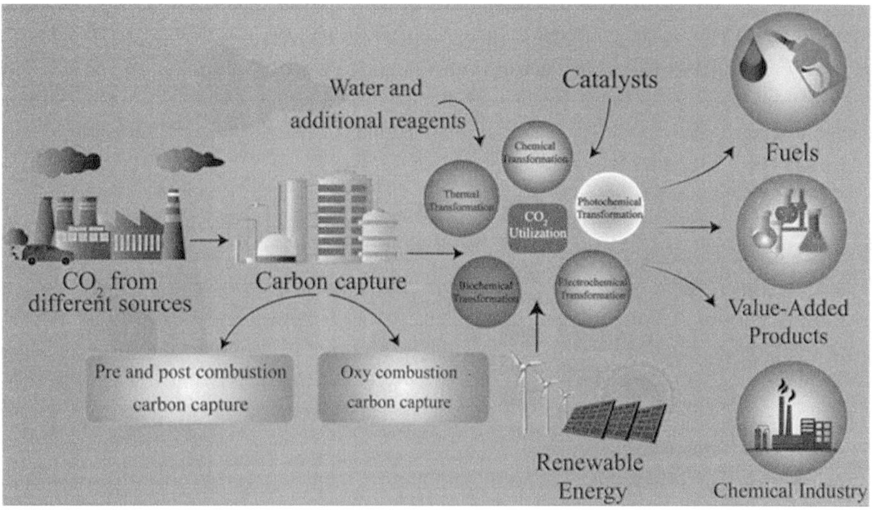

Figure 6.2 Sources of CO_2 emissions and how it is used to create goods with additional value. Reproduced with authorization from [14].

The key points given below explain why we need to convert CO_2 into valuable products:

6.4.1 Economic Challenges

The transformation of CO_2 into usable goods creates economic possibilities by creating new markets and enterprises. Sectors such as tackle environmental challenges. The terms are Chemical Capture and Use (CCU) and CCS stand for Carbon Capture and Storage can foster innovation, provide employment, and propel economic growth. By completing the carbon cycle, CO_2 conversion aids in the shift to a circular economy. CO_2 can be gathered and turned into goods rather than being released into the atmosphere as trash, encouraging the reuse and recycling of carbon resources.

6.4.2 Resistance to Climate Change

Using CO_2 in valued products promotes resilience to the effects of climate change by cutting emissions and promoting sustainable practices. By using this method, the effects of global warming on ecosystems, communities, and economies are mitigated. Mitigating the consequences of climate change can be achieved by the transformation of CO_2 into useful goods, hence reducing greenhouse gas emissions. We can efficiently store carbon that might otherwise contribute to global warming and climate instability by capturing and using CO_2.

6.4.3 Managing Resources

CO_2 is a lucrative commodity and may be turned from a problem to an asset. It was formerly thought of as a waste product. In support of sustainable resource management, this strategy encourages resource efficiency and lessens dependency on limited natural resources. The market is becoming more interested in sustainable goods and materials. Businesses can supply this demand while lowering their carbon impact by turning CO_2 into valuable products like chemicals, food products, building materials, and synthetic fuels.

6.5 CONVERSION OF CO_2 USING NANOMATERIALS

Utilizing nanomaterials for the conversion of CO_2 represents a groundbreaking approach to combating climate change and advancing sustainable practices. Nanomaterials, due to their unique properties at the nanoscale exhibit remarkable catalytic capabilities that make them highly effective in driving CO_2 conversion processes. These materials can be engineered with precise control over their composition, structure, and surface properties to optimize catalytic performance (Figure 6.3) [14]. Nanomaterials serve as catalysts in various CO_2 conversion reactions such as photocatalysis, electrocatalysis, and thermal catalysis. These

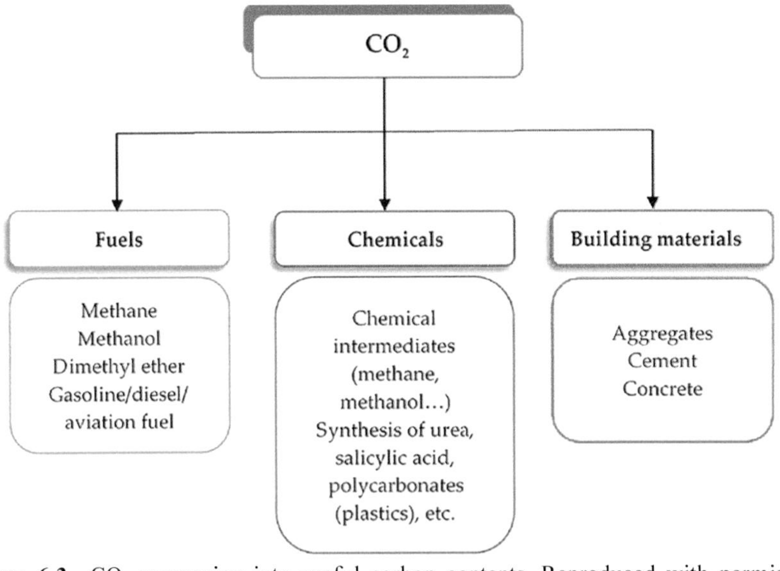

Figure 6.3 CO_2 conversion into useful carbon contents. Reproduced with permission from [15].

processes offer versatility and efficiency across different energy inputs. Photocatalytic nanomaterials harness solar energy to initiate CO_2 conversion reactions by providing a renewable and environmentally friendly route to carbon utilization. Electrocatalytic nanomaterials enable the electrochemical reduction of CO_2 converting it into valuable chemicals and fuels using electricity from renewable sources.

Additionally, nanomaterials can facilitate the direct conversion of carbon dioxide into high-value added products such as CH_4, CH_3OH, and HCOOH through heterogeneous catalysis. By harnessing the unique properties of nanomaterials CO_2 conversion technologies provide a viable approach to reducing greenhouse gas emissions, promoting sustainable resource utilization, and driving the shift towards a future that is carbon neutral. Ongoing research and development in this domain are crucial for unleashing the complete capabilities of nanomaterial-enabled CO_2 conversion and realizing its widespread implementation on a global scale.

6.5.1 Fuels

In recent years, the Earth has faced significant threats from worldwide climate change, mostly attributed to the escalating energy crisis and the rise in global warming resulting from fossil fuels' depletion. The escalating number of people on the planet, the amount of transportation used, and the rate of industrialization have all increased the demand for fossil fuels, consequently elevating CO_2 emissions into the atmosphere [15]. This surge in CO_2 emissions stems primarily from the combustion of fossil fuels, which releases substantial amounts of this greenhouse gas, thereby contributing to the escalation of atmospheric CO_2 concentrations. Addressing these pressing challenges necessitates innovative approaches. Among them, the transformation of coal dioxide (CO_2) into usable carbon based fuels through electrochemistry, utilizing renewable electricity, emerges as a promising avenue (Figure 6.4) [16]. Consequently, there is an urgent need for the development of technologies aimed to mitigate the impact of global warming, it is crucial to decrease the buildup of CO_2 in the atmosphere caused by this combustion. Carbon capture and utilization presents a promising and sustainable solution, serving as an substitute to carbon capture and storage. This strategy not only offers a potential solution to environmental degradation but also presents a means to mitigate the energy crisis [5]. The main way to this transition is the development of robust technologies capable of efficiently converting small, low energy molecules such as H_2O, N_2, and CO_2 into chemicals and fuels, relying solely on sustainable energy sources, with solar energy being of particular importance. Researchers have devoted considerable effort over the past three decades to evaluating various methods for the reduction of CO_2 into valuable products [17]. These methods encompass a combination of photocatalytic, electrocatalytic, and photoelectrochemical approaches. By harnessing renewable energy, particularly solar energy, these methodologies hold the promise of transforming CO_2 from a pollutant into a valuable resource, thereby simultaneously addressing the challenges posed by climate change and the energy crisis. Although projecting the market for CO_2 applications in the future is difficult, there are certain applications and domains in which CO_2 can be used efficiently. This involves the resource intensive process of reacting CO_2 with hydrogen to produce fuels that contain carbon and are easier to handle and use than hydrogen alone. Aviation fuels, petrol, methanol, and methane are among the illustrative examples.

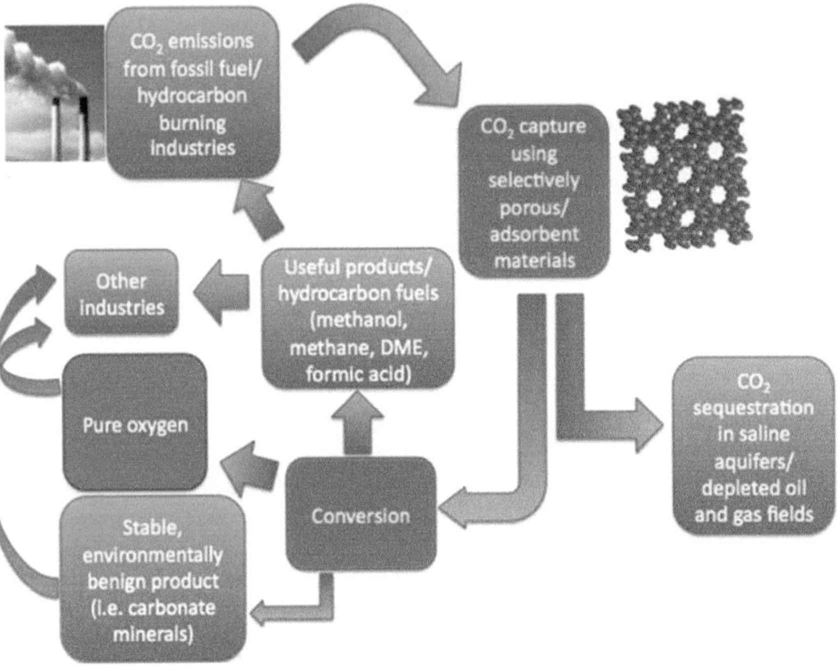

Figure 6.4 Schematic illustration of the chemical recycling of CO_2 for a sustainable economy. Reproduced with permission from [17].

Zhang et al. synthesized a novel 2D hierarchical structure composed of Pd/SnO_2, featuring ultrathin Pd nanosheets with SnO_2 nanoparticles partially covering them (Figure 6.5) [18]. This innovative design facilitates the transmission of multiple electrons, which is essential for the efficient electroreduction of CO_2 into CH_3OH. Harnessing electricity from renewable sources, this electrochemical conversion process contributes to the development of an artificial carbon cycle. By adjusting the coverage of SnO_2 on the Pd nanosheets, researchers achieved optimized selectivity and long-term stability. The exceptional selectivity towards CH_3OH is attributed to several factors: firstly, the Pd/SnO_2 nanosheets exhibit outstanding CO_2 adsorption capacity; secondly, the partially exposed Pd sites on the nanosheets facilitate the adsorption of CO* intermediates for subsequent electroreduction; and thirdly, the presence of Pd–O–Sn interfaces aid in the additional reduction of CO* intermediates to CH_3OH.

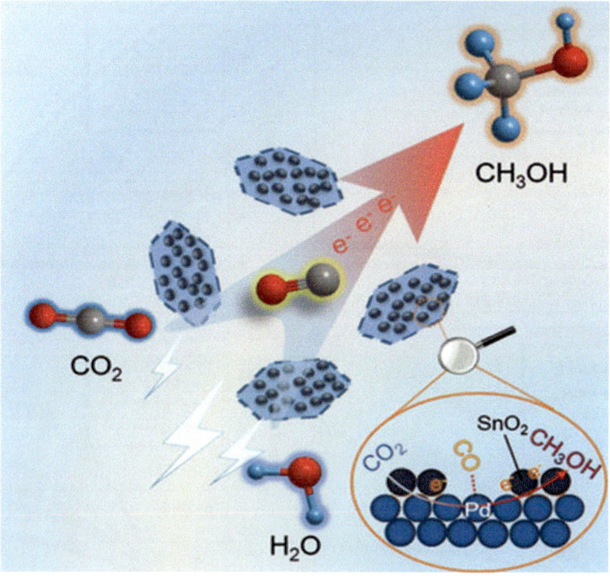

Figure 6.5 Pictorial representation of the CO_2 is converted electrochemically. Reproduced with authorization from [19].

Similarly, Renones et al. explored the impact of indium doping on the staging of TiO_2 as a photocatalyst for CO_2 reduction using water as an electron donor (Figure 6.6) [19]. Additionally, they investigated how the deposition of gold nanoparticles these effects are amplified on the semiconductor surface, potentially creating photocatalytic devices with improved efficiency. To achieve this, catalysis with varying indium concentrations was produced through a straightforward wet biochemical method commencing from anatase TiO_2. Subsequently, further modification in a selected indium doped sample was done by decorating it with silver using an impregnation method. The results demonstrated that indium doped catalysts exhibit increased selectivity towards the highly reduced product CH_4 compared to undoped titania, which predominantly produces CO in photocatalytic CO_2 reduction tests. The addition of silver decoration plays a crucial role in enhancing performance. Initially, the decoration results in the production of silver oxides, that are subsequently converted to metallic Ag particles by the methods' usage of UV radiation. These nanoparticles effectively reduce electron hole recombination using photoexcited electron extraction from Titania's conduction band, thus serving as electron transfer centers. This process significantly increases the selectivity towards methanol and propane in CO_2 photocatalytic reduction.

Nhu-Nang Vu et al. synthesized diminished Perovskite nanosheets $HCa_2Ta_3O_{10}$ incorporating Pt and Cu aimed at utilizing sunlight to convert carbon dioxide with water vapor into valuable fuels (Figure 6.7) [20]. These nanosheets are derived and produced layered perovskite $CsCa_2Ta_3O_{10}$ by liquid acoustic exfoliation and tetra butyl ammonium ion exchange. It results in a high specific surface area exceeding 200 $m^2\ g^{-1}$. Assessing their photocatalytic efficacy in the presence of sun rays in

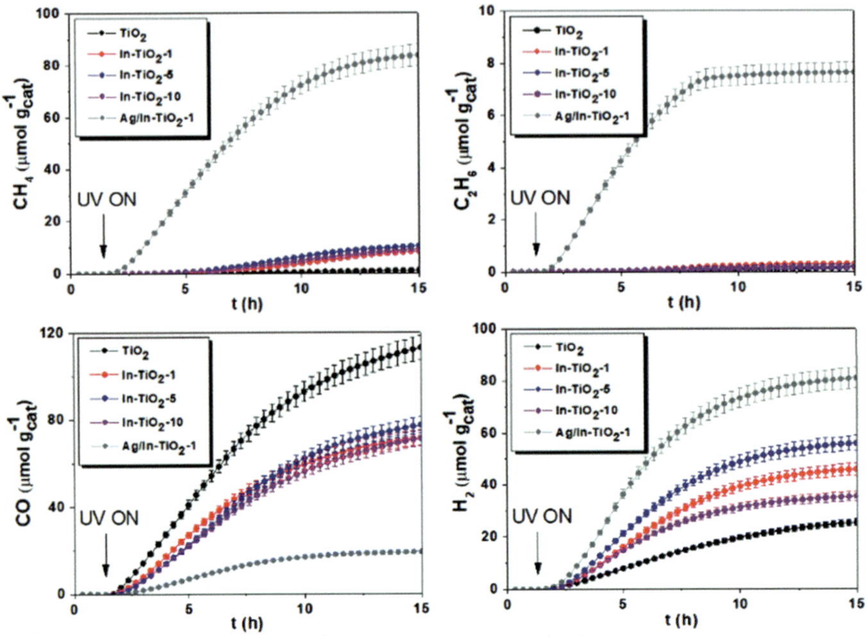

Figure 6.6 Total output of the primary photocatalytic products CO_2 reduction using the various manufactured catalysts. Reproduced with permission from [20].

the form of saturation water vapor for CO_2 photoreduction reveals significantly enhanced photoactivity in the reduced nanosheets compared to their nonreduced counterparts, owing to their unique composition. When platinum is present, the hydrogen treatment causes notable Ta^{+4} and oxygen vacancies, and enhances visible light absorption. Moreover, the incorporation of electron-hole separation is improved by CuO nanoparticles, forming a p-n junction. This enhancement is allocated to strengthen CO_2 adsorption and stabilization of C_1 intermediates, facilitating C-C interaction to generate C_2 compounds, such as ethanol. CH_2OH and CH_3OH form at 113 and 7.4 μmol g^{-1} h^{-1}, respectively, in the presence of CuO nanoparticles. as opposed to 125.9 μmol g^{-1} h^{-1} for ethanol by itself when they are not present.

Li Gong et al. used self-consistent and periodic DFT (Density Functional Theory) simulations to theoretically assess the catalytic efficiency and underlying processes of NiFe-based tiny clusters in the hydrogenating of CO_2 to create formic acid and CO. According to their theoretical research, NiFe nanoclusters show promise for both selectivity and catalytic activity when it comes to reducing acidosis to formic acid. According to the suggested pathway, bicarbonate preferentially reacts with hydrogen atoms that have been adsorbed from H_2 within the Nanoclusters of NiFe alloy, particularly at its carbon atom location. Additionally, their simulations show that NiFe alloy tiny clusters with an uncovered Fe atom on the Ni cluster surface perform better than those with Fe doping in the Ni cluster corner, as evidenced by reduced energy barriers.

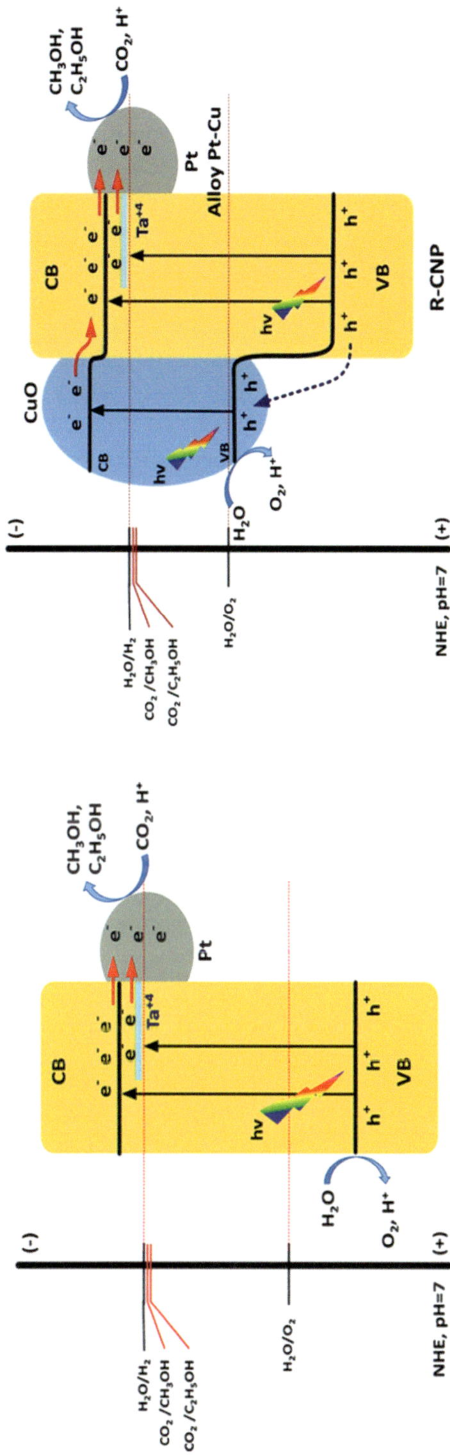

Figure 6.7 Schematic diagram illustrating the mechanism of CO_2 photoreduction on the R-Pt-CNP system. Reproduced with permission from [21].

This suggests the significance of surface composition in influencing catalytic activity and efficiency in this reaction [21].

Zhang et al. investigated the carboxylate a cyclization of propargylic amino acids with CO_2 under mild circumstances, two pyridine-based organic covalent frameworks with rich porosity (ideal for the controlled manufacture of silver nanoparticles) showed outstanding catalytic activity. Characterization through experimentation techniques reveals that in addition to providing silver with anchoring sites, pyridinic-N in the frameworks carefully directs nanoparticles of silver into various pores and controls their sizes. The accuracy and stability of the catalytic process are greatly influenced by this control. In particular, under mild circumstances, the Ag@2,6-FPP-TAPT catalyst with ultrasmall nanoparticles (only 1.35 nm) demonstrates remarkable catalytic efficiency and durability in the carboxylate cyclization of propargylic amino acids with CO_2 to create oxazolidinones. It performs better than Ag@3,5-FPP-TAPT. highlighting the significance of nanoparticle size and distribution within the COF structure [22].

Mahmud et al. investigated the effectiveness of defective ZnO nanoparticles as locations for activation and adsorption of carbon dioxide (CO_2) in photocatalytic processes (Figure 6.8) [24]. Using a straightforward precipitation hydrothermal process with a concentration of NaOH as the precipitating agent, these nanoparticles were created. Being the only controlled variable, the heating treatment during synthesis induced a rough surface on the ZnO nanoparticles, effectively removing any attached inorganic molecule. Among the tested concentrations, the 1:4 ZnO ratio showed the best methanol (CH_4) yield (1.3×10^{-1} μmol), demonstrating a threefold increase relative to commercialized ZnO (4.8×10^{-2} μmol). Additionally,

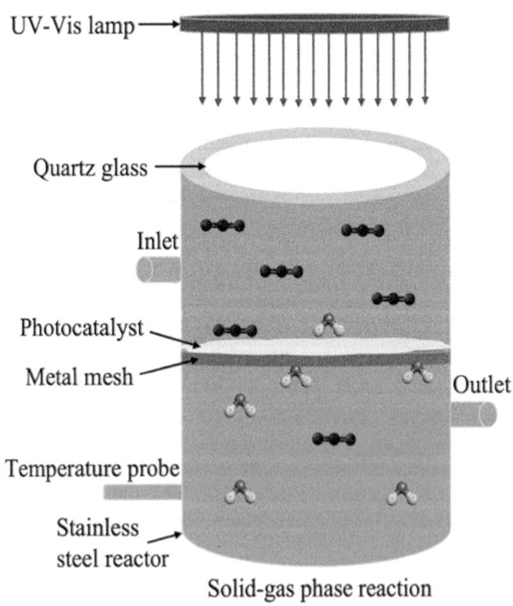

Figure 6.8 CO_2 reduction through photocatalysts in a batch reactor setup. Reproduced with permission from [24].

the versatility of the 1:4 ZnO nanoparticles was confirmed through a stability test consisting of four cycles, disclose only a 7.7% reduction in performance after the fourth cycle. This underscores the robustness and efficiency of the photocatalyst for multiple uses.

6.5.2 Chemicals

In today's scientific community, the primary objective is to transform greenhouse gases like carbon dioxide into valuable commodities such as fuels or chemicals, aiming to mitigate their adverse impacts (Figure 6.9) [6], Recognizing the environmental consequences of GHG emissions, the Paris Agreement was signed by 195 countries on December 13, 2015. This agreement mandates the necessity of continuous efforts to halt the increase in global average temperature at 1.5°C and keep it below 2°C beyond pre industrial levels. As such, it is crucial to manage CO_2 emissions effectively while also meeting global energy demands. Moreover, carbon capture, storage, and utilization technologies emerge as paramount solutions in the quest to combat global warming and address energy shortages. Carbon dioxide and methane [23], the primary components of biogas, despite being greenhouse gases, these substances can be utilized as flexible resources for the manufacturing of valuable fuels and chemicals. Conversion technologies can significantly lower GHG emissions through constant process enhancement and cooperation with the public and private sectors. The by- products of these transformations can also be used as sustainable energy and fuel sources. The growing public and corporate emphasis on clean technology within the chemical industry underscores the significant attention directed toward mitigating emissions of volatile organic compounds [24]. The rise in atmospheric CO_2 levels has sparked significant concern, prompting ongoing efforts in carbon sequestration. One promising avenue for CO_2 sequestration involves its photocatalytic conversion into hydrocarbons with solar irradiation offering an ideal solution [25]. The carbon and oxygen found in CO_2 have great potential as fossil fuel replacements in the production of vast assortment of chemicals and products available from synthetic rubber to fibers and plastics. The conversion of carbon dioxide into methanol and methane stands out as the most advanced technological method available [26].

Luan et al. investigated the concurrent Methylene Blue's (MB) dye breakdown alongside CO_2 reduction was conducted through a two-compartment electrochemical device in which carbon nanotubes are utilized, which are grown on carbon fiber (CNTs/CFM) electrodes as both the cathode and the anode in an electrochemical process. Numerous operational parameters, including applied currents, starting MB concentration (between 5 and 20 mg L^{-1}), and backing ions ($KHCO_3$ and H_2SO_4) (ranging from 10 to 100 mA) were scrutinized for their impact on MB degradation. The findings underscored the crucial effect of delivered voltage and initial MB focus, while supporting fluids exerted a small role in MB decomposition. Conversely, electrolytes exhibited a more substantial influence on the performance of CO_2 reduction, affecting the creation and quality of the things that are produced. The ideal electrochemical setup yielded an EC

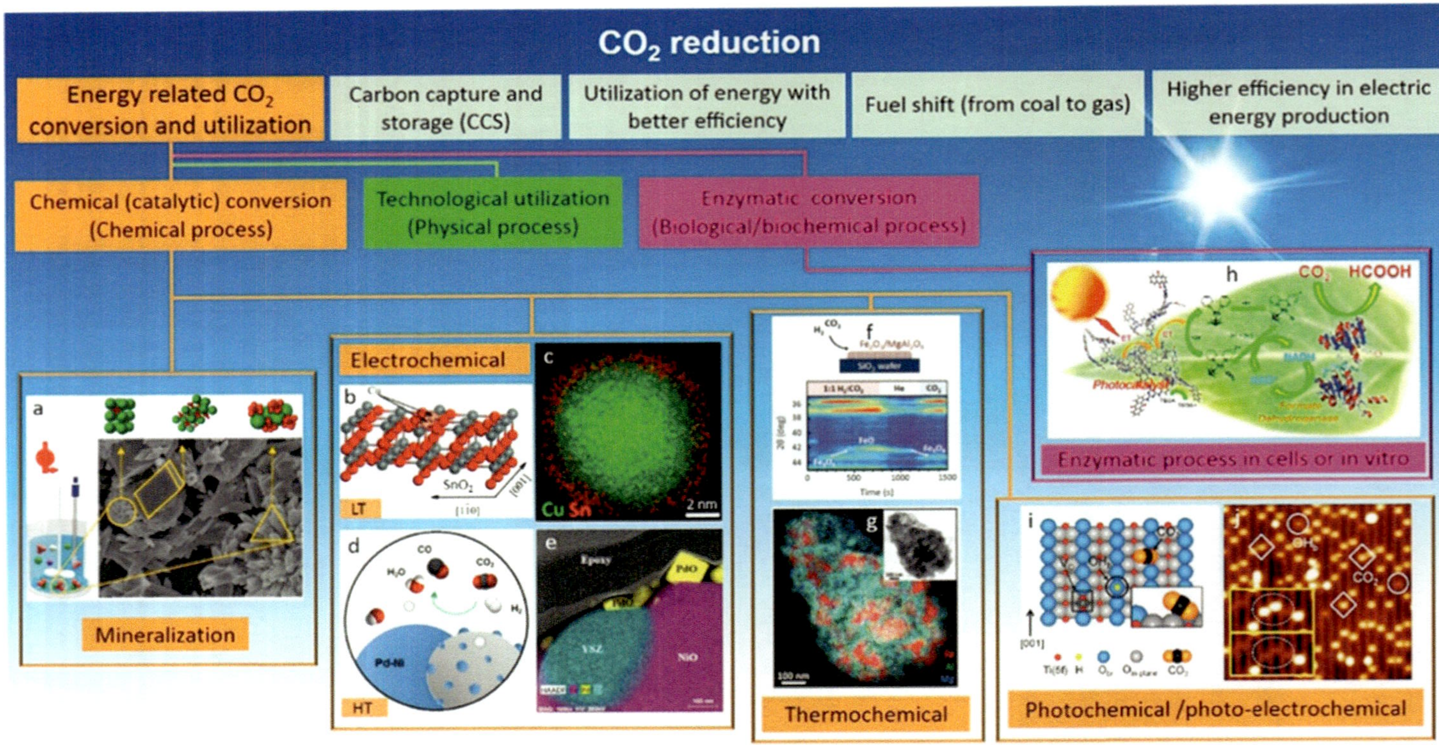

Figure 6.9 Schematic illustration of chemical reduction pathways for CO_2. Reproduced with permission from [25].

degrading rate of 72% for MB at a starting concentration in the anodized cell of 10 mg L^{-1}. This was achieved by using 0.1 M $KHCO_3$ as the electrolyte, a 50 mA induced voltage in the anodized cell, and a CO_2 solution that was saturated in the cathodic cell. Moreover, the cathodic cell produces 11.3 mg L^{-1} of oxalic acid, 67 mM cm^{-2} H_2, and 4.7 mM cm^{-2} CO. was achieved. This study underscores the promising potential of CNTs/CFM as dual-function electrodes for EC dye oxidation and CO_2 reduction to yield valuable compounds [27].

Song et al. studied a novel nanostructured catalyst composed of a prevalent element for the rapid electric synthesis of ethanol from CO_2. This catalyst operates in water, at room temperature and pressure, and exhibits exceptional performance, with high selection (84%) and Faradaic efficiency (63% at 1.2 V against RHE). The catalyst is notable for having Cu nanoparticles backed by an extremely textured film of carbon nanospike that has been doped with N. It is devoid of noble metallic elements and other costly and uncommon components. An initial mechanism is provided by calculations using Density Functional Theory (DFT) and electrical analysis [28].

It suggests that active areas on both the Cu nanostructures and the nanospikes simultaneously facilitate the electrolytic reduction of carbon cyanide dimer to ethanol. This research represents a significant advancement in the field, offering a promising pathway toward sustainable CO_2 utilization for the production of valuable chemicals.

Pathak et al. investigated TiO_2 nanoparticles that are added to Nafion membrane films, which made it easier to coat the nanoparticles with silver later. This process did not affect the dispersion of the nanoparticles or the optical transparency of catalytic films. Methanol was the main reduction product as a result of notable improvements in the photocatalytic conversion of CO_2 brought about by this silver covering. These catalytic sheets may also be recycled for further photocatalytic processes and demonstrate both chemical and photochemical stability. For a variety of photocatalysis applications, the same coating method may be used to prepare additional metal coated semiconductor nanoparticles that are uniformly distributed within optically transparent ionomer membrane films [29].

Alhaddad et al. determined a straightforward method for preparing visible light $ZnMn_2O_4$ (ZMO) tiny rods as photographic catalysts are reported (Figure 6.10) [33]. To enhance efficient respiration of CO_2 into methanol (MeOH), a moderate amount of Pt nanoparticle (0.5–2.0 wt%) is attached to these ZMO nanorods. After Pt was added, the synthesized materials' surface qualities and structure significantly improved, according to analysis. Moreover, Pt deposition promoted photo charge separation and improved visible light absorption, which caused the bandgap of the 1.5% Pt/ZMO composite to decrease to 1.98 eV. At a concentration of 2.4 g L^{-1}, this optimized Pt/ZMO photocatalyst demonstrated an impressive specific photoreduction of CO_2 to MeOH, achieving 1906 μmol/g in just 9 hours. The produced photocatalyst additionally showed sustainable recyclability, continuing to function after five reuse cycles. The inhibition of recombination and enhancement of mobility are responsible for Pt/ZMO's increased photocatalytic activity. The potential of spinel-structured photocatalysts assisted by less noble metals for the production of renewable fuels is highlighted by this work.

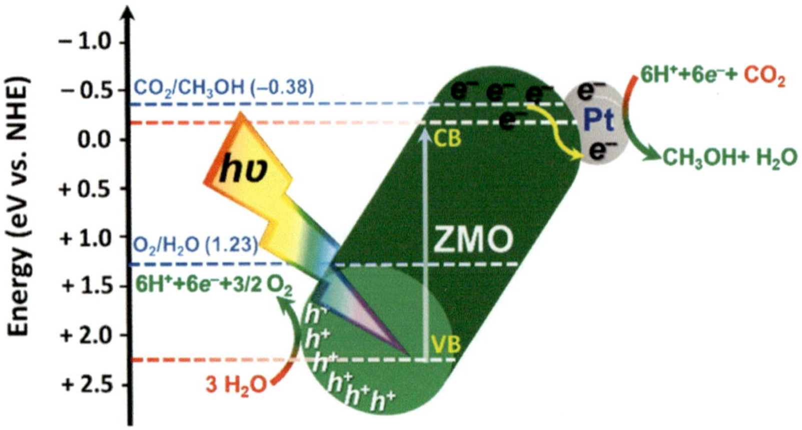

Figure 6.10 The photocatalytic method of Pt/ZMO photocatalyst ratio of 1.5% is used at the specific photoreduction of CO_2 to MeOH. Reproduced with permission from [33].

Rawool et al. investigated utilizing magnesium (both nanoparticle and bulk forms) to use water as the only water source in order to transform CO_2 into methane, ethanol, and formic acid. Remarkably, at 300 K and 1 bar, CO_2 could be converted in a matter of minutes, whether it was pure or directly from the air, without the need for external thermal, optical, or electric energy. Nonetheless, because water and magnesium oxide reacted more favorably than CO_2 and water did, oxygen was the elementary result. The favorable actions of mineral magnesium, basic aluminum carbonates, CO_2, and water were necessary for this CO_2 transition; the absence of any of these elements stopped the conversion process. Magnesium ions were converted into hydroxide of magnesium and carbonate during the CO_2 conversion process, which may be regenerated. The low temperature trials also suggest that this CO_2-to-fuel conversion method might be used on the surface of Mars, where there is an abundance of CO_2, water (ice), and magnesium. The procedure as a whole is not catalytic, but it is a step in the direction of a sustainable CO_2 utilization plant [34].

Ola et al. developed methods to convert atmospheric CO_2 into fuel without the need for external energy input at pressure and temperature at room temperature is a pressing challenge (Figure 6.11) [30]. Hence, there is a critical need for CO_2 conversion techniques operating under ambient conditions. In this study, the author explores the use of magnesium, an abundant and affordable element in both nanoparticle and bulk ways to produce formic acid, methanol, and methane from CO_2. The only source of hydrogen needed for this reaction is the water process. Remarkably, CO_2 conversion occurs rapidly even from air without requiring additional thermal, photo, or electric energy inputs. However, hydrogen predominates due to the favored reaction of water with magnesium. This conversion process relies on the interaction of basic magnesium carbonate, magnesium, CO_2, and water, the absence of any of these components inhibits the process. During CO_2 conversion, magnesium is transformed into magnesium hydroxide and carbonate, potentially regenerative substances. These experiments hint at the possibility of applying this process on Mars that converts CO_2 into fuel by utilizing the planet's resources of CO_2, ice (ice), and magnesium on its surface. Although non catalytic, this process represents a significant step towards a sustainable CO_2 utilization strategy and holds the potential for initiating A Martian civilization powered by magnesium.

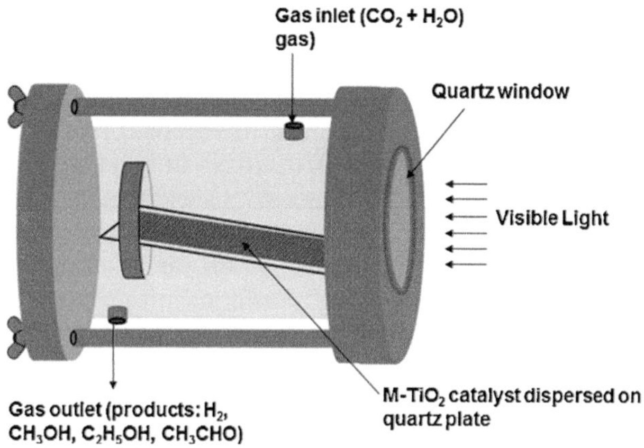

Figure 6.11 Representation of photoreactor to convert atmospheric CO_2 into fuel. Reproduced with permission from [30].

6.5.3 Building Materials

A comprehensive investigation of various approaches for reducing the release of greenhouse gases is essential to address the existing crisis of global warming. These strategies include the increasing use of renewable energy sources, technological improvements, and innovative building designs. Carbon Capture and Utilization is an established approach for reducing anthropogenic CO_2 emissions mostly from power stations and cement factories which are two of the main sources of industrial carbon dioxide emissions worldwide. It will be changed into mineral carbonates by CO_2 which are thermodynamically stable. The development of distinct phases of calcium and magnesium carbonates is reliant on several parameters including reaction routes, temperature, the CO_2 partial pressure, the composition and concentration of the reactants, and the reaction fluids [31]. Phases can be achieved by using specific structures and chemistry. There are practical ways to capture and store CO_2 by using natively natural silicate minerals like the mineral olivine (Mg_2SiO_4), serpent ($Mg_3(OH)_4(Si_3O_5)$), limestone ($CaSiO_3$), and alkali industrial by products. Achieving the full potential of CO_2 as an economically viable and ecologically sustainable carbon source requires resolving issues related to CO_2 activation, designing appropriate reactor designs, and effectively integrating processes. There is a current wave of interest in the study of organic-inorganic hybrid nanocomposites, with a wide range of research teams and real-world applications producing a large amount of data [32]. Personalizing the qualities of nanomaterials to maximize their performance is made possible by utilizing this. Composites that are created at the molecular or nanoscale level by a well-balanced combination of organic and inorganic ingredients are known as hybrid materials. Covalent Triazine Frameworks (CTFs) are a highly regarded class of contemporary nanoporous materials that exhibit remarkable performance [33]. These structures, which are distinguished by aromatic triazine bonds, are meticulously constructed by covalently joining light elements like hydrogen, nitrogen, and carbon. Among its noteworthy characteristics are high porosity, enhanced nitrogen content, and strong chemical and thermal stability. The process of utilizing CO_2 in a concrete production can involve substituting water with this gas and is known as CO_2 curing. Additionally, CO_2 has the potential to be utilized as a primary component in the composition of concrete, encompassing cement and building materials [34]. Tiny particles utilized in building materials known as construction aggregates, can be synthesized by blending CO_2 with industrial or waste by products originating from power plant operations. Examples of such by products include coal fly ash and iron slag, which would otherwise be disposed of in landfills. While the use of CO_2 in cement manufacturing is still relatively new, one of the most established and promising uses of CO_2 utilization is CO_2-cured concrete. The market for CO_2 utilization is expected to develop, especially in industries like building materials and as a carbon feedstock for chemicals and fuels. Still, several critical criteria impact the potential market for CO_2-derived products, including competitiveness, scalability, climate benefits, and the visibility of related technology [33].

6.6 SUMMARY

The recent trends for the transformation of CO_2 into useful materials through a variety of processes have been outlined with special emphasis on some representative studies. The management and mitigation of CO_2 is a primary necessity for a sustainable global economy and environment. To maintain that, nanomaterials-based conversion of CO_2 can play a vital role. Nanomaterials can be used with variable composition, structure, and surface properties to optimize and enhance catalytic performance. Nanomaterials may serve as catalysts in photocatalysis, electrocatalysis, and thermal catalysis processes for CO_2 conversion. This promising industrial advancement motivates the research community to collaborate further in scaling up other significant catalytic pathways for CO_2 conversion.

REFERENCES

[1] Johnson, J.M.F., Franzluebbers, A.J., Weyers, S.L. and Reicosky, D.C. 2007. Agricultural opportunities to mitigate greenhouse gas emissions. Environmental Pollution. 150(1): 107–124.

[2] Hepburn, C., Adlen, E., Beddington, J., Carter, E.A., Fuss, S., Mac Dowell, N., et al. 2019. The technological and economic prospects for CO_2 utilization and removal. Nature. 575(7781): 87–97. http://dx.doi.org/10.1038/s41586-019-1681-6.

[3] Huang, C.H. and Tan, C.S. 2014. A review: CO_2 utilization. Aerosol and Air Quality Research. 14(2): 480–499.

[4] Hendriks, I.E., Duarte, C.M., Olsen, Y.S., Steckbauer, A., Ramajo, L., Moore, T.S., et al. 2015. Biological mechanisms supporting adaptation to ocean acidification in coastal ecosystems. Estuarine, Coastal and Shelf Science. 152: A1–A8. http://dx.doi.org/10.1016/j.ecss.2014.07.019.

[5] Kandathil, V. and Manoj, N. 2023. Advances in CO_2 utilization employing anisotropic nanomaterials as catalysts: A review. Frontiers in Chemistry. 11: 1175132.

[6] Zheng, Y., Zhang, W., Li, Y., Chen, J., Yu, B., Wang, J., et al. 2017. Energy related CO_2 conversion and utilization: advanced materials/nanomaterials, reaction mechanisms and technologies. Nano Energy. 40: 512–539.

[7] Gupta, R., Mishra, A., Thirupathaiah, Y. and Chandel, A.K. 2024. Biochemical conversion of CO_2 in fuels and chemicals: Status, innovation, and industrial aspects. Biomass Conversion and Biorefinery. 14(3): 3007–3030. https://doi.org/10.1007/s13399-022-02552-8.

[8] Li, T., Baležentis, T., Makutėnienė, D., Streimikiene, D. and Kriščiukaitienė, I. 2016. Energy-related CO_2 emission in European Union agriculture: Driving forces and possibilities for reduction. Applied Energy. 180: 682–694.

[9] Bierwirth, P.N. 2018. Carbon dioxide toxicity and climate change: a major unapprehended risk for human health. Web Published: ResearchGate. 10. https://www.researchgate.net/publication/311844520_Long-term_carbon_dioxide_toxicity_and_climate_change_a_major_unapprehended_risk_for_human_health.

[10] Karnauskas, K.B., Miller, S.L. and Schapiro, A.C. 2020. Fossil fuel combustion is driving indoor CO_2 toward levels harmful to human cognition. GeoHealth, 4(5): e2019GH000237.

[11] Visser, P.M., Verspagen, J.M., Sandrini, G., Stal, L.J., Matthijs, H.C., Davis, T.W., et al. 2016. How rising CO_2 and global warming may stimulate harmful cyanobacterial blooms. Harmful Algae. 54: 145–159. http://dx.doi.org/10.1016/j.hal.2015.12.006.

[12] Zhang, T., Zhang, W., Yang, R., Liu, Y. and Jafari, M. 2021. CO_2 capture and storage monitoring based on remote sensing techniques: A review. Journal of Cleaner Production. 281: 124409. https://doi.org/10.1016/j.jclepro.2020.124409.

[13] Alok, A., Shrestha, R., Ban, S., Devkota, S., Uprety, B. and Joshi, R. 2022. Technological advances in the transformative utilization of CO_2 to value-added products. Journal of Environmental Chemical Engineering. 10(1): 106922. https://doi.org/10.1016/j.jece.2021.106922.

[14] Bratovčić, A. and Tomašić, V. 2023. Design and development of photocatalytic systems for reduction of CO_2 into valuable chemicals and fuels. Processes. 11(5): 1433.

[15] Sundar, D., Liu, C.H., Anandan, S. and Wu, J.J. 2023. Photocatalytic CO_2 conversion into solar fuels using carbon-based materials—A review. Molecules. 28(14): 5383.

[16] Damiani, D., Litynski, J.T., McIlvried, H.G., Vikara, D.M. and Srivastava, R.D. 2012. The US department of Energy's R&D program to reduce greenhouse gas emissions through beneficial uses of carbon dioxide. Greenhouse Gases: Science and Technology. 2(1): 9–16.

[17] Rizzato, L., Cavazzani, J., Osti, A., Scavini, M. and Glisenti, A. 2023. Cu-Doped $SrTiO_3$ Nanostructured Catalysts for CO_2 Conversion into Solar Fuels Using Localised Surface Plasmon Resonance. Catalysts. 13(10): 1377.

[18] Zhang, W., Qin, Q., Dai, L., Qin, R., Zhao, X., Chen, X., et al. 2018. Electrochemical reduction of carbon dioxide to methanol on hierarchical Pd/SnO_2 nanosheets with abundant Pd–O–Sn interfaces. Angewandte Chemie International Edition. 57(30): 9475–9479.

[19] Reñones, P., Fresno, F., Oropeza, F.E. and Gorni, G. 2022. Structural and electronic insight into the effect of indium doping on the photocatalytic performance of TiO_2 for CO_2 conversion. Journal of Materials Chemistry A. 10(11): 6054–6064.

[20] Vu, N.N., Nguyen, C.C., Kaliaguine, S. and Do, T.O. 2017. Reduced Cu/Pt–$HCa_2Ta_3O_{10}$ perovskite nanosheets for sunlight-driven conversion of CO_2 into valuable fuels. Advanced Sustainable Systems. 1(9): 1700048

[21] Gong, L., Chen, J.J. and Mu, Y. 2017. Catalytic CO_2 reduction to valuable chemicals using NiFe-based nanoclusters: A first-principles theoretical evaluation. Physical Chemistry Chemical Physics. 19(41): 28344–28353.

[22] Zhang, W., Qin, Q., Dai, L., Qin, R., Zhao, X., Chen, X., et al. 2018. Electrochemical reduction of carbon dioxide to methanol on hierarchical Pd/SnO_2 nanosheets with abundant Pd–O–Sn interfaces. Angewandte Chemie International Edition. 57(30): 9475–9479.

[23] Zain, M.M. and Mohamed, A.R. 2018. An overview on conversion technologies to produce value added products from CH_4 and CO_2 as major biogas constituents. Renewable and Sustainable Energy Reviews. 98: 56–63. https://doi.org/10.1016/j.rser.2018.09.003

[24] Yaashikaa, P.R., Kumar, P.S., Varjani, S.J. and Saravanan, A. 2019. A review on photochemical, biochemical and electrochemical transformation of CO_2 into value-added products. Journal of CO_2 Utilization. 33: 131–147. https://doi.org/10.1016/j.jcou.2019.05.017

[25] Alper, E. and Orhan, O.Y. 2017. CO_2 utilization: Developments in conversion processes. Petroleum. 3(1): 109–126. http://dx.doi.org/10.1016/j.petlm.2016.11.003.

[26] Saleh, T.A. 2022. Nanomaterials and hybrid nanocomposites for CO_2 capture and utilization: environmental and energy sustainability. RSC Advances. 12(37): 23869–23888.

[27] Luan, N.H., Yang, Y.T. and Chang, C.F. 2022. Electrochemical degradation of methylene blue accompanied with the reduction of CO_2 by using carbon nanotubes grown on carbon fiber electrodes. Sustainable Environment Research. 32(1): 13.

[28] Song, Y., Peng, R., Hensley, D.K., Bonnesen, P.V., Liang, L., Wu, Z., et al. 2016. High-selectivity electrochemical conversion of CO_2 to ethanol using a copper nanoparticle/N-doped graphene electrode. Chemistry Select. 1(19): 6055–6061.

[29] Pathak, P., Meziani, M.J., Castillo, L. and Sun, Y.P. 2005. Metal-coated nanoscale TiO_2 catalysts for enhanced CO_2 photoreduction. Green Chemistry. 7(9): 667–670.

[30] Ola, O. and Maroto-Valer, M.M. 2015. Transition metal oxide based TiO_2 nanoparticles for visible light induced CO2 photoreduction. Applied Catalysis A: General. 502: 114–121. http://dx.doi.org/10.1016/j.apcata.2015.06.007.

[31] Rasul, M.G. and Arutla, L.K.R. 2020. Environmental impact assessment of green roofs using life cycle assessment. Energy Reports. 6: 503–508

[32] Kerenkan, A.E., Ello, A.S., Echchahed, B. and Do, T.O. 2016. Synthesis of mesoporous tungsten oxide/γ-alumina and surfactant-capped tungsten oxide nanoparticles and their catalytic activities in oxidative cleavage of oleic acid. International Journal of Chemical Reactor Engineering. 14(4): 899–907.

[33] Gadikota, G., Fricker, K., Jang, S.H. and Park, A.H.A. 2015. Carbonation of silicate minerals and industrial wastes and their potential use as sustainable construction materials. *In*: Advances in CO_2 Capture, Sequestration, and Conversion. 295–322. American Chemical Society.

[34] Jamil, R., Tomé, L.C., Mecerreyes, D. and Silvester, D.S. 2021. Emerging ionic polymers for CO_2 conversion to cyclic carbonates: an overview of recent developments. Australian Journal of Chemistry. 74(11): 767–777.

Chapter 7

Electrochemical Conversion of Carbon Using Nanomaterials

Kailash Devi[1], Sajjad Hussain[2], Sandeep Kaur[2], Akash Katoch*[2] and Deepika Jamwal*[1]

[1]Department of Chemistry, University Institute of Science, Chandigarh University, Gharuan, Mohali, Punjab, India-140413

[2]Centre for Nanoscience and Nanotechnology, University Institute of Science, Panjab University, Chandigarh, India-160014

7.1 INTRODUCTION

Rising carbon dioxide (CO_2) levels in the atmosphere are a major contributor to climate change. CO_2 emissions from the combustion of fossil fuels have dramatically increased since the industrial revolution. Developing methods for the electrochemical conversion of CO_2 into value-added products provides an opportunity to reduce atmospheric CO_2 levels while transforming this greenhouse gas into sustainable fuels and chemicals [1, 2].

The electrochemical reduction of CO_2 utilizes electricity to drive non spontaneous chemical reactions that can convert CO_2 into compounds like carbon monoxide, formic acid, methane, ethylene, or alcohol. This approach can store intermittent renewable electricity in the form of chemical bonds while recycling CO_2 [3, 4]. However, traditional CO_2 electrocatalysis suffers from high energy costs and low conversion rates or product yields. This is primarily due to the stability of the CO_2 molecule and the multiple reaction pathways that are possible depending on the electrocatalyst used. By using nanomaterials to rationally

*For Correspondence: jamwaldeepika@gmail.com; katochakash@gmail.com

DOI: 10.1201/9781003492184-7

build selective and energy efficient CO_2 reduction catalysts, these problems can be effectively addressed. Materials that have at least one dimension in the 1–100 nm size range frequently display unique optical, electrical, or chemical characteristics [5, 6]. These characteristics are mostly caused by high surface area to volume ratios and quantum phenomena that emerge at the nanoscale. These characteristics can be used to strategically modify nanomaterial-based electrocatalysts in order to maximize CO_2 adsorption orientations and stabilize important reaction intermediates in order to increase product selectivity [7].

The rigorous control given by bottom up chemical synthesis techniques has provided the way to ground breaking advancements in the field of electrochemical CO_2 reduction. Through the growth and nucleation stages, these techniques enable scientists to control nanomaterial's size, shape, facets, and composition down to the atomic level [5–8]. Researchers may construct catalysts with previously unattainable efficiency by precisely altering these properties, increasing their potential to accelerate CO_2 conversion processes [9, 10]. Furthermore, a different approach to shaping nanomaterials with certain orientations and revealing highly active areas essential for electrocatalysis is using top down lithographic processing methods. By carefully arranging each atom to optimize catalytic activity, this nanoscale manipulation raises the efficiency of CO_2 conversion reactions even more [4–11]. The introduction of localized charge effects by nanomaterials with many grain boundaries, flaws, interfaces, or twisted configurations is crucial in activating CO_2 molecules and promoting subsequent conversion stages. These intrinsic properties increase the electrochemical process's overall efficiency and encourage CO_2 activation. Although bimetallic catalysts and nanostructured metals have shown potential in reducing CO_2, carbon-based nanomaterials like graphene and carbon nanotubes have clear benefits. Their remarkable stability over an extensive pH range, large surface area, adjustable electrical characteristics, and economical nature made them highly desirable options for electrocatalytic uses [12, 13]. Emerging hybrid catalysts that combine nitrogen-doped carbon nanotubes or graphene nanosheets with metal nanoparticles have become a strong strategy for improving catalytic performance [14]. In CO_2 conversion processes, the synergistic effects of carbon based supports and metal nanoparticles result in enhanced catalytic activity and selectivity.

Though tremendous progress has been made, more understanding of the molecular mechanisms controlling product selectivity over these nanoelectrocatalysts is still required [15, 16]. The key to understanding these complex dynamics lies in the combination of computational modeling and advanced spectroscopic characterization techniques. Researchers can improve catalyst design and reaction conditions to produce desired product outcomes with unmatched precision by clarifying the underlying processes at work. Fundamentally, the development of nano-electrocatalysts for CO_2 conversion is being driven by the synergistic integration of top-down fabrication techniques, bottom-up synthesis processes, and sophisticated characterization tools working together. The pursuit of effective and selective CO_2 consumption is accelerated by sustained innovation and interdisciplinary cooperation, opening the door to a sustainable future [4–17].

7.2 CARBON EMISSIONS AND CLIMATE CHANGE

7.2.1 The Impact of Anthropogenic Carbon Emissions on Climate Change

The climate of the Earth is an intricately balanced system driven by the flow of energy from the Sun and dynamic exchanges between the atmosphere, land, and oceans. This energy balance has been disrupted over the past century by rising emissions of carbon dioxide and other greenhouse gases due to human activities, notably the combustion of fossil fuels and large scale deforestation [18, 19]. The elevated concentrations of atmospheric carbon are inducing measurable changes to the Earth's climate system including increased temperatures, altered precipitation patterns, and more extreme weather events [1, 20, 21].

7.2.2 Sources and Sinks of Carbon Emissions

Carbon dioxide is a naturally occurring gas present in the Earth's atmosphere essential for supporting life by trapping heat. Historically over millions of years before human influence, the main source of atmospheric CO_2 was volcanic outgassing which was balanced by the sequestration of carbon through gradual geological processes like rock weathering as well as biological carbon fixation into sediments and fossil organic matter. This carbon cycle maintained CO_2 levels between 180 and 280 parts per million (ppm) for nearly one million years up to the 19th century [22, 23]. Furthermore, the steam engine marks a turning moment in human history when we started using fossil fuels extensively, which resulted in a huge amount of carbon dioxide emissions. Knowing the history of the steam engine helps us to identify the causes of the climatic problems we have now and emphasizes how critical it is to switch to more environmentally friendly energy sources [24, 25]. The invention of the steam engine marking the onset of the industrial revolution in the late 18th century ushered in large scale extraction and burning of coal as an energy source to power machinery, initiating the significant addition of anthropogenic carbon emissions. The internal combustion engine, industrial manufacturing, increasing use of cement, and land-use changes like deforestation further escalated CO_2 emissions throughout the 20th century. Current atmospheric CO_2 levels have exceeded 415 ppm, far surpassing natural historical concentrations [26–28]. Reports by the Intergovernmental Panel on Climate Change estimate carbon dioxide emissions from fossil fuel usage and industry to have risen by over 16 times between the years 1900 to 2010 [11].

Such enormous quantities of carbon emissions are overwhelming natural carbon sequestration mechanisms leading to a measurable increase of CO_2 in the atmosphere by about 40% since pre-industrial times [29]. Ocean waters are the most significant carbon sink absorbing nearly 30% of total anthropogenic emissions thus far and helping mitigate some global warming. However, both the upper and deep ocean waters are becoming more acidic due to higher dissolved CO_2 levels which impacts calcifying marine organisms [30, 31]. Terrestrial ecosystems particularly forests also serve as an important biological sink. Carbon dioxide (CO_2) emissions are significantly increased by deforestation and changes in land use, in addition to the enormous contribution from the burning of fossil fuels [21]. These activities lead to the decomposition of soil and the destruction of biomass, which releases stored carbon into the atmosphere and accounts for more than 10% of total emissions. Deforestation results in the loss of carbon sinks like forests and wetlands and is mostly caused by urbanization, industrialization, and agricultural development. The carbon held in these ecosystems is released when trees are cut down and vegetation is removed, which raises the atmospheric concentration of CO_2. Additionally, the delicate balance of carbon storage and release is disturbed by changes in land use, such as the conversion of natural habitats into agricultural fields or urban areas [32, 33].

7.2.3 Observed Warming and Severe Events

Direct temperature recordings over the past century paired with paleoclimate proxies on atmospheric composition reveal the extent of recent climate change attributable to anthropogenic forcing, often termed the 'hockey stick' effect due to the sharp inflection in warming. Global surface temperatures have increased by 1°C relative to pre-industrial baselines caused by heightened atmospheric greenhouse gas concentrations preventing efficient heat radiation from the Earth back into space [34, 35]. Multiple independently produced temperature reconstructions eliminate any possibility of natural climate factors driving the observed warming.

The Intergovernmental Panel on Climate Change (IPCC) special report Global Warming of 1.5°C provides extensive evidence correlating human activity with nearly all aspects of climate change based on thousands of independent studies analyzing real world data sets across various parameters. Analyses definitively conclude that human influence has led to substantial warming of the Earth's surface, with each decade successively warmer than all preceding decades [36, 37].

Besides a persistent rise in global temperatures, human induced climate change has also augmented the intensity and frequency of heatwaves, altered spatial and temporal precipitation levels globally leading to intense droughts or flooding, and increased the rate of ice melt covering Greenland and Antarctic land mass resulting in accelerated sea level rise. Recent attribution studies have confirmed that the severity of damaging extreme weather events occurring presently like heatwaves, intense hurricanes, and flooding have been markedly exacerbated by anthropogenic emissions. Immediate steps to curtail emissions and remove excess carbon from the atmosphere are vital to prevent catastrophic climate change [38, 39].

7.3 ELECTROCHEMICAL CO_2 REDUCTION

7.3.1 Electrochemical Conversion of CO_2 to Value-added Products

Among the various carbon mitigation strategies being explored, the electrochemical reduction of CO_2 has emerged as a promising approach to convert this greenhouse gas into useful fuels and chemicals. This technique utilizes electricity to power non-spontaneous chemical reactions that can transform CO_2 into compounds like carbon monoxide, formic acid, methane, ethylene, or alcohol [40–42]. The process provides a means to store intermittent renewable electricity in the form of chemical bonds while recycling CO_2 emissions to valuable products. As shown in below Figure 7.1.

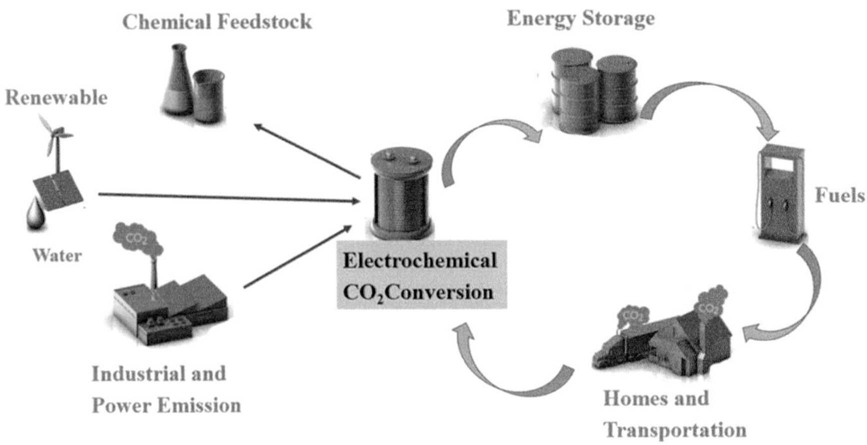

Figure 7.1 Electrochemical conversion of CO_2 to valuable products.

Though promising, there are substantial thermodynamic obstacles to the electrochemical conversion of CO_2. Significant challenges are presented by the intrinsic stability of CO_2 molecules and the enormous amounts of energy needed to break their chemical bonds [43, 44].

7.3.2 Challenges for CO_2 Conversion

However, the electrochemical conversion of CO_2 faces significant thermodynamic and kinetic challenges. The linear geometry and symmetry of the CO_2 molecule makes it extremely stable. Breaking the C=O double bond to bend and activate CO_2 requires a high energy input. The first one-electron transfer step to form the CO_2 anion radical is also highly energy intensive with an estimated potential of –1.9 V vs the hydrogen electrode. Furthermore, the bent radical intermediate can participate in multiple downstream proton coupled electron transfer pathways forming a variety of possible products [45, 46].

This multiplicity of reactions coupled with the sluggish activation of CO_2 leads to poor energy efficiency and low conversion rates during electrocatalysis. For perspective, fossil fuel-derived hydrocarbons have a combustion energy density of around 40 MJ/kg whereas the heat of formation for CO_2 is below 0.4 MJ/kg. Functionalizing CO_2 into fuels or chemicals with even a fraction of hydrocarbon energy density requires substantial energy input. Identification and selective stabilization of key intermediates are necessary to obtain meaningful yields [47, 48].

Product Spectrum: Depending on the electrocatalyst morphology and surrounding chemical environment, CO_2 can be transformed into several compounds. In aqueous electrolytes, the one-electron reduced $CO_2^{\bullet-}$ readily reacts with a proton to form either carbon monoxide (CO) or formic acid (HCOOH) through a hydrated formaldehyde intermediate [49]. Furthermore, the multi electron, multi proton reactions can produce other oxygenates like methanol, ethanol, n-propanol, or ethylene glycol. With non-aqueous solvents and additional hydrogen sources, more reduced C1 or C2 hydrocarbons like methane, ethylene, or ethane and longer chain alcohols are also possible [50].

However, the production of highly reduced hydrocarbons or alcohols involves large overpotentials driving the need for selective and energy efficient catalysts and process configurations. Beyond chemical selectivity, the crystal facet, exposed active sites, adsorption geometry, and intermediate stabilization capability of catalysts play a key role in determining favorable reactions. Nanomaterials provide potential pathways to rationally engineer certain desired properties to improve electrocatalytic performance through careful morphological and compositional modulation during synthesis [51–53].

Reactor design: In addition to the catalyst itself, the electrolytic cell reactor design and operating conditions can significantly influence product distribution and process efficiency. Flow reactors ensure optimal mass transport of gaseous CO_2 to the electrode interface enhancing the availability of dissolved CO_2 for electrochemical conversion reactions [54, 55]. The separation distance between the cathode and anode needs to be optimized to balance cell resistance and product crossover effects. Managing co-reactants like water or protons through strategic feeding or membrane isolation along with applying optimal potential ranges helps steer selectivity [56].

The basic parts of a typical CO_2 electrolyzer setup are as follows: an anode for the Oxygen Evolution Reaction (OER), which consumes OH or produces protons (H^+) and electrons (e^-) an electrolyte for ion conduction, CO_2 dissolution, and transport to the cathode, an ion-exchange membrane or porous diaphragm for electrode separation; and a voltage source to drive electron transfer from an anode to a cathode. The following are important phases in the reduction of CO_2:

(1) Mass transfer of CO_2 from gas to electrolyte; (2) transport of dissolved CO_2 to the interface between the cathode and electrolyte; (3) absorption of CO_2 at the cathode surface; and (4) dissociation of adsorbed CO_2 into intermediates such as *COOH, *CO, *CHO, and *COH [57], (5) electron transfer from a cathode catalyst to intermediates. (6) Product desorption from the electrode, and (7) product migration from the cathode/electrolyte interface to bulk gas or liquid phases. As shown in below Figure 7.2.

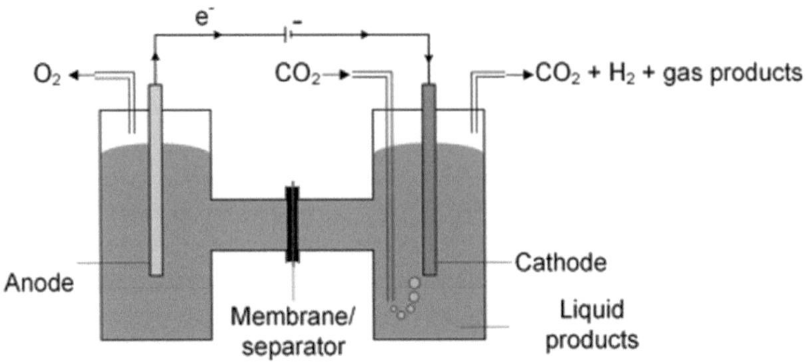

Figure 7.2 Diagram of a laboratory electrochemical H-cell (reactor) that reduces CO_2 at the cathode to produce gas (CO) and liquid (methanol and formic acid) at the anode while oxidizing water at the cathode. Reprinted with permission, Copyright: 2020, Journal of Material Chemistry A, RSC.

There is also growing interest in photo electrocatalytic configurations using semiconductor interfaces to directly leverage solar irradiation to activate CO_2 while overcoming intrinsic thermodynamic barriers to photocatalysis alone. Rational reactor engineering guided by mechanistic computational models and real time spectroscopic analysis will be vital to complement advances in nano-electrocatalysts to drive benchmark performances [58].

Techno economic sustainability: While the field has expanded rapidly over the past decade, considerable work remains in addressing fundamental selectivity challenges, enhancing conversion rates, and scaling up devices to meet commercial metrics. Recent technoeconomic analyses taking into account capital and operating expenses estimate electrochemically derived ethylene carbonate from CO_2 using nanocatalysts could be economically viable provided conversion rates above 50% are consistently achieved. Process intensification by continuous reactant feeding, smart reactor design, and stacked cell configurations need to be explored within a holistic framework factoring in sustainability benchmarks and life cycle impacts to realize widespread feasibility [59].

In summary, while electrochemical CO_2 reduction holds significant promise as a carbon capture and utilization pathway, deliverables have presently been modest outside laboratory settings. Concerted interdisciplinary efforts bringing together advances in nanomaterial synthesis, computational modeling, reactor engineering, and systems level optimization will be essential to address inherent energy intensity and selectivity challenges for commercially relevant processes [8, 60].

7.4 ROLE OF NANOMATERIALS IN ELECTROCHEMICAL CO_2 CONVERSION

Compared to bulk materials, the remarkable small size of nanomaterials (100–200 nm) has drawn a lot of attention in the research community and is the reason for their enhanced functionality over their bulk counterparts. In this sense, a potential way to address both the global energy problem and climate change at the same time is to directly convert CO_2 into renewable organic fuels and other feedstocks using nanomaterials [61, 62].

The two main particle effects responsible for these characteristics are the increased relative surface area and the effects of quantum mechanics. First, there are more active sites available for electrochemical reactions at the solid fluid interface due to the high surface area to volume ratio of nanomaterials. This property speeds up CO_2 adsorption rates, makes activation stages easier, and stabilizes important intermediates. Furthermore, the increased surface curvature and unsaturated bonding of nanomaterials enhance their reactivity with CO_2 molecules, encouraging specific transitions that lead to the creation of desired products. As such, nanostructuring offers a chance to customize the shape of the electrocatalyst to suit particular reaction pathways [63–65].

Nanostructured materials provide potential strategies to address key challenges in electrochemical CO_2 reduction regarding activation barriers, intermediate stabilization, and competing side reactions [66].

7.5 PROPERTIES OF NANOMATERIALS

Nanomaterials refer to substances or structures that have at least one physical dimension in the nanoscale range. For reference, the width of a human hair is about 80,000 nm while a sheet of paper is 100,000 nm thick. Individual atoms measure around 0.1 to 0.5 nm, therefore nanomaterials comprise just tens to thousands of atoms across their smallest axis. Nevertheless, the minute size confers astonishingly unique and useful properties befitting their classification as a distinct class of materials straddling the quantum realm between isolated atoms and bulk matter [67].

At the most basic physical level, the geometric attribute of nanomaterials that shifts exponentially compared to larger bulk is the surface area to volume ratio. As any material transitions from macro to micro and further down to nanoscale , the percentage of atoms located at the surface increases markedly relative to the total number of atoms contained within the volume. This enormously raises the exposure of reactive surface atoms available to interact at solid-fluid interfaces, conferring advantages like enhanced catalytic activity and control over surface-derived phenomena [68].

Additionally, the nanoscale dimensions become comparable to various intrinsic physical lengths associated with photons, electrons, and phonons leading to observable quantum effects that alter nanomaterial behavior substantially. Optical excitation wavelengths, electronic de Broglie waves, charge transfer lengths, and phonon scattering pathways share dimension similarities allowing precise tuning of associated interactions through nanomaterial size modulations with profound impacts technologically [69].

7.5.1 Optical and Electronic Properties

The optical absorption and emission patterns of metals and semiconductors are radically modified in nanomaterials due to abrupt changes in an electronic structure compared to bulk. Quantum confinement effects emerge as the electronic motion gets restricted across one or more physical axis within nanoscale volumes comparable to de Broglie wavelengths. Consequently, continuous valence and conduction energy bands transform into discrete, quantized electronic levels exhibiting molecule-like properties [70].

The bandgap widens, enabling band structure engineering through precise size control allowing customized optical response from the ultraviolet, visible to infrared spectra. Additionally, the density of electronic states fluctuates substantially which radically alters the accompanied phenomena like electron confinement lengths and also drastically modulate the nanoscale transistor channels and interconnects underlying modern electronics. Charge carriers also exhibit altered mobilities depending on nanoparticle composition and morphology.

7.5.2 Enhanced Active Surface Sites

Nanosizing to particles or thin films with higher surface area-to-volume ratios amplifies available active sites for surface adsorption and electrocatalytic reactions. The increased number of under coordinated atoms and abundant grain boundaries serve as anchoring points for CO_2 activation through its bent radical anion transition state [71].

7.5.3 Modulated Electronic Properties

The emergent quantum properties in nanomaterials significantly alter redox potentials, band energetics, work functions and electronic densities of states relative to bulk metals. Through control of size, shape and morphology these can be judiciously tuned to match energies of transition states and intermediates involved during CO_2 electroreduction and hydrogenation steps enhancing catalytic activity [72, 73]. Edge enriched hollow nickel sulfide nanocubes augment alcohol selectivity achieving 57% ethylene glycol yields by optimizing adsorption dynamics. Deliberate band shifting drives selective C–C coupling avoiding competitive hydrogen production pathways [74].

7.5.4 Intermediate Confinement

The hollow interiors, porous walls and confined spaces in nanoreactors provide a protected environment for the stabilization of highly transient reactive intermediates. Preventing rapid surface desorption or electrolyte quenching of partially reduced radicals enhances the likelihood of further reduction events translating to higher product yields [75].

Dual-shelled core-shell copper nanoparticles deliver pressurized CO_2 gas directly into the hollow interior while permeable shells provide mass transport regulation attaining localized enrichment and favorable conversion kinetics otherwise unattainable [76].

7.5.5 Exposed Facets

The abundance of low coordinated atoms on vertices, edges and corners of nanomaterials allows selective amplification of particular crystal facets like (100), (110) or (211) surfaces. Differential facet reactivities induce orientation preferences during CO_2 adsorption and activation leading to directed reaction channeling improving selectivity akin to enzymes [77]. Overall engineering nanoscale architectures through compositional, electronic structure and morphological modulations facilitate unprecedented tuning possibilities to overcome inherent challenges around CO_2 conversion through enhanced activation, optimized localization and tailored selectivity unprecedented in bulk catalysts [78].

7.5.6 Emergence of Quantum Effects

Quantum mechanical effects become prominent at nanoscale levels, changing electronic characteristics. This effect, which is especially noticeable in metal nanoparticles, results in band structure narrowing and shifting of electrical potential energy levels relative to the bulk. Discrete electronic states produced over constrained dimensions act like molecules, enabling intentional control of the electron density, conductivity, charge transfer rates, and localization behaviors of nanomaterials. Changes in composition, size, and form act as artificial levers to manipulate these attributes, which are critical factors influencing the electrocatalytic activity [79, 80].

7.5.7 Synergy from Combined Attributes

Increased surface area and emerging electrical features of nanomaterials provide synergistic benefits that allow for accurate electrocatalyst characteristic optimization to address CO_2 reduction issues. Increased interface active site counts reduce energy barriers and speed up activation processes; on the other hand, variable surface reactivity and regulated electrical potentials enhance product selectivity. Nanomaterials have great potential to advance electrochemical CO_2 conversion technologies toward more selective and efficient processes by using these combined properties [81, 82].

7.5.8 Mechanical Characteristics

Mechanical properties including hardness, yield strength, and ductility also improve considerably for nanomaterials over bulk owing to reduced defect density and high-volume fractions of grain boundaries that hinder dislocation pileups.

This allows designs to achieve superior strength but lower weight ratios. The high surface area additionally amplifies surface elastic responses [96].

The extremely large fraction of surface atoms also impacts compatibility and interactions with surrounding chemical species by exposing more reactive sites and anchoring points facilitating the development of sensor platforms. The combination of tunable electrical, optical, magnetic, and catalytic traits paired with adjustable mechanical strength and compatibility enables precise and previously unattainable customization of material performance indicators empowering disruptive technologies [83–86].

7.6 STRATEGIES FOR SYNTHESIS OF NANOMATERIALS

The exceptional degree of control required over material size, shape, and composition essential to manifest useful nanoscale properties necessitated the development of novel techniques integrating chemistry, physics, and engineering for tailored synthesis and fabrication. Broadly, approaches involve either topdown methods aimed at carving, etching, milling, or lithographically printing patterns at the nanoscale. Or bottomup techniques focused on atom-by-atom or molecular building block assemblies using self-organization and biomimicry (Fig. 7.3) [87–90].

Continued innovation across these diverse strategies further supplements existing processes facilitating scalable, reliable, and high precision manufacture to fully leverage the promise of nanomaterials commercially across industries ranging from electronics, computing, energy, healthcare, and sustainability.

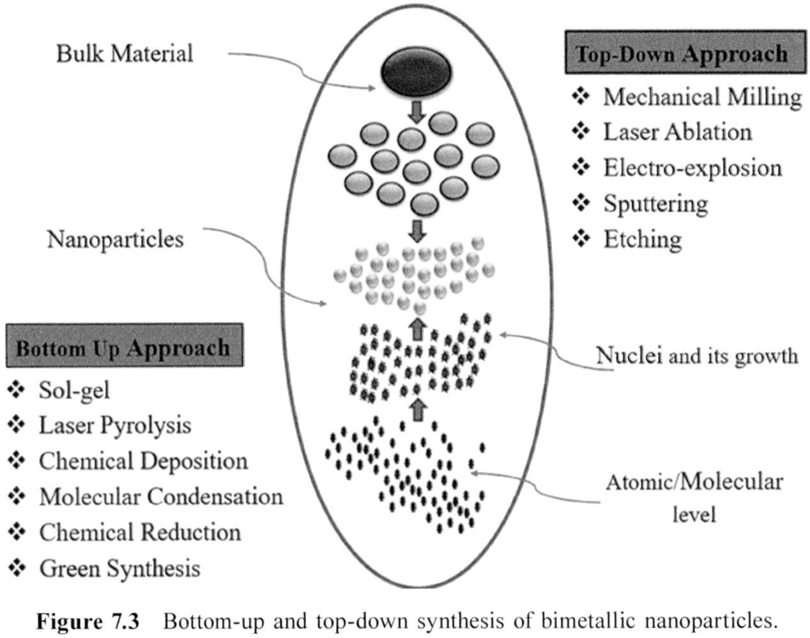

Figure 7.3 Bottom-up and top-down synthesis of bimetallic nanoparticles.

7.6.1 Top-Down Methods for Fabricating of Nanomaterials

Top-down nanofabrication refers to a set of processes that carve, etch, imprint, or deposit materials starting from larger thin films or bulk substrates and scaling down precisely to achieve nano-sized structures and devices. These approaches allow excellent control over nanoparticle dimensions, placement, and orientation leveraging advanced instrumentation and imaging capabilities. Top-down methods dominate the manufacturing of electronic integrated circuits and optoelectronic devices underpinning modern computing. The rich toolbox of nanolithography and printing techniques developed by the electronics industry can be adapted for the scalable synthesis of designed nanocatalysts for electrochemical CO_2 conversion and other applications as well [91, 92].

7.6.1.1 Photolithography

Photolithography involves selectively exposing parts of a light-sensitive polymer resin coating on a substrate to UV light or other radiation. The exposed regions undergo chemical amplification that either promotes dissolution away in a developer solvent or polymerization and cross-linking strengthening areas masking the substrate beneath. By using patterned masks, shining light to induce differentiation across the resist layer in the desired geometry allows the successive transfer of those resolution patterns via etching or material deposition steps [93, 94, 104].

Photolithography resolution has been scaled down multiple times over decades enabling continued miniaturization of transistors down to a few tens of nanometers in size for the current electronics fabrication. For electrocatalysis applications, lithographic methods can be adopted to produce periodic nanopillar arrays, nanodisks, nanocone forests, and nanoelectrodes with tuned spacing and densities on conductive substrates providing customized active area and sites for CO_2 adsorption and activation [95].

7.6.1.2 Electron Beam Lithography

Rather than using an optical mask and a large illumination area, electron beam lithography focuses a narrow beam of electrons across the resist in a scanning fashion to imprint customer patterns serially reaching resolutions below 10 nm. The electron doses change solvent solubility for positive resists while prompting chain scissioning in negative resists aiding dissolution. Metropolitan scanning patterns allow rapid writing across large area substrates. This technique has been leveraged to create nanoporous films, nanodots, nanowire arrays, and graphene nanoribbons with reproducible dimensions and affordable throughput at the research scale [96, 97].

7.6.1.3 Ion Beam Techniques

Energetic ions can also be utilized for direct nanopatterning. Focused ion beam lithography scans a narrow ion beam akin to electron lithography but drives sputtering and milling allowing subtractive fabrication. Atomic layer precision down to sub 10 nm dimensions is feasible. Dual beam systems integrating electron and ion beams advance patterning with *in-situ* imaging. Higher rate plasma etch processes like reactive ion etching create directional sidewalls aiding 3D nanostructure definition. Metal oxide nanotubes, nanocolumns, and nanorod arrays assist with preferential CO_2 adsorption configurations and intermediate stabilization [98, 99].

7.6.1.4 Nanoimprinting Lithography

Mechanical deformation embossing into thermoplastic polymers or photo-curable resins provides a high resolution, parallel alternative. Smooth mold masters are pressed into heated polymer films deforming and retaining patterns lifting off. Renewed molds allow reuse lowering costs for volume production. Reported resolutions reach sub-5 nm for some polymer systems. Imprinting grants 3D control unattainable in other top-down processes, useful for designing ordered mesoporous architectures to optimize reactant transport through tailored pore sizes and high internal surface areas [100].

7.6.1.5 Self-assembly Patterning

Guided organization leveraging spontaneous assembly driven by surface minimization allows versatile nanoscale patterning extending beyond standalone photolithography. Microphase separated block copolymer system combined with graphoepitaxy using lithographic trench templates induce alignment of microdomains granting guided self-assembly. Block lengths tune repeating domain periodicity from 5–50 nm readily. Alternating brush-like lamellae, cylinder or other morphologies translate into useful masks for selective infiltration or directed etch transfer [101, 102].

These collective set of top-down processes cater to both research development allowing rapid prototype iteration for mechanistic studies as well as production ready systems for scaled synthesis of tailored nanocatalysts and reactor configurations to realize application targets for carbon dioxide electroreduction and beyond.

7.6.1.6 Hybrid Approaches

Integrating bottom-up self-assembly with top-down lithography provides combined advantages of spontaneous organization guided across larger areas. Precisely deposited nanocrystals confined within lithographic channels or graphoepitaxy directed block copolymer blends exemplify this synergy by dividing extended constructs into configurable nanoscale subunits for customizability.

As advances continue, marrying directions of top-down sculpture with bottom-up construction will usher flexibility mimicking nature's engineering elegance across dimensions for transformative nanomanufacturing schemes tackling global challenges.

7.6.2 Bottom-up Chemical Approaches for Synthesis of Nanomaterials

Fundamental techniques in the synthesis of nanomaterials, bottom-up methods provide exact control over atom or molecular assembly to create nanostructures from the bottom up. These techniques entail taking separate parts and methodically arranging them into the desired form in order to construct complicated structures. In contrast to top-down strategies, which entail disassembling large materials into smaller parts, bottom-up strategies use physical or chemical reactions to create nanomaterials from atomic or molecule building blocks. The capacity of bottom-up synthesis to yield nanomaterials with customized features, including size, shape, composition, and functionality, is one of its main benefits. For a number of applications in industries like electronics, photonics, catalysis, medicine, and energy storage, this degree of control is essential [103, 104].

7.6.2.1 Sol-gel Method

One particularly adaptable soft chemical technique that is widely used to produce ceramics, glass, and metal oxides is the sol-gel process. Ceramic and glass materials can be created with this approach in a number of forms, such as spherical forms, thin film coatings, ceramic fibers, and microporous inorganic membranes. As starting materials, metal alkoxides or organometallic inorganic salts are frequently used in the sol-gel process. These precursors go through a series of hydrolysis and polycondensation reactions in this approach, which results in the development of a colloidal suspension or sol. The system is changed from a liquid "sol" state often colloidal to a solid "gel" phase during the sol-gel process. The gel is then dried and calcined at different temperatures to produce metal oxide nanopowders. It is possible to control the final materials' morphology, form, and textural characteristics by using the sol-gel method. The sol-gel approach has a number of advantages over high-temperature processes, including the ability to create metastable materials and improve product purity and compositional uniformity at moderate temperatures. Furthermore, this method affects the shape of the particles during the chemical transformation from the molecular precursor to the final oxidative network. The sol-gel method has been used by several research groups to synthesize iron oxide and alumina nanoparticles, which have been previously published [105, 106].

7.6.2.2 Hydrothermal Method

Usually, hydrothermal synthesis is carried out in an autoclave, a pressurized vessel in which an aqueous solution is used to facilitate the reaction. The autoclave's interior can get hotter above the boiling point of water and achieve vapor saturation pressure. This technique is commonly used to produce metal oxide nanoparticles, which are obtained by treating the peptized precipitates of a metal precursor with water hydrothermally. It regulates solution composition and reaction conditions to give control over surface chemistry, crystalline phase, grain size, and particle morphology. Whereas precursors in Chemical Vapor Deposition (CVD) are introduced in a gaseous state, precursors in hydrothermal synthesis begin as solid-form raw materials. The process is mostly atomistically based and frequently requires only a few small chemical events. Methods like as evaporation and sputtering have been developed to eliminate growth species from the source or target. Whereas sputtering is the process of removing atoms or molecules from a solid target by the impact of gaseous ions (plasma), evaporation is the thermal removal of growth species from the source. Deposits can have thicknesses ranging from angstroms to millimeters, providing a wide range of applications [107, 108].

7.6.2.3 Gas Phase Method

Gas phase techniques are very beneficial for effectively creating thin films. A chemical or physical process can be used to create such a synthesis. One popular industrial method that may swiftly coat huge regions is chemical vapor deposition, or CVD. Metal oxides are created in CVD via gas phase precursor breakdown or chemical reactions. While Physical Vapor Deposition (PVD) and CVD are comparable techniques, PVD starts with raw materials or precursors in a solid state, whereas CVD introduces precursors in a gaseous state. Both methods work atomistically and usually entail few chemical reactions. Methods for evaporating and sputtering growth species have been developed and are used to remove them from sources or targets. Growth species are thermally removed from the source during evaporation, whereas atoms or molecules are dislodged from solid targets during sputtering by the impact of gaseous ions (plasma). Deposits can have thicknesses ranging from angstroms to millimeters, providing a wide range of use [109, 110].

7.7 ADVANCED ELECTROCATALYST FOR CO_2 REDUCTION

To increase the efficiency and selectivity of CO_2 conversion into useful products like carbon monoxide (CO), methane (CH_4), ethylene (C_2H_4), and other hydrocarbons or oxygenates, advanced electrocatalysts for CO_2 reduction are essential. There are some fundamental features to consider for the efficient electrocatalysts in CO_2 reduction [111, 112].

High catalytic activity in the electrocatalyst is necessary to encourage CO_2 reduction at low overpotentials, allowing for efficient conversion even at low voltages. Additionally, the ability to selectively target specific products is crucial for optimizing the production of valuable compounds while reducing the production of unwanted ones. Surface characteristics may affect a product's selectivity. A sophisticated electrocatalyst must be long lasting and robust in the severe electrochemical conditions that arise during CO_2 reduction. It must also enable quick electron transfer between the CO_2 molecules and the electrode surface in order to maximize catalytic activity. Electron transfer kinetics can be enhanced by optimizing the electrical structure and surface shape of the catalyst [113, 114]. Controlling the electrocatalyst's catalytic characteristics, such as the binding energies of reaction intermediates and reaction routes, can improve process efficiency by allowing for better distribution of products. However, scalability and affordability are key components of advanced electrocatalysts intended for large-scale industrial CO_2 reduction systems. Using inexpensive, readily available materials from the environment can lower production costs and improve economic viability. Also to maximize process performance and efficiency, the electrocatalyst's compatibility with various reactor configurations and electrode designs must be taken into account. It is important to customize catalysts to fit certain reactor geometries and operating environments. To maintain safety and lessen environmental effect, the electrocatalyst synthesis process should use as few poisonous or ecologically hazardous ingredients as possible. Prioritizing sustainable manufacturing techniques and green synthesis technologies are necessary. By taking these factors into account, scientists can create sophisticated electrocatalysts for CO_2 reduction that work better and are more efficient, which will help create sustainable energy conversion technologies and slow down global warming [115, 116]. For CO_2 conversion, a number of advanced electrocatalysts have been studied with the goal of improving activity, selectivity, and stability.

7.7.1 Metallic Catalysts

As electrocatalysts for CO_2 reduction, metals like gold (Au) [117], silver (Ag) [118], copper (Cu), [119] and tin (Sn) [120] have demonstrated potential. These metals' surface characteristics and crystal structures can determine how active and selective they are toward particular products.

According to recent developments in nanoparticles (NPs) synthesis, assessment of possibly enhanced reaction kinetics has been made possible by the controlled surface area and surface morphology attained. The electrochemical conversion of CO_2 into hydrocarbons on Cu NPs or into CO on gold-based NPs/clusters serves as an example metal catalyst. Recently, a novel type of Au nanostructured catalyst has been utilized to demonstrate great selectivity for catalyzing CO_2 reduction to CO: its Faraday Efficiency (FE) was 96% at 0.35 V (vs. Normal Hydrogen Electrode (NHE)) with current densities ranging from 2 to 4 mA/cm^2. This catalyst was created by anodizing and electro-reducing an Au electrode [121, 122].

A simple hydrothermal approach produced high surface area tin oxide (SnO_2) nanocrystals that demonstrated outstanding electrocatalytic activity for CO_2 reduction to formate. At overpotentials as low as 340 mV, these new nanostructured tin (Sn) catalysts have the ability to selectively reduce CO_2 to formate. Maximum FE of 493% for formate generation with high stability and current densities of 410 mA/cm^2 on graphene supports has been achieved in aqueous $NaHCO_3$ solutions. This compromise between CO_2 and the nanoscale Sn surface, followed by kinetic activation toward protonation and further reduction, may be the source of the remarkable reactivity toward CO_2 reduction attained here [123].

7.7.2 Carbon Nanotubes

Diverse synthetic routes yield multi-walled concentric, few-wall archetypes, or single walled entities with diameters tunable 1–100 nm and lengths ranging from tens of microns sufficient for electronic connections. Compositionally, predominantly carbon incarnations demonstrate advantages for stability and conductivity excelling certain metals for electrocatalysis. Furthermore, tractable carbon nanotubes (CNTs) functionalization and structural hijacking tactics greatly bolster catalytic prowess positioning them as versatile supports renovating conventional materials constraints across myriad reaction chemistries including carbon dioxide conversion.

Conceptually, CNT architectures stem from dimensionality constraints imposed on versatile hexagonal carbon-carbon sp^2 hybridized planar graphene sheets which stack into various dimensional allotropes graphite, fullerenes, nanotubes, nanoscrolls, graphene, etc. each with intriguing characteristics. For example, while graphite stacks into overlapping flatland grids and fullerenes wrap into zero-dimensional spheres, carbon nanotubes elongate a single atom thick graphene sheet seamlessly into a tubular morphology translating exceptional in plane durability and electrical mobility into an axial one-dimensional conduit just nanometers wide but micrometers long.

Based on orientation angles of folded graphene, CNTs classify as either metallic armchair confirmation with hexagons circling round, semiconducting zigzag with hexagons lining axially, or chiral hybrids. This structural DNA imprints signature optoelectronic, thermal, and mechanical fingerprints with impressive values benefiting from quantum confinement perpendicular to the axial tube direction. Technology visions leveraging CNTs thus center around customizable 1D platform scaffoldings [124–127].

Individual single walled CNTs demonstrate ballistic electron transfer exceeding conventional metals. Layered multiwalled morphologies ease connectivities fabricating conductive pathways for circuitry, chemical sensors, and electrode interfaces with effective surface area multiplication promoting reaction rates useful for electrocatalysis. The intrinsic hollow cavity allows housing guest atoms, NPs, or concentric inner tubes forming encapsulated 1D reactors with tunable permeability. [128–130].

Integrating inorganic nanoparticles like copper, zinc, iron, indium, or quantum dots with CNTs promotes beneficial charge and multi-electron transfer relays assisted via tightly coupled interfaces critical in multi-step electrochemical CO_2 conversion reactions. Non-covalent pi-pi stacking attachments using pyrene ligands aid simpler nanoparticle assembly overcoming intrinsic inertness of nanotube sidewalls, while covalent linkages through oxidative acid processing enable robust permanent conjugation like amination salinization, thiolation, etc., although disrupting electronic mobility in certain cases [131, 132].

Micro-mesoporous composite networks with CNT inclusions facilitate diffusion ensuring optimal reactant access to anchored catalysts which amplifies productivity spatial decoupling and also prevents reverse reactions or other losses. Hierarchical CNT-nanoparticle hybrid constructed electrodes outperform conventional gas diffusion configurations for selective CO_2 electroreduction. Although scalability and stability remain ongoing challenges. Despite incredible progress, CNT catalyst promise remains predominantly untapped [133, 134].

Looking ahead, interfacing designed multi scale molecular assemblies mimicking enzymatic active sites on modular CNT scaffolds can potentially realize mimicking Nature's masterful catalytic prowess for transformative carbon recycling technologies aiding sustainable processes benefiting society globally.

7.7.3 Graphene

Graphene refers to a single atomic layer of sp^2-bonded carbon atoms tightly packed into a honeycomb lattice structurally constituting the fundamental building block underpinning graphitic materials of all other dimensionalities. First isolated in 2004 by mechanical exfoliation, this extraordinarily thin 2D material exhibits exceptional electronic, thermal, optical, and mechanical properties superior to conventional metals or semiconductors owing to synergistic quantum confinement and pi-cloud delocalization effects making graphene an ideal candidate for –next-generation electronics, photonics, energy storage/conversion applications including electrochemical CO_2 reduction.

Band structure fundamentals: Conceptually, the cross-sectional structure of graphene comprises two equivalent triangular sub-lattices of carbon atoms bonded together through in-plane sigma and perpendicular pi-orbital hybridizations. This geometric arrangement translates to electronic properties where valence and conductance bands derived from localized sigma orbitals lie far apart in energy while pi and pi* bands responsible for electronic transports overlap at the Dirac K points [135, 136].

Consequently, pristine graphene exhibits zero bandgap semi metallic behaviors with linear energy dispersion relations enabling manipulation of electronic densities and conductivities spanning metal to semi-metal through bandstructure engineering, doping, and external field effects, unlike traditional parabolic dispersion in metals and semiconductors. Furthermore, minimized backscattering from equivocal sublattices allows ballistic charge carrier mobilities promising innovations in electronics and photonics.

Synthetic approaches: Primarily, graphene synthesis involves top-down exfoliation from bulk graphite into single sheets or bottom-up recombinant growth processes on catalytic substrates. Micromechanical cleavage most reliably produces the highest crystalline quality graphene but lacks scalable throughput. Solution processing like sonication assisted liquid phase exfoliation grants mass quantities albeit with more defects impacting electronic properties [137, 138].

Epitaxial approaches on transition metal substrates using chemical vapor deposition permit large area single-crystal graphene sheets with moderate defects amenable for roll-to-roll transfer to device structures. Additionally, solid-state transformations of SiC, metal carbides under high temperature reordering assistance of small molecules or thermal shock waves propagate graphene layers over relevant substrates providing alternate fabrication pathways as the field continues to mature rapidly.

Graphene catalysis fundamentals: The exotic electronic structure allows exceptional optical adsorption and high intrinsic charge mobility even at room temperatures. Combined with superior mechanical stiffness and thermal conductivity graphene offers advantages over metal catalysts plagued with aggregation and carbon deposition issues during harsh electrochemical operations. Nitrogen or other heteroatom doping provides a further lever to tune catalytic interaction potentials and band energetics favorable for selective CO_2 conversion [139, 140].

Furthermore, the high theoretical specific surface area per gram rivals' porous supports accommodating abundant anchoring sites when appropriately functionalized for decorating metal, metal oxide, and organometallic nanoparticles critical for multi-electron reduction events necessary to catalyze CO_2 electrolysis. Molecular preorientation of epoxide and hydroxyls also assists CO_2 capture near electropositive sites promoting activation. Overall graphene fosters excellent synergistic hybridization opportunities transforming limitations of conventional catalysts for efficient carbon capture and electroreduction.

Practical implementations: Graphene electrodes with copper nanoparticle decoration exhibit marked improvements for ethylene and ethanol productions at substantially lower overpotentials than pure copper. Detailed mechanistic studies confirm cooperative effects between graphene band structure and copper interfaces facilitate key proton transfer steps and intermediate stabilization uniquely activating CHO– like species after initial CO_2 reduction which otherwise readily desorb on pure copper surfaces [141, 142].

Likewise, defect engineered graphene triple bonded to individually size-controlled sub-nanometric gold clusters displayed over 95% Faradaic efficiencies for selective CO production at 300 Mv lower overpotentials markedly outperforming extended gold catalysts. Strong metal support interactions through graphene and electronic structure effects aid these advances. Further doping with nitrogen and sulfur produced diverse oxygenate selectivity controllable by cluster size and band bending effects. The development of recent advance catalyst have been shown in Table 7.1.

Table 7.1 Performance of nanostructured catalysts for CO_2 reduction

Catalyst	**Major Product**	**Morphology**	**Faradaic Efficiency (%)**	**Current Density (mA/cm^2)**	**Ref.**
Sn/SnO_x	HCOOH	NP	19	–0.6	[143]
Au	CO	NP	60	0.3	[117]
Au	CO	NP	90	_	[144]
Cu	CO	NP	25	47.5	[119]
Au	CO	NP	20	400	[145]
MoS_2	CO	NS	98	–65	[146]
Ag/C	CO	NP	84.4	10	[147]
Cu	CH_4	NW	55	–6	[148]
N-doped Carbon	CO	NT	80	1	[149]
Co phthalocyanine/CNT	CO	NT	92	9	[132]
N, Fe-decorated CNT	CO	NT	69	–	[131]

Catalyst	Major Product	Morphology	Faradaic Efficiency (%)	Current Density (mA/cm^2)	Ref.
Pt	$HCOO^-$	NP	100	10	[150]
Au or Pd	CO	Nanoneedle	95	22	[151]
NiGa	C_2H_6	Multi-phases	1.3	100	[152]
Au	CO	Nanoneedles	95	22	[151]
$Cu/Ni(OH)_2$	CO	NS	92	4.3	[153]
CuO	CO	NW	_	100	[154]
SnO_2	CO	Wire-in-tube	63	7.6	[120]
Bi	HCOOH	NS	100	24	[155]
Vo-rich ZnO	CO	NS	83	–16.1	[156]
A-Ni-NG	CO	Single atoms	97	22	[157]
Au/PE	CO	NP/ Nanoporous Structures	92	25.5	[158]
Nb_4N_5–NO/NC	CO	NP/NS	91	–1	[159]
Bi_2S_3-Bi_2O_3	HCOOH	NS	70.8	6.14	[160]
In–In_2S_3	HCOOH	NS	76	70.3	[161]
B-doped graphene (G)	HCOOH	NT	66	1.4	[162]
Sn sheets confined in G	HCOOH	NP	89	22.1	[163]
N-doped G(NG)	HCOOH	NP	73	7.5	[164]
Sb nanosheets/G	HCOOH	NS	84	8	[165]
Cu NPs/NG	C_2H_5OH	NP	63	1.7	[166]
Au NPs/GNR	CO	NP	92	5	[167]
Fe/NG	CO	NP	80	2.5	[168]
Cu/TiO_2/NG	C_2H_5OH	NP	43.6	1.43	[169]
Ag/NG/carbon	C_2H_5OH	NP	85.2	0.31	[170]
Cu_2O/ZnO/G	PrOH	NP	30	8	[171]

NP-Nanoparticles, NS-Nanosheets, Nanowires-NW, Nanotubes-NT

7.8 ELECTROCHEMICAL PROCESSES AND MECHANISM

7.8.1 Electrochemical CO_2 Conversion to CO, HCOOH, CH_4, CH_3OH

Electrochemically reducing carbon dioxide can produce a range of compounds based on the catalyst, electrolyte, and operating conditions used. The possible products include carbon monoxide, formic acid, hydrocarbons like methane or ethylene, and alcohols such as methanol or ethanol. Maximizing selectivity and yield for specific target products is essential for efficient processes [172].

7.8.1.1 Carbon Monoxide (CO)

CO forms through an initial one electron transfer to CO_2 generating a radical anion that readily converts to CO by protonation from water:

$$CO_2 + e^- \rightarrow CO_2^{\bullet-}$$

$$CO_2^{\bullet-} + H^+ \rightarrow CO + OH^-$$

Overall:

$$CO_2 + e^- + H^+ \rightarrow CO + OH^-$$

As the simplest one carbon reduction product requiring minimal hydrogen input, CO forms with relatively high selectivity on metallic catalysts like silver, zinc, and gold. By tuning particle size and morphology influences on intermediate binding strengths, some control over CO selectivity is possible. While CO is an important industrial feedstock for further conversions, the low carbon number limits recoverable energy density per CO_2 molecule. Nevertheless, in certain chemical production contexts, CO can offer easier purification relative to more complex hydrocarbons or alcohols that require additional separation [173].

7.8.1.2 Formic Acid (HCOOH)

Formic acid (HCOOH) remains one of the simplest reductions requiring one proton coupled electron transfer. Formate salts represent chemical storage possibilities for renewable electricity. Selectivity control stems from the relative ease of hydrating the CO_2 radical anion to format an acidifying formaldehyde like intermediate that readily converts to formic acid by protonation:

$$CO_2 + e^- \rightarrow CO_2^{\bullet-}$$
$$CO_2^{\bullet-} + H_2O \rightarrow HCOO^- + H^+$$
$$HCOO^- + H^+ \rightarrow HCOOH$$

Overall:

$$CO_2 + 2e^- + 2H^+ \rightarrow HCOOH$$

Limiting hydrogen sources through the use of protic ionic liquids and carbon nanotube based catalysts have achieved high ~90% formic acid selectivity at modest overpotentials [174].

7.8.1.3 Hydrocarbons Like Methane (CH_4)

More reduced C1 or C2 hydrocarbon formation like methane, ethylene, and ethane requires additional protons and electrons coupled at much higher overpotentials driving hydrogen evolution as a competing side reaction:

$$CO_2 + 8e^- + 8H^+ \rightarrow CH_4 + 2H_2O$$

Preventing the back reaction of hydrocarbons like CH_4 to CO intermediates also remains challenging. These energetics constraints limit most catalyst hydrocarbon efficiencies to below 50% [175].

7.8.1.4 Alcohols Like Methanol (CH_3OH)

The production of more oxidized C1 or C2 alcohols like methanol, ethanol, and n-propanol through electrochemical CO_2 reduction faces both selectivity and efficiency challenges. Multi-step proton coupled transfers faced with competing single electron pathways towards CO limits most alcohol productions in water to below 20% currently:

$$CO_2 + 6e^- + 6H^+ \rightarrow CH_3OH + H_2O$$

Particle size effects to modulate key transition state energies on copper nanoparticles have achieved ~60% ethylene glycol selectivity. Ionic liquids and advanced cell configurations also improve toward desired alcohol formation but often use expensive precious metal catalysts like gold-palladium posing commercial viability issues requiring further development of optimal catalyst-electrolyte systems [176].

In summary, producing fuels or easily separable compounds selectively using CO_2 electrolysis remains a great challenge requiring interdisciplinary advances pairing smart electrocatalysts and reactor designs guided by advanced spectroscopic investigations and computational modeling to elucidate underlying mechanisms for targeted product optimizations.

7.8.2 Mechanistic Pathways

The electrochemical conversion of CO_2 is a complex multi-step process involving electron transfer, proton transfer, and intermediate reactions. Elucidating the mechanistic pathways and identifying rate-limiting steps is vital for designing selective and efficient catalyst systems.

1. ***Initial Activation Steps:*** The first step in aqueous media involves one-electron transfer to CO_2 to form an activated $CO_2^{\bullet-}$ radical anion:

   $$CO_2 + e^- \rightarrow CO_2^{\bullet-}$$

   The bent radical configuration enables attack by nucleophiles like H_2O. Potential pathways involve protonation to CO or reaction with water to form carboxylic acids:

   $$CO_2^{\bullet-} + H^+ \rightarrow CO + OH-$$

   $$CO_2^{\bullet-} + H_2O \rightarrow HCOO^- + H^+$$

   $$HCOO^- + H^+ \rightarrow HCOOH$$

   Over metal catalysts, CO release competes kinetically vs further reduction. Modulating particle size and morphology that tune intermediate binding can help steer selectivity between CO and other multi carbon products.

2. ***C1 vs C2 Pathways:*** The C1 pathway involves step-wise proton coupled multi-electron transfers producing methane, methanol, or formic acid with each sequential addition:

$$CO_2 + H^+ + e^- \rightarrow {}^{\bullet}COOH$$
$$COOH + H^+ + e^- \rightarrow {}^{\bullet}CHO$$
$$CHO + H^+ + e^- \rightarrow {}^{\bullet}CH_2OH$$
$$CH_2OH + H^+ + e^- \rightarrow CH_3OH$$

In contrast, C^{2+} products require coupling of partially reduced intermediates:

$$2(CO_2 + e^-) \rightarrow 2(CO_2^{\bullet -})$$
$$CO_2^{\bullet -} + {}^{\bullet}CO_2^- \rightarrow \text{Oxalate}$$
$$\text{Oxalate} + 6e^- + 6H^+ \rightarrow \text{Ethylene glycol}$$

Ethylene, ethanol, and n-propanol form through similar radical couplings. Discriminating between single, double, or triple C–C bonds governs selectivity splitting between 1C or 2C products.

3. ***Intermediate Stabilization:*** The high reactivity of radicals formed -post-CO_2 activation drives quicker desorption before further reductions can occur which limits yields. Functional groups on the catalyst surface may stabilize these intermediates by hydrogen bonding or valence adjustments [177, 178].

   Metal-oxide interfaces on copper catalysts assist in stronger binding of CHx and OHx groups improving selectivity. Ionic liquids also stabilize some intermediates by solvation effects. Further research into intermediate characterization using advanced *in-situ* techniques and computational modeling will provide greater details on prevailing mechanistic pathways.

   Overall, the competing dynamics between activation barriers, intermediate stability, and managing proton-electron transfers govern product distributions which can be systematically modulated through nanocatalyst and electrolyte design paired with mechanistic insights.

4. ***Rate Limiting Steps:*** The overall efficiency of converting CO_2 into value added fuels and chemicals is governed by the kinetics of the slowest steps in the mechanistic pathway. Understanding these rate-determining barriers is a key to designing improved catalysts and processes [179].

   First Electron Transfer: The initial one-electron transfer to bend and activate CO_2 is widely accepted as one of the most significant rate-limiting steps. This is driven by the thermodynamic inertness of the linear CO_2 molecule ($\Delta G = +104$ kJ/mol). This energy demand correlates with the observed high overpotentials.

   $$CO_2 + e^- \rightarrow [CO_2^{\bullet -}]^* \text{ (High overpotential} \sim 0.5\text{–}1 \text{ V)}$$

   Additionally, electrode interfaces stabilize the bent $[CO_2^{\bullet -}]^*$ radical transiently allowing competitive secondary reactions to occur instead of desired multi-electron reductions. Rapid surface desorption regenerates unreacted CO_2 limiting overall yield. Modulating particle size, morphology, and pore structures may better confine reactive intermediates.

Subsequent proton-coupled transfers: Following the initial electron uptake and activation through the $[CO_2^{\bullet -}]^*$ radical anion, a series of proton coupled electron transfers are required to further reduce CO_2 into hydrocarbons, oxygenates, or alcohols. Each sequential protonation represents an additional energy barrier

$$[CO_2^{\bullet -}]^* + H^+ + e^- \rightarrow [HCO_2^{\bullet}]^*$$

$$[HCO_2^{\bullet}]^* + H^+ + e^- \rightarrow [H_2CO]^*$$

$$[H_2CO]^* + H^+ + e^- \rightarrow [H_3COH]^*$$

Kinetic control over these Proton Coupled Electron Transfer (PCET) steps fine-tunes intermediate species permitting some selectivity between C1 vs C2 or oxygenate vs reduced hydrocarbon pathways [176].

5. ***Hydrogen Evolution Activity:*** The high concentration of protons in aqueous electrolytes causes intense competition from the hydrogen evolution -half-reaction. This thermodynamically favored proton reduction limits many CO_2 electrocatalysts which also inadvertently make active hydrogen evolution co-catalysts leading to lowered efficiencies.

$$2H^+ + 2e^- \rightarrow H_2\ (0\ V)$$

Using non-aqueous solvents or solid proton conducting membranes to modulate proton availability has improved hydrocarbon selectivitiesby over 40% by suppressing hydrogen side reactions.

The collective energy barriers spanning initial CO_2 activation coupled with subsequent reduction potentials for C–C or C–O bond formation compete kinetically against faster non-productive proton reduction warranting extensive mechanistic examinations to identify ideal catalyst-electrolyte pairs for selective CO_2 electrolysis.

7.9 CONCLUSION

The electrochemical conversion of CO_2 into fuels or value added chemicals provides a promising strategy for carbon capture and utilization that can simultaneously mitigate greenhouse gas emissions while also storing intermittent renewable electricity in the form of chemical bonds. However, traditional catalytic processes face fundamental challenges regarding the stabilization of key reaction intermediates and competing side reactions due to the incredible stability of the CO_2 molecule.

Leveraging Nanomaterials: Engineering catalysts and processes at the nanoscale provides unprecedented opportunities to overcome these limitations through deliberate modulation of physicochemical properties inaccessible to bulk materials. The high density of surface-active sites lowers activation barriers while quantum confinement effects allow tuning of band energetics to match binding energies and stabilization of transitional states. Hollow interiors and porous architectures facilitate reactant enrichment and intermediate stabilization. Exposed facets, abundant grain boundaries and compositional strains further optimize activity and selectivity.

Recent Progress: Significant progress has taken place demonstrating enhanced electrocatalytic conversion of CO_2 selectively to CO, hydrocarbons and alcohols using nano-engineered metals, bimetallics, carbon-based and hybrid nanomaterials. Strategic material configurations targeting identified rate limiting steps such as initial electron transfer activation, proton coupled reductions and intermediate stabilization have achieved promising improvements in conversion efficiencies and selectivity in laboratory studies.

Remaining Challenges: However, most reports have been limited to proof-of-concept stages under model conditions with simple electrolytes. Practical processes must contend with additional factors like electrode fouling, poor reactant transport, variability in renewable electricity supplies etc. which can degrade nanomaterial effectiveness over long-term operations. Scalability and reliable large scale manufacturing of designed nanocatalysts remains an ongoing challenge. Device and reactor engineering aspects also require further innovation to meet commercial targets.

Future Outlook: In summary, while nanomaterials have clearly opened promising directions for electrochemical CO_2 valorization, considerable multidisciplinary advances bridging fundamental catalysis design with applied engineering across reactor and process scales will be essential to enable translatable technologies. Combined progress in precision nanomaterial synthesis, advanced spectroscopic characterization, computational modeling and modular systems engineering within an integrative framework can potentially transform carbon capture and conversion measures to mitigate climate change and accelerate sustainable chemical production.

The exciting innovations in electrochemical CO_2 reduction with nanomaterials over the past decade suggest we are still only scratching the surface of possibilities. Rapidly evolving tools for atomic-level structure tuning guided by molecular scale reaction insights promise to reveal enhanced catalytic phenomena facilitating customized optimization of targeted product lines at previously inaccessible selectivity, yield and energy efficiency metrics leading to commercially viable processes benefitting society globally.

7.10 ACKNOWLEDGMENT

Deepika Jamwal acknowledges financial support from Shastri Indo-Canadian Institute for Shastri Research Student Fellowship (SRSF).

REFERENCES

[1] Birdja, Y.Y., Pérez-Gallent, E., Figueiredo, M.C., Göttle, A.J., Calle-Vallejo, F. and Koper, M.T. 2019. Advances and challenges in understanding the electrocatalytic

conversion of carbon dioxide to fuels. Nature Energy. 4(9): 732–745. doi:10.1038/s41560-019-0450-y.

[2] Ganesh, I. 2016. Electrochemical conversion of carbon dioxide into renewable fuel chemicals–The role of nanomaterials and the commercialization. Renewable and Sustainable Energy Reviews. 59: 1269–1297. doi:10.1016/j.rser.2016.01.026.

[3] Garrett, T.J. 2011. Are there basic physical constraints on future anthropogenic emissions of carbon dioxide? Climatic Change. 104: 437–455. doi:10.1007/s10584-009-9717-9.

[4] Ješić, D., Jurković, D.L., Pohar, A., Suhadolnik, L. and Likozar, B. 2021. Engineering photocatalytic and photoelectrocatalytic CO_2 reduction reactions: Mechanisms, intrinsic kinetics, mass transfer resistances, reactors and multi-scale modelling simulations. Chemical Engineering Journal. 407: 126799. doi:10.1016/j.cej.2020.126799.

[5] Lu, Q. and Jiao, F. 2016. Electrochemical CO_2 reduction: Electrocatalyst, reaction mechanism, and process engineering. Nano Energy. 29: 439–456. doi:10.1016/j.nanoen.2016.04.009.

[6] Harison, M. 2024. Review of: 'Nanostructures are materials or structures that have at least one dimension between 1 and 100 nanometers have. *Qeios* [Preprint]. doi:10.32388/f7hy6f.

[7] Jones, M.W., Peters, G.P., Gasser, T., Andrew, R.M., Schwingshackl, C., Gütschow, J., et al. 2023. National contributions to climate change due to historical emissions of carbon dioxide, methane, and nitrous oxide since 1850. Scientific Data. 10(1): 155. doi:10.1038/s41597-023-02041-1.

[8] Zhou, Y., Wang, K., Zheng, S., Cheng, X., He, Y., Qin, W., et al. 2024. Advancements in electrochemical CO_2 reduction reaction: A review on CO_2 mass transport enhancement strategies. Chemical Engineering Journal. 150169. doi:10.1016/j.cej.2024.150169.

[9] Elhenawy, S.E.M., Khraisheh, M., AlMomani, F. and Walker, G. 2020. Metal-organic frameworks as a platform for CO2 capture and chemical processes: Adsorption, membrane separation, catalytic-conversion, and electrochemical reduction of CO_2. Catalysts. 10(11): 1293. doi:10.3390/catal10111293.

[10] CO_2 conversion and utilization. 2002. Focus on Catalysts. 2002(8): 8. doi:10.1016/s1351-4180(02)00857-7.

[11] Yadav, T.P., Yadav, R.M. and Singh, D.P. 2012. Mechanical milling: A top down approach for the synthesis of nanomaterials and nanocomposites. Nanoscience and Nanotechnology. 2(3): 22–48. doi:10.5923/j.nn.20120203.01.

[12] Garg, B., Bisht, T. and Ling, Y.C. 2016. Graphene-based nanomaterials: Versatile catalysts for carbon-carbon bond forming reactions. Current Organic Chemistry. 20(15): 1547–1566. doi:10.2174/1385272820666151208212302.

[13] Gregorczyk, K. and Knez, M. 2016. Hybrid nanomaterials through molecular and atomic layer deposition: Top down, bottom up, and in-between approaches to new materials. Progress in Materials Science. 75: 1–37. doi:10.1016/j.pmatsci.2015.06.004.

[14] Choi, E.Y., Kim, M.H. and Kim, C.K. 2018. Nitrogen doped graphene-carbon nanotubes and nitrogen doped nano-onion hybrids as high-performance catalysts for oxygen reduction reaction. *In*: Electrochemical Society Meeting Abstracts. 233(5): 662–662. The Electrochemical Society, Inc. doi:10.1149/ma2018-01/5/662.

[15] Olajire, A.A. 2018. Recent progress on the nanoparticles-assisted greenhouse carbon dioxide conversion processes. Journal of CO_2 Utilization. 24: 522–547. doi:10.1016/j.jcou.2018.02.012.

[16] Adegoke, K.A. and Maxakato, N.W. 2023. Electrocatalytic CO_2 conversion on metal-organic frameworks derivative electrocatalysts. Journal of CO_2 Utilization. 69: 102412. doi:10.1016/j.jcou.2023.102412.

[17] Wang, H. 2022. Nanostructure@metal-organic frameworks (MOFs) for catalytic carbon dioxide (CO_2) conversion in photocatalysis, electrocatalysis, and thermal catalysis. Nano Research. 15(4): 2834–2854. doi:10.1007/s12274-021-3984-9.

[18] Smith, M.R. and Myers, S.S. 2018. Impact of anthropogenic CO_2 emissions on global human nutrition. Nature Climate Change. 8(9): 834–839. doi:10.1038/s41558-018-0253-3.

[19] Removal of carbon dioxide from flue gases from fossil fuel combustion. 2003. Fuel and Energy Abstracts. 44(4): 260. doi:10.1016/s0140-6701(03)82188-4.

[20] Hansen, G. and Stone, D. 2016. Assessing the observed impact of anthropogenic climate change. Nature Climate Change. 6(5): 532–537. doi:10.1038/nclimate2896.

[21] Paraschiv, S. and Paraschiv, L.S. 2020. Trends of carbon dioxide (CO_2) emissions from fossil fuels combustion (coal, gas and oil) in the EU member states from 1960 to 2018. Energy Reports. 6: 237–242. doi:10.1016/j.egyr.2020.11.116.

[22] Copard, Y., Di-Giovanni, C., Martaud, T., Albéric, P. and Olivier, J.E. 2006. Using Rock-Eval 6 pyrolysis for tracking fossil organic carbon in modern environments: implications for the roles of erosion and weathering. Earth Surface Processes and Landforms. 31(2): 135–153. doi:10.1002/esp.1319.

[23] He, Y. and Song, W. 2022. Analysis of the impact of carbon trading policies on carbon emission and carbon emission efficiency. Sustainability. 14(16): 10216. doi:10.3390/su141610216.

[24] BRZEŻAŃSKI, M. 2007. Carbon dioxide emissions in the aspect of applied engine fuels. Combustion Engines. 131(4): 62–67. doi:10.19206/ce-117319.

[25] Sun, X., Zhao, P., Liang, X., Jing, G. and Chen, G. 2023. Effects of Steam-Injection Strategies on Performance and Emission Characteristics of Marine Engines. Journal of Energy Engineering. 149(3): 04023009. doi:10.1061/jleed9.eyeng-4743.

[26] Kubik, A., Turoń, K., Folęga, P. and Chen, F. 2023. CO_2 emissions—evidence from internal combustion and electric engine vehicles from car-sharing systems. Energies. 16(5): 2185. doi:10.3390/en16052185.

[27] Nivesrangsan, P., Steel, J.A. and Reuben, R.L. 2007. Source location of acoustic emission in diesel engines. Mechanical Systems and Signal Processing. 21(2): 1103–1114. doi:10.1016/j.ymssp.2005.12.010.

[28] O'Connor, J.P. 2020. Modeling of atmospheric carbon dioxide (CO_2) concentrations as a function of fossil-fuel and land-use change CO_2 emissions coupled with oceanic and terrestrial sequestration. Climate. 8(5): 61. doi:10.3390/cli8050061.

[29] Shan, S., Li, Y., Zhang, Z., Zhu, W. and Zhang, T. 2023. Identification of Key Carbon Emission Industries and Emission Reduction Control Based on Complex Network of Embodied Carbon Emission Transfers: The Case of Hei-Ji-Liao, China. International Journal of Environmental Research and Public Health. 20(3): 2603. doi:10.3390/ijerph20032603.

[30] Ito, A. and Kawamiya, M. 2010. Potential impact of ocean ecosystem changes due to global warming on marine organic carbon aerosols. Global Biogeochemical Cycles. 24(1). doi:10.1029/2009gb003559.

[31] Northcott, D., Sevadjian, J., Sancho-Gallegos, D.A., Wahl, C., Friederich, J. and Chavez, F.P. 2019. Impacts of urban carbon dioxide emissions on sea-air flux and ocean acidification in nearshore waters. PloS One. 14(3): e0214403. doi:10.1371/journal.pone.0214403.

[32] Bernold M. Hanson. 1994. The Burning of Fossil Fuels and Carbon Dioxide: Abstract. AAPG Bulletin. 78. doi:10.1306/a260007f-171b-11d7-8645000102c1865d.

[33] Ding, Y. and Li, F. 2017. Examining the effects of urbanization and industrialization on carbon dioxide emission: evidence from China's provincial regions. Energy. 125, 533–542. doi:10.1016/j.energy.2017.02.156.

[34] Solomon, S., Plattner, G.K., Knutti, R. and Friedlingstein, P. 2009. Irreversible climate change due to carbon dioxide emissions. Proceedings of the National Academy Of Sciences. 106(6): 1704–1709. doi:10.1073/pnas.0812721106.

[35] Global atmospheric greenhouse gas concentrations reach record levels. 2015. Physics Today [Preprint]. doi:10.1063/pt.5.029367.

[36] The war has produced more planet-warming gases than the annual emissions of 20 countries... 2024. üha [Preprint]. doi:10.32739/uha.id.44236.

[37] Podobnikar, T. 2007. Visualisations of the Human Impacts on the Earth's Surface. Acta Carsologica. 36(2). doi:10.3986/ac.v36i2.192.

[38] OA, O. 2018. Climate Change and its Effect on the Global Economy and Security: A Call for more Robust Climate Finance, Prevention of Climate Finance Against Corrupt Spending and Review of Articles 9(1): (3) & (4) of the Paris Agreement and 12(8) of Kyoto Protocol to the United Nations Framework Convention on Climate Change. Environment Pollution and Climate Change. 02(03). doi:10.4172/2573-458x.1000157.

[39] Colucci, R.R. and Guglielmin, M. 2019. Climate change and rapid ice melt: Suggestions from abrupt permafrost degradation and ice melting in an alpine ice cave. Progress in Physical Geography: Earth and Environment. 43(4): 561–573. doi:10.1177/0309133319846056.

[40] Ramírez-Valencia, L.D., Bailón-García, E., Carrasco-Marín, F. and Perez-Cadenas, A.F. 2021. From CO_2 to value-added products: a review about carbon-based materials for electro-chemical CO_2 conversion. Catalysts. 11(3): 351. doi:10.3390/catal11030351.

[41] Hu, B., Guild, C. and Suib, S.L. 2013. Thermal, electrochemical, and photochemical conversion of CO_2 to fuels and value-added products. Journal of CO_2 Utilization. 1: 18–27. doi:10.1016/j.jcou.2013.03.004

[42] Yusuf, N., Almomani, F. and Qiblawey, H. 2023. Catalytic CO_2 conversion to C1 value-added products: Review on latest catalytic and process developments. Fuel. 345. 128178. doi:10.1016/j.fuel.2023.128178.

[43] Wickramasinghe, S., Wang, J., Morsi, B. and Li, B. 2021. Carbon dioxide conversion to nanomaterials: Methods, applications, and challenges. Energy and Fuels. 35(15): 11820–11834. doi:10.1021/acs.energyfuels.1c01533.

[44] Dan, D. 2020. Electrochemistry: Electrochemical cell, thermodynamic and kinetic aspects. Journal of Marine Science Research and Oceanography. 5(2). doi:10.33140/jmsro.05.02.01.

[45] Ma, S. and Kenis, P.J. 2013. Electrochemical conversion of CO_2 to useful chemicals: current status, remaining challenges, and future opportunities. Current Opinion in Chemical Engineering. 2(2): 191–199. doi:10.1016/j.coche.2013.03.005.

[46] Simões, A.L.A., Rodrigues, J.S., Araujo, M.H. and Vieira, S.S. 2022. The chemistry involved in CO_2 conversion: Challenges and opportunities. Revista Virtual de Química. 14(3): 468–483. doi:10.21577/1984-6835.20220091.

[47] Kafi, M., Sanaeepur, H. and Ebadi Amooghin, A. 2023. Grand Challenges in CO_2 Capture and Conversion. Journal of Resource Recovery. 1. doi:10.52547/jrr.2302-1007.

[48] Galadima, A. and Muraza, O. 2019. Catalytic thermal conversion of CO_2 into fuels: Perspective and challenges. Renewable and Sustainable Energy Reviews. 115: 109333. doi:10.1016/j.rser.2019.109333.

[49] Upadhyay, P.K. and Srivastava, V. 2016. Carbon sequestration: hydrogenation CO_2 to formic acid. Present Environment and Sustainable Development, (2): 14–34. doi:10.1515/pesd-2016-0022.

[50] Mousavi, Z., Pirdashti, M., Rostami, A.A. and Dragoi, E.N. 2020. Thermophysical properties analysis of poly (ethylene glycol) 600+ methanol, ethanol, 1-propanol, and 2-propanol binary liquid mixtures. International Journal of Thermophysics. 41(2): 19. doi:10.1007/s10765-019-2600-7.

[51] Hongxuan, W. and Zhen, Z. 2005. The role of the catalysts with highly dispersed and isolated active sites in the selective oxidation of light hydrocarbons. Progress in Natural Science. 15(12): 1066–1072. doi:10.1080/10020070512331343186.

[52] Honciuc, M. and Honciuc, A. 2024. Morphological Design and Synthesis of Nanoparticles. Nanomaterials. 14(4): 360. doi:10.3390/nano14040360.

[53] Perduca, M. 2022. Synthesis, properties and applications of polymeric nanomaterials. Nanomaterials. 12(24): 4385. doi:10.3390/nano12244385.

[54] Vennekoetter, J.B., Sengpiel, R. and Wessling, M. 2019. Beyond the catalyst: How electrode and reactor design determine the product spectrum during electrochemical CO_2 reduction. Chemical Engineering Journal. 364: 89–101. doi:10.1016/j.cej.2019.01.045.

[55] Wan, Q. and Ke, C. 2022. CO_2 Electrochemical Reduction to CO: From Catalysts, Electrodes to Electrolytic Cells and Effect of Operating Conditions. In Advanced Electrochemical Materials in Energy Conversion and Storage. 297–328. CRC Press. doi:10.1201/9781003133971-12.

[56] Kasherman, D. and Skyllas-Kazacos, M. 1988. Effects of anode-cathode distance on the cell potential and electrical bath resistivity in an aluminium electrolysis cell with a sloping TiB_2 composite cathode. Journal of Applied Electrochemistry. 18(6): 863–868. doi:10.1007/bf01016043.

[57] Garg, S., Li, M., Weber, A.Z., Ge, L., Li, L., Rudolph, V., et al. 2020. Advances and challenges in electrochemical CO_2 reduction processes: an engineering and design perspective looking beyond new catalyst materials. Journal of Materials Chemistry A. 8(4): 1511–1544. doi:10.1039/c9ta13298h.

[58] Karanjkar, D.S., S. Chatterji, None Shimi S.L, and Amod Kumar. 2014. Real Time Simulation and Analysis of Maximum Power Point Tracking (MPPT) Techniques for Solar Photo-Voltaic System. doi:10.1109/raecs.2014.6799656.

[59] Karimi, I.A. and Kawi, S. 2016. Technoeconomic perspectives on sustainable CO_2 capture and utilization. Environmental Science and Pollution Research. 23: 22223–22225. doi:10.1007/s11356-016-7838-z.

[60] Zhu, Y., Yu, J., Fan, Z. and Zhang, Z. 2021. 2D Materials for electrochemical carbon dioxide reduction. *In*: Nanomaterials for CO_2 Capture, Storage, Conversion and Utilization. 183–196. Elsevier. doi:10.1016/b978-0-12-822894-4.00006-x.

[61] Zhang, Z. and Wang, R. 2022. Editorial for the special issue: Advanced nanomaterials for electrochemical energy conversion and storage. Nanomaterials. 12(20): 3579. doi:10.3390/nano12203579.

[62] Mathur, S. 2003. Molecule derived nanomaterials: Chemical concepts for composition, morphology and particle size control. MRS Online Proceedings Library (OPL). 788, L1–6. doi:10.1557/proc-788-l1.6.

[63] Freund, H.J. and Messmer, R.P. 1986. On the bonding and reactivity of CO_2 on metal surfaces. Surface Science. 172(1): 1–30. doi:10.1016/0039-6028(86)90580-7.

[64] Lu, Q. and Jiao, F. 2016. Electrochemical CO_2 reduction: Electrocatalyst, reaction mechanism, and process engineering. Nano Energy. 29: 439–456. doi:10.1016/j.nanoen.2016.04.009.

[65] Liu, Y., Tang, H., Zhou, Y. and Lin, B.L. 2024. Improved catalytic performance of CO_2 electrochemical reduction reaction towards ethanol on chlorine-modified Cu-based electrocatalyst. Nano Research. 17(5): 3761–3768. doi:10.1007/s12274-023-6301-y.

[66] Rothenberg, G. 2023. A realistic look at CO_2 emissions, climate change and the role of sustainable chemistry. Sustainable Chemistry for Climate Action. 2: 100012. doi:10.1016/j.scca.2023.100012.

[67] Chen, W., Pei, J., He, C.T., Wan, J., Ren, H., Wang, Y., et al. 2018. Single tungsten atoms supported on MOF-derived N-doped carbon for robust electrochemical hydrogen evolution. Advanced Materials. 30(30): 1800396. doi:10.1002/adma.201800396.

[68] Liu, P., Chen, B., Liang, C., Yao, W., Cui, Y., Hu, S., et al. 2021. Tip-enhanced electric field: a new mechanism promoting mass transfer in oxygen evolution reactions. Advanced Materials. 33(9): 2007377. doi:10.1002/adma.202007377.

[69] Ling, T., Jaroniec, M. and Qiao, S.Z. 2020. Recent progress in engineering the atomic and electronic structure of electrocatalysts via cation exchange reactions. Advanced Materials. 32(46): 2001866. doi:10.1002/adma.202001866.

[70] Jamwal, D., Hussain, S., Rana, D., Katoch, A. and Mehta, S.K. 2022. Mixed transition metal oxide nanostructures based supercapacitors. Encyclopedia of Energy Storage. 4, 566–580. doi: 10.1016/B978-0-12-819723-3.00097-4.

[71] Xu, L., Papanikolaou, K.G., Lechner, B.A., Je, L., Somorjai, G.A., Salmeron, M., et al. 2023. Formation of active sites on transition metals through reaction-driven migration of surface atoms. Science. 380(6640): 70–76. doi:10.1126/science.add0089.

[72] da Silva Freitas, W., D'Epifanio, A. and Mecheri, B. 2021. Electrocatalytic CO_2 reduction on nanostructured metal-based materials: Challenges and constraints for a sustainable pathway to decarbonization. Journal of CO_2 Utilization. 50: 101579. doi:10.1016/j.jcou.2021.101579.

[73] Li, L., Huang, Y. and Li, Y. 2020. Carbonaceous materials for electrochemical CO_2 reduction. EnergyChem. 2(1): 100024. doi:10.1016/j.enchem.2019.100024.

[74] Clegg, F., Breen, C. and Khairuddin. 2014. Synergistic and competitive aspects of the adsorption of poly (ethylene glycol) and poly (vinyl alcohol) onto Na-bentonite. The Journal of Physical Chemistry B. 118(46): 13268–13278. doi:10.1021/jp507772t.

[75] Gupta, P., Coon, P.A., Koehler, B.G. and George, S.M. 1991. Desorption product yields following Cl_2 adsorption on Si (111) 7×7: Coverage and temperature dependence. Surface Science. 249(1–3): 92–104. doi:10.1016/0039-6028(91)90835-g.

[76] Soomro, F.A., Zaib, A., Haq, R.U. and Sheikholeslami, M. 2019. Dual nature solution of water functionalized copper nanoparticles along a permeable shrinking

cylinder: FDM approach. International Journal of Heat and Mass Transfer. 129: 1242-1249. doi:10.1016/j.ijheatmasstransfer.2018.10.05

[77] Halicioğlu, T. 1976. Computer simulation of adsorption of Cu atoms on W (100) and W (110) surfaces. Journal of Crystal Growth. 33(2): 212–214. doi:10.1016/0022-0248(76)90045-2.

[78] Li, F., Wang, Q., Liu, G., Lv, W. and Zhou, Z. 2023. Crystal facet-dependent product selectivity of CO_2 reduction reaction on indium revealed by first-principles calculations. Surface Science. 735: 122333. doi:10.1016/j.susc.2023.122333.

[79] Rana, D., Hussain, S., Mehta, S.K., Jamwal, D. and Katoch, A. 2022. Metal Chalcogenide Nanomaterials Based Supercapacitors. Encyclopedia of Energy Storage. 4: 599–607. doi: 10.1016/B978-0-12-819723-3.00024-X.

[80] Yang, J., Li, W., Wang, D. and Li, Y. 2020. Electronic metal–support interaction of single-atom catalysts and applications in electrocatalysis. Advanced Materials. 32(49): 2003300. doi:10.1002/adma.202003300.

[81] Zhao, G., Rui, K., Dou, S.X. and Sun, W. 2020. Boosting electrochemical water oxidation: the merits of heterostructured electrocatalysts. Journal of Materials Chemistry A. 8(14): 6393–6405. doi:10.1039/d0ta00708k.

[82] Han, C., Li, W., Liu, H.K., Dou, S. and Wang, J. 2019. Design strategies for developing non-precious metal based bi-functional catalysts for alkaline electrolyte based zinc–air batteries. Materials Horizons. 6(9): 1812–1827. doi:10.1039/c9mh00502a.

[83] Xu, J., Lian, Z., Wei, B., Li, Y., Bondarchuk, O., Zhang, N., et al. 2020. Strong electronic coupling between ultrafine iridium–ruthenium nanoclusters and conductive, acid-stable tellurium nanoparticle support for efficient and durable oxygen evolution in acidic and neutral media. ACS Catalysis. 10(6): 3571–3579. doi:10.1021/acscatal.9b05611.

[84] Lagadec, M.F. and Grimaud, A. 2020. Water electrolysers with closed and open electrochemical systems. Nature Materials. 19(11): 1140–1150. doi:10.1038/s41563-020-0788-3.

[85] Yu, L., Deng, D. and Bao, X. 2020. Chain mail for catalysts. Angewandte Chemie International Edition. 59(36): 15294–15297. doi:10.1002/anie.202007604.

[86] Tseng, T.C., Urban, C., Wang, Y., Otero, R., Tait, S.L., Alcamí, M., et al. 2010. Charge-transfer-induced structural rearrangements at both sides of organic/metal interfaces. Nature Chemistry. 2(5): 374–379. doi:10.1038/nchem.591.

[87] Wang, D., Li, Q., Han, C., Xing, Z. and Yang, X. 2019. Single-atom ruthenium based catalyst for enhanced hydrogen evolution. Applied Catalysis B: Environmental. 249: 91–97. doi:10.1016/j.apcatb.2019.02.059.

[88] Qiu, H.J., Ito, Y., Cong, W., Tan, Y., Liu, P., Hirata, A., et al. 2015. Nanoporous graphene with single-atom nickel dopants: an efficient and stable catalyst for electrochemical hydrogen production. Angewandte Chemie International Edition. 54(47): 14031–14035. doi:10.1002/anie.201507381.

[89] Liang, C., Zou, P., Nairan, A., Zhang, Y., Liu, J., Liu, K., et al. 2020. Exceptional performance of hierarchical Ni–Fe oxyhydroxide@ NiFe alloy nanowire array electrocatalysts for large current density water splitting. Energy and Environmental Science. 13(1): 86–95. doi:10.1039/c9ee02388g.

[90] Deng, D., Novoselov, K.S., Fu, Q., Zheng, N., Tian, Z. and Bao, X. 2016. Catalysis with two-dimensional materials and their heterostructures. Nature Nanotechnology. 11(3): 218–230. doi:10.1038/nnano.2015.340.

[91] Liu, M., Pang, Y., Zhang, B., De Luna, P., Voznyy, O., Xu, J., et al. 2016. Enhanced electrocatalytic CO_2 reduction via field-induced reagent concentration. Nature. 537(7620): 382–386. doi:10.1038/nature19060.

[92] Liu, D., Li, X., Chen, S., Yan, H., Wang, C., Wu, C., et al. 2019. Atomically dispersed platinum supported on curved carbon supports for efficient electrocatalytic hydrogen evolution. Nature Energy. 4(6): 512–518. https://doi.org/10.1038/s41560-019-0402-6

[93] Jiang, H., Hou, Z. and Luo, Y. 2017. Unraveling the mechanism for the sharp-tip enhanced electrocatalytic carbon dioxide reduction: the kinetics decide. Angewandte Chemie International Edition. 56(49): 15617–15621. doi:10.1002/anie.201708825.

[94] Fang, S., Zhu, X., Liu, X., Gu, J., Liu, W., Wang, D., et al. 2020. Uncovering near-free platinum single-atom dynamics during electrochemical hydrogen evolution reaction. Nature Communications. 11(1): 1029. doi:10.1038/s41467-020-14848-2.

[95] Wang, Y., Pan, Y., Zhu, L., Yu, H., Duan, B., Wang, R., et al. 2019. Solvent-free assembly of Co/Fe-containing MOFs derived N-doped mesoporous carbon nanosheets for ORR and HER. Carbon. 146: 671–679. doi:10.1016/j.carbon.2019.02.002.

[96] Jia, Y., Zhang, L., Du, A., Gao, G., Chen, J., Yan, X., et al. 2016. Defect graphene as a trifunctional catalyst for electrochemical reactions. Advanced Materials. 28(43): 9532–9538. doi:10.1002/adma.201602912.

[97] Zhu, Y., Murali, S., Cai, W., Li, X., Suk, J.W., Potts, J.R., et al. 2010. Graphene and graphene oxide: synthesis, properties, and applications. Advanced Materials. 22(35): 3906–3924. doi:10.1002/adma.201001068.

[98] Tang, T., Jiang, W.J., Liu, X.Z., Deng, J., Niu, S., Wang, B., et al. 2020. Metastable rock salt oxide-mediated synthesis of high-density dual-protected M@ NC for long-life rechargeable zinc–air batteries with record power density. Journal of the American Chemical Society. 142(15): 7116–7127. doi:10.1021/jacs.0c01349.

[99] He, C., Zhang, Y., Zhang, Y., Zhao, L., Yuan, L.P., Zhang, J., et al. 2020. Molecular evidence for metallic cobalt boosting CO_2 electroreduction on pyridinic nitrogen. Angewandte Chemie. 132(12): 4944–4949. doi:10.1002/anie.201916520.

[100] Zhang, J., Jiang, W.J., Niu, S., Zhang, H., Liu, J., Li, H., et al. 2020. Organic small molecule activates transition metal foam for efficient oxygen evolution reaction. Advanced Materials. 32(11): 1906015. doi:10.1002/adma.201906015.

[101] Yin, P., Yao, T., Wu, Y., Zheng, L., Lin, Y., Liu, W., et al. 2016. Single cobalt atoms with precise N-coordination as superior oxygen reduction reaction catalysts. Angewandte Chemie. 128(36): 10958–10963. doi:10.1002/anie.201604802.

[102] Wang, J., Huang, Z., Liu, W., Chang, C., Tang, H., Li, Z., et al. 2017. Design of N-coordinated dual-metal sites: a stable and active Pt-free catalyst for acidic oxygen reduction reaction. Journal of the American Chemical Society. 139(48): 17281–17284. doi:10.1021/jacs.7b10385.

[103] Qiao, Y., Yuan, P., Hu, Y., Zhang, J., Mu, S., Zhou, J., et al. 2018. Sulfuration of an Fe–N–C catalyst containing FexC/Fe species to enhance the catalysis of oxygen reduction in acidic media and for use in flexible Zn–air batteries. Advanced Materials. 30(46): 1804504. doi:10.1002/adma.201804504.

[104] Yang, Q., Yang, C.C., Lin, C.H. and Jiang, H.L. 2019. Metal–organic-framework-derived hollow N-doped porous carbon with ultrahigh concentrations of single Zn atoms for efficient carbon dioxide conversion. Angewandte Chemie. 131(11): 3549–3553. doi:10.1002/ange.201813494.

[105] Chen, Y., Ji, S., Wang, Y., Dong, J., Chen, W., Li, Z., et al. 2017. Isolated single iron atoms anchored on N-doped porous carbon as an efficient electrocatalyst for the oxygen reduction reaction. Angewandte Chemie International Edition. 56(24): 6937–6941. doi:10.1002/anie.201702473.

[106] Park, C., Koo, W.T., Chong, S., Shin, H., Kim, Y.H., Cho, H.J., et al. 2021. Confinement of Ultrasmall Bimetallic Nanoparticles in Conductive Metal–Organic Frameworks via Site-Specific Nucleation. Advanced Materials. 33(38): 2101216. doi:10.1002/adma.202101216.

[107] Ao, X., Zhang, W., Li, Z., Li, J.G., Soule, L., Huang, X., et al. 2019. Markedly enhanced oxygen reduction activity of single-atom Fe catalysts via integration with Fe nanoclusters. ACS Nano. 13(10): 11853–11862. doi:10.1021/acsnano.9b05913.

[108] Wang, T., Kou, Z., Mu, S., Liu, J., He, D., Amiinu, I.S., et al. 2018. 2D dual-metal zeolitic-imidazolate-framework-(ZIF)-derived bifunctional air electrodes with ultrahigh electrochemical properties for rechargeable zinc–air batteries. Advanced Functional Materials. 28(5): 1705048. doi:10.1002/adfm.201705048.

[109] Chen, Z., Wu, R., Liu, Y., Ha, Y., Guo, Y., Sun, D., et al. 2018. Ultrafine Co nanoparticles encapsulated in carbon-nanotubes-grafted graphene sheets as advanced electrocatalysts for the hydrogen evolution reaction. Advanced Materials. 30(30): 1802011. doi:10.1002/adma.201802011.

[110] Schwab, G.M. 1970. Chemical effects at the solid/solid phase boundary. Journal of Colloid and Interface Science. 34(3): 337–342. doi:10.1016/0021-9797(70)90192-x.

[111] Tributsch, H., Koslowski, U.I. and Dorbandt, I. 2008. Experimental and theoretical modeling of Fe-, Co-, Cu-, Mn-based electrocatalysts for oxygen reduction. Electrochimica Acta. 53(5): 2198–2209. doi:10.1016/j.electacta.2007.09.027.

[112] Campbell, C.T. 2012. Electronic perturbations. Nature Chemistry. 4(8): 597–598. doi:10.1038/nchem.1412.

[113] Hu, P., Huang, Z., Amghouz, Z., Makkee, M., Xu, F., Kapteijn, F., et al. 2014. Electronic metal–support interactions in single-atom catalysts. Angewandte Chemie International Edition. 53(13): 3418–3421. doi:10.1002/anie.201309248.

[114] Van Deelen, T.W., Hernández Mejía, C. and de Jong, K.P. 2019. Control of metal-support interactions in heterogeneous catalysts to enhance activity and selectivity. Nature Catalysis. 2(11): 955–970. doi:10.1038/s41929-019-0364-x.

[115] Ro, I., Resasco, J. and Christopher, P. 2018. Approaches for understanding and controlling interfacial effects in oxide-supported metal catalysts. ACS Catalysis. 8(8): 7368–7387. doi:10.1021/acscatal.8b02071.

[116] Shi, J.L., Wu, J.H., Zhao, X.J., Xue, X.L., Gao, Y.F., Guo, Z.X., et al. 2016. Substrate co-doping modulates electronic metal–support interactions and significantly enhances single-atom catalysis. Nanoscale. 8(46): 19256–19262. doi:10.1039/c6nr04292a.

[117] Chen, Y., Li, C.W. and Kanan, M.W. 2012. Aqueous CO_2 reduction at very low overpotential on oxide-derived Au nanoparticles. Journal of the American Chemical Society. 134(49): 19969–19972. doi:10.1021/ja309317u.

[118] Kim, C., Eom, T., Jee, M.S., Jung, H., Kim, H., Min, B.K., et al. 2017. Insight into electrochemical CO_2 reduction on surface-molecule-mediated Ag nanoparticles. ACS Catalysis. 7(1): 779–785. doi:10.1021/acscatal.6b01862.

[119] Reske, R., Mistry, H., Behafarid, F., Roldan Cuenya, B. and Strasser, P. 2014. Particle size effects in the catalytic electroreduction of CO_2 on Cu nanoparticles. Journal of the American Chemical Society. 136(19): 6978–6986. doi:10.1021/ja500328k.

[120] Fan, L., Xia, Z., Xu, M., Lu, Y. and Li, Z. 2018. 1D SnO_2 with wire-in-tube architectures for highly selective electrochemical reduction of CO_2 to C1 products. Advanced Functional Materials. 28(17): 1706289. doi:10.1002/adfm.201706289.

[121] Chung, M.W., Cha, I.Y., Ha, M.G., Na, Y., Hwang, J., Ham, H.C., et al. 2018. Enhanced CO_2 reduction activity of polyethylene glycol-modified Au nanoparticles prepared via liquid medium sputtering. Applied Catalysis B: Environmental. 237: 673–680. doi:10.1016/j.apcatb.2018.06.022.

[122] Xie, Y. and Dimitrov, N. 2020. Nanoporous Au-Cu-Pt As a Superior Electro-Catalyst for the Methanol Oxidation Reaction. *In*: Electrochemical Society Meeting Abstracts Prime. 49: 3741–3741. The Electrochemical Society, Inc. doi:10.1149/ma2020-02493741mtgabs.

[123] Zhang, S., Kang, P. and Meyer, T.J. 2014. Nanostructured tin catalysts for selective electrochemical reduction of carbon dioxide to formate. Journal of the American Chemical Society. 136(5): 1734–1737. doi:10.1021/ja4113885.

[124] Ahmadi, M., Mistry, H. and Roldan Cuenya, B. 2016. Tailoring the catalytic properties of metal nanoparticles via support interactions. The Journal of Physical Chemistry Letters. 7(17): 3519–3533. doi:10.1021/acs.jpclett.6b01198.

[125] Li, Z., Wu, R., Zhao, L., Li, P., Wei, X., Wang, J., et al. 2021. Metal-support interactions in designing noble metal-based catalysts for electrochemical CO_2 reduction: Recent advances and future perspectives. Nano Research. 1–15. doi:10.1007/s12274-021-3363-6.

[126] Pan, C.J., Tsai, M.C., Su, W.N., Rick, J., Akalework, N.G., Agegnehu, A.K., et al. 2017. Tuning/exploiting strong metal-support interaction (SMSI) in heterogeneous catalysis. Journal of the Taiwan Institute of Chemical Engineers. 74: 154–186. doi:10.1016/j.jtice.2017.02.012.

[127] Lou, Y., Xu, J., Zhang, Y., Pan, C., Dong, Y. and Zhu, Y. 2020. Metal-support interaction for heterogeneous catalysis: from nanoparticles to single atoms. Materials Today Nano. 12: 100093. doi:10.1016/j.mtnano.2020.100093.

[128] Tributsch, H., Koslowski, U.I. and Dorbandt, I. 2008. Experimental and theoretical modeling of Fe¯, Co¯, Cu¯, Mn-based electrocatalysts for oxygen reduction. Electrochimica Acta. 53(5): 2198–2209. doi:10.1016/j.electacta.2007.09.027.

[129] Deng, D., Novoselov, K.S., Fu, Q., Zheng, N., Tian, Z. and Bao, X. 2016. Catalysis with two-dimensional materials and their heterostructures. Nature Nanotechnology. 11(3): 218–230. doi:10.1038/nnano.2015.340.

[130] Liu, M., Pang, Y., Zhang, B., De Luna, P., Voznyy, O., Xu, J., et al. 2016. Enhanced electrocatalytic CO_2 reduction via field-induced reagent concentration. Nature. 537(7620): 382–386. doi:10.1038/nature19060.

[131] Pan, F., Zhao, H., Deng, W., Feng, X. and Li, Y. 2018. A novel N, Fe-Decorated carbon nanotube/carbon nanosheet architecture for efficient CO_2 reduction. Electrochimica Acta. 273: 154–161. doi:10.1016/j.electacta.2018.04.047.

[132] Zhang, X., Wu, Z., Zhang, X., Li, L., Li, Y., Xu, H., et al. 2017. Highly selective and active CO_2 reduction electrocatalysts based on cobalt phthalocyanine/carbon nanotube hybrid structures. Nature Communications. 8(1): 14675. doi:10.1038/ncomms14675.

[133] Shi, J.L., Wu, J.H., Zhao, X.J., Xue, X.L., Gao, Y.F., Guo, Z.X., et al. 2016. Substrate co-doping modulates electronic metal–support interactions and significantly enhances single-atom catalysis. Nanoscale. 8(46): 19256–19262. https://doi.org/10.1039/c6nr04292a

[134] Fang, S., Zhu, X., Liu, X., Gu, J., Liu, W., Wang, D., et al. 2020. Uncovering near-free platinum single-atom dynamics during electrochemical hydrogen evolution reaction. Nature Communications. 11(1): 1029. https://doi.org/10.1038/s41467-020-14848-2

[135] Hu, P., Huang, Z., Amghouz, Z., Makkee, M., Xu, F., Kapteijn, F., et al. 2014. Electronic metal–support interactions in single-atom catalysts. Angewandte Chemie International Edition. 53(13): 3418–3421. https://doi.org/10.1002/anie.201309248

[136] Wang, Y., Pan, Y., Zhu, L., Yu, H., Duan, B., Wang, R., et al. 2019. Solvent-free assembly of Co/Fe-containing MOFs derived N-doped mesoporous carbon nanosheets for ORR and HER. Carbon. 146: 671–679. https://doi.org/10.1016/j.carbon.2019.02.002

[137] Van Deelen, T.W., Hernández Mejía, C. and de Jong, K.P. 2019. Control of metal-support interactions in heterogeneous catalysts to enhance activity and selectivity. Nature Catalysis. 2(11): 955–970. https://doi.org/10.1038/s41929-019-0364-x

[138] Jia, Y., Zhang, L., Du, A., Gao, G., Chen, J., Yan, X., et al. 2016. Defect graphene as a trifunctional catalyst for electrochemical reactions. Advanced Materials. 28(43): 9532–9538. https://doi.org/10.1002/adma.201602912

[139] Ro, I., Resasco, J. and Christopher, P. 2018. Approaches for understanding and controlling interfacial effects in oxide-supported metal catalysts. ACS Catalysis. 8(8): 7368–7387. https://doi.org/10.1021/acscatal.8b02071

[140] Zhu, Y., Murali, S., Cai, W., Li, X., Suk, J.W., Potts, J.R., et al. 2010. Graphene and graphene oxide: synthesis, properties, and applications. Advanced Materials. 22(35): 3906–3924. https://doi.org/10.1002/adma.201001068

[141] Ahmadi, M., Mistry, H. and Roldan Cuenya, B. 2016. Tailoring the catalytic properties of metal nanoparticles via support interactions. The Journal of Physical Chemistry Letters. 7(17): 3519–3533. https://doi.org/10.1021/acs.jpclett.6b01198

[142] Tang, T., Jiang, W.J., Liu, X.Z., Deng, J., Niu, S., Wang, B., et al. 2020. Metastable rock salt oxide-mediated synthesis of high-density dual-protected M@ NC for long-life rechargeable zinc–air batteries with record power density. Journal of the American Chemical Society. 142(15): 7116–7127. https://doi.org/10.1021/jacs.0c01349

[143] Chen, Y. and Kanan, M.W. 2012. Tin oxide dependence of the CO_2 reduction efficiency on tin electrodes and enhanced activity for tin/tin oxide thin-film catalysts. Journal of the American Chemical Society. 134(4): 1986–1989. doi:10.1021/ja2108799.

[144] Zhu, W., Michalsky, R., Metin, O., Lv, H., Guo, S., Wright, C.J., et al. 2013. Monodisperse Au nanoparticles for selective electrocatalytic reduction of CO_2 to CO. Journal of the American Chemical Society. 135(45): 16833–16836. doi:10.1021/ja409445p.

[145] Mistry, H., Reske, R., Zeng, Z., Zhao, Z.J., Greeley, J., Strasser, P., et al. 2014. Exceptional size-dependent activity enhancement in the electroreduction of CO_2 over Au nanoparticles. Journal of the American Chemical Society. 136(47): 16473–16476. doi:10.1021/ja508879j.

[146] Asadi, M., Kumar, B., Behranginia, A., Rosen, B.A., Baskin, A., Repnin, N., et al. 2014. Robust carbon dioxide reduction on molybdenum disulphide edges. Nature Communications. 5(1): 4470. doi:10.1038/ncomms5470.

[147] Kim, C., Jeon, H.S., Eom, T., Jee, M.S., Kim, H., Friend, C.M., et al. 2015. Achieving selective and efficient electrocatalytic activity for CO_2 reduction using immobilized silver nanoparticles. Journal of the American Chemical Society. 137(43): 13844–13850. doi:10.1021/jacs.5b06568.

[148] Raciti, D., Livi, K.J. and Wang, C. 2015. Highly dense Cu nanowires for low-overpotential CO_2 reduction. Nano Letters. 15(10): 6829–6835. doi:10.1021/acs.nanolett.5b03298.

[149] Wu, J., Yadav, R.M., Liu, M., Sharma, P.P., Tiwary, C.S., Ma, L., et al. 2015. Achieving highly efficient, selective, and stable CO_2 reduction on nitrogen-doped carbon nanotubes. ACS Nano. 9(5): 5364–5371. doi:10.1021/acsnano.5b01079.

[150] Min, X. and Kanan, M.W. 2015. Pd-catalyzed electrohydrogenation of carbon dioxide to formate: high mass activity at low overpotential and identification of the deactivation pathway. Journal of the American Chemical Society. 137(14): 4701–4708. doi:10.1021/ja511890h.

[151] Liu, M., Pang, Y., Zhang, B., De Luna, P., Voznyy, O., Xu, J., et al. 2016. Enhanced electrocatalytic CO_2 reduction via field-induced reagent concentration. Nature. 537(7620): 382–386. doi:10.1038/nature19060.

[152] Torelli, D.A., Francis, S.A., Crompton, J.C., Javier, A., Thompson, J.R., Brunschwig, B.S., et al. 2016. Nickel–gallium-catalyzed electrochemical reduction of CO_2 to highly reduced products at low overpotentials. ACS Catalysis. 6(3): 2100–2104. doi:10.1021/acscatal.5b02888.

[153] Dai, L., Qin, Q., Wang, P., Zhao, X., Hu, C., Liu, P., et al. 2017. Ultrastable atomic copper nanosheets for selective electrochemical reduction of carbon dioxide. Science Advances. 3(9): e1701069. doi:10.1126/sciadv.1701069.

[154] Zhang, Q., Huang, L., Kang, S., Yin, C., Ma, Z., Cui, L., et al. 2017. CuO/Cu_2O nanowire arrays grafted by reduced graphene oxide: synthesis, characterization, and application in photocatalytic reduction of CO_2. doi:10.1039/c7ra07310k.

[155] Han, N., Wang, Y., Yang, H., Deng, J., Wu, J., Li, Y., et al. 2018. Ultrathin bismuth nanosheets from in situ topotactic transformation for selective electrocatalytic CO_2 reduction to formate. Nature Communications. 9(1): 1320. doi:10.1038/s41467-018-03712-z.

[156] Geng, Z., Kong, X., Chen, W., Su, H., Liu, Y., Cai, F., et al. 2018. Oxygen vacancies in ZnO nanosheets enhance CO_2 electrochemical reduction to CO. Angewandte Chemie. 130(21): 6162–6167. doi:10.1002/ange.201711255.

[157] Yang, H.B., Hung, S.F., Liu, S., Yuan, K., Miao, S., Zhang, L., et al. 2018. Atomically dispersed Ni (i) as the active site for electrochemical CO_2 reduction. Nature Energy. 3(2): 140–147. doi:10.1038/s41560-017-0078-8.

[158] Li, J., Chen, G., Zhu, Y., Liang, Z., Pei, A., Wu, C.L., et al. 2018. Efficient electrocatalytic CO_2 reduction on a three-phase interface. Nature Catalysis. 1(8): 592–600. doi:10.1038/s41929-018-0108-3.

[159] Fu, J., Bao, H., Liu, Y., Mi, Y., Qiu, Y., Zhuo, L., et al. 2020. Oxygen doping induced by nitrogen vacancies in Nb_4N_5 enables highly selective CO_2 reduction. Small. 16(2): 1905825. doi:10.1002/smll.201905825.

[160] Yang, X., Deng, P., Liu, D., Zhao, S., Li, D., Wu, H., et al. 2020. Partial sulfuration-induced defect and interface tailoring on bismuth oxide for promoting electrocatalytic CO_2 reduction. Journal of Materials Chemistry A. 8(5): 2472–2480. doi:10.1039/c9ta11363k.

[161] Yuan, X., Luo, Y., Zhang, B., Dong, C., Lei, J., Yi, F., et al. 2020. Decoration of In nanoparticles on In_2S_3 nanosheets enables efficient electrochemical reduction of CO_2. Chemical Communications. 56(30): 4212–4215. doi:10.1039/c9cc10078d.

[162] Sreekanth, N., Nazrulla, M.A., Vineesh, T.V., Sailaja, K. and Phani, K.L. 2015. Metal-free boron-doped graphene for selective electroreduction of carbon dioxide to

formic acid/formate. Chemical Communications. 51(89): 16061–16064. doi:10.1039/c5cc06051f.

[163] Lei, F., Liu, W., Sun, Y., Xu, J., Liu, K., Liang, L., et al. 2016. Metallic tin quantum sheets confined in graphene toward high-efficiency carbon dioxide electroreduction. Nature Communications. 7(1): 12697. doi:10.1038/ncomms12697.

[164] Wang, H., Chen, Y., Hou, X., Ma, C. and Tan, T. 2016. Nitrogen-doped graphenes as efficient electrocatalysts for the selective reduction of carbon dioxide to formate in aqueous solution. Green Chemistry. 18(11): 3250–3256. doi:10.1039/c6gc00410e.

[165] Li, F., Xue, M., Li, J., Ma, X., Chen, L., Zhang, X., et al. 2017. Unlocking the electrocatalytic activity of antimony for CO_2 reduction by two-dimensional engineering of the bulk material. Angewandte Chemie. 129(46): 14910–14914. doi:10.1002/anie.201710038.

[166] Song, Y., Peng, R., Hensley, D.K., Bonnesen, P.V., Liang, L., Wu, Z., et al. 2016. High-selectivity electrochemical conversion of CO_2 to ethanol using a copper nanoparticle/N-doped graphene electrode. Chemistry Select. 1(19): 6055–6061. doi:10.1002/slct.201601169.

[167] Rogers, C., Perkins, W.S., Veber, G., Williams, T.E., Cloke, R.R. and Fischer, F.R. 2017. Synergistic enhancement of electrocatalytic CO_2 reduction with gold nanoparticles embedded in functional graphene nanoribbon composite electrodes. Journal of the American Chemical Society. 139(11): 4052–4061. doi:10.1021/jacs.6b12217.

[168] Zhang, C., Yang, S., Wu, J., Liu, M., Yazdi, S., Ren, M., et al. 2018. Electrochemical CO_2 reduction with atomic iron-dispersed on nitrogen-doped graphene. Advanced Energy Materials. 8(19): 1703487. doi:10.1002/aenm.201703487.

[169] Yuan, J., Yang, M.P., Hu, Q.L., Li, S.M., Wang, H. and Lu, J.X. 2018. Cu/TiO_2 nanoparticles modified nitrogen-doped graphene as a highly efficient catalyst for the selective electroreduction of CO_2 to different alcohols. Journal of CO_2 Utilization. 24: 334–340. doi:10.1016/j.jcou.2018.01.021.

[170] Lv, K., Fan, Y., Zhu, Y., Yuan, Y., Wang, J. and Zhang, Q. 2018. Elastic Ag-anchored N-doped graphene/carbon foam for the selective electrochemical reduction of carbon dioxide to ethanol. Journal of Materials Chemistry A. 6(12): 5025–5031. doi:10.1039/c7ta10802h.

[171] Geioushy, R.A., Khaled, M.M., Alhooshani, K., Hakeem, A.S. and Rinaldi, A. 2017. $Graphene/ZnO/Cu_2O$ electrocatalyst for selective conversion of CO_2 into n-propanol. Electrochimica Acta. 245: 456–462. doi:10.1016/j.electacta.2017.05.185.

[172] Li, Z., Wu, R., Zhao, L., Li, P., Wei, X., Wang, J., et al. 2021. Metal-support interactions in designing noble metal-based catalysts for electrochemical CO_2 reduction: Recent advances and future perspectives. Nano Research. 1–15. https://doi.org/10.1007/s12274-021-3363-6

[173] Liu, Z., Qian, J., Zhang, G., Zhang, B. and He, Y. 2024. Electrochemical CO_2-to-CO conversion: A comprehensive review of recent developments and emerging trends. Separation and Purification Technology. 330: 125177. doi:10.1016/j.seppur.2023.125177.

[174] Proietto, F., Schiavo, B., Galia, A. and Scialdone, O. 2018. Electrochemical conversion of CO_2 to HCOOH at tin cathode in a pressurized undivided filter-press cell. Electrochimica Acta. 277: 30–40. doi:10.1016/j.electacta.2018.04.159.

[175] Liu, D.J. 2022. Electrochemical conversion of CO_2 to long-chain hydrocarbons. Joule. 6(9): 1978–1980. doi:10.1016/j.joule.2022.08.012.

[176] Gonçalves, M.R., Gomes, A., Condeço, J., Fernandes, R., Pardal, T., Sequeira, C.A.C., et al. 2010. Selective electrochemical conversion of CO_2 to C_2 hydrocarbons. Energy Conversion and Management. 51(1): 30–32. doi:10.1016/j.enconman.2009.08.002.

[177] He, C., Zhang, Y., Zhang, Y., Zhao, L., Yuan, L.P., Zhang, J., et al. 2020. Molecular evidence for metallic cobalt boosting CO_2 electroreduction on pyridinic nitrogen. Angewandte Chemie. 132(12): 4944–4949. https://doi.org/10.1002/anie.201916520

[178] Lou, Y., Xu, J., Zhang, Y., Pan, C., Dong, Y. and Zhu, Y. 2020. Metal-support interaction for heterogeneous catalysis: from nanoparticles to single atoms. Materials Today Nano. 12: 100093. https://doi.org/10.1016/j.mtnano.2020.100093

[179] Savéant, J.M. 2012. Electrochemical approach to proton-coupled electron transfers: recent advances. Energy and Environmental Science. 5(7): 7718–7731. doi:10.1002/chin.201310270.

Chapter 8

Photocatalytic Conversion of CO_2 Using Nanomaterials

Garima Sharma, Gaurav Bhargava and Rupesh Kumar*
Department of Chemistry, I.K. Gujral Punjab Technical University, Kapurthala, Punjab, India-144603

8.1 INTRODUCTION AND SIGNIFICANCE

Climate change states the long-term changes in temperature and weather patterns over time. Historically, variations in the solar cycle have been the primary drivers of such fluctuations. However, since the 1800s, human activities, particularly the combustion of fossil fuels and coal, have emerged as significant catalysts for these changes.

In our atmosphere, increase in carbon dioxide levels disrupts the balance of energy flow, leading to a notable rise in global temperatures. This shift triggers a cascade of consequences, including the rapid melting of polar ice caps and the conversion of fertile land into barren wastelands, potentially exacerbating sea level rise and compromising habitable spaces for humanity in the near future. The depletion of natural resources and the subsequent release of harmful gases predominantly stem from human activities on a large scale. Despite our continued reliance on fossil fuels, their combustion releases greenhouse gases, posing a significant threat to current living conditions. Notably, carbon dioxide emitted from the burning of hydrocarbon fuels like coal, natural gas, and oil stands out as the primary culprit among anthropogenic greenhouse gases, highlighting the urgent need for mitigation measures.

*For Correspondence: rupesh.manak@gmail.com

DOI: 10.1201/9781003492184-8

At present, greenhouse gas concentrations have reached unparalleled heights, surpassing levels not witnessed in the past 2 million years, and the trend of emissions shows no signs of abating. This surge in greenhouse gases has led to an approximate increase of 1.1 degrees Celsius in the Earth's temperature since the late 1800, [1] making the most recent decade (2011–2020) notably the warmest on record. The global mean temperature for the period 2011–2020 was recorded at 1.10 ± 0.12°C above the average for the period from 1850 to 1900.

It is crucial to acknowledge that climate change encompasses more than merely rising temperatures. It initiates a domino effect, influencing numerous interconnected facets of the ecosystem. Consequently, changes in one component inevitably cascade through others, shaping the intricate storyline of climate change [2].

Carbon dioxide (CO_2) exhibits potential as a carbon feedstock to generate value added products through different catalytic protocols. The capacity of carbon dioxide (CO_2) to function as a chemical precursor has garnered significant research interest across various disciplines, particularly in the realms of photocatalysis [3, 4], biocatalysis [5, 6], electrochemistry [7, 8], thermochemistry [9, 10], and photothermal [11, 12] conversion of CO_2. The technology of carbon dioxide capture, utilization, and storage (CCUS) has garnered significant interest in recent years [13]. However, owing to the inherent thermodynamic stability and kinetic inertness of carbon dioxide, its chemical conversion demands substantial energy input. Therefore, the development of efficient catalysts is imperative to activate carbon dioxide and mitigate the reaction energy barrier [14].

The field of nanomaterials has garnered significant attention from researchers owing to the unique properties exhibited by materials at the nanoscale, which are absent in their micro-sized counterparts. The distinctiveness of nanoscale properties primarily stems from surface effects and quantum size effects [15]. These effects result in the breakdown of many thermodynamic laws observed at the macroscopic scale. The thermodynamic phenomenon, such as phase transitions, typically manifest as collective behavior . However, at the nanoscale, the high surface-to-volume ratio significantly influences the thermodynamic behavior of nanomaterials [16]. The atomic interactions occurring at the surfaces of nanomaterials diverge significantly from those observed at macroscopic scales. Consequently, this disparity in atomic behavior leads to notable distinctions in the mechanical properties between nanoscale and macroscopic materials [17, 18]. Moreover, nanoscale effects exert a substantial influence not only on the physical properties but also on the chemical properties of materials. Particularly noteworthy are their impacts on catalytic and biochemical reactions. One contributing factor to the variance in catalytic activity between nanoscale and macroscopic materials is the occurrence of polarization changes in the bond with the adsorbate [15].

Photocatalysis provides numerous advantages and versatile applications, including cost-effectiveness and ease of use in energy production, environmental remediation, and CO_2 reduction. This process initiates through light absorption by a solid material, typically semiconductor photocatalysts, which are commonly perceived as facilitating reactions at their solid surfaces. An ideal photocatalyst should exhibit non-toxicity, stability, high photoactivity, and affordability. The redox potential of the H_2O/OH^- pair, which ideally falls inside the semiconductor band gap area, is another essential component in the disintegration of organic molecules. A wide range of semiconductors with suitable band gap energies for photocatalysis are available, such as $SrTiO_3$, TiO_2, ZnO, WO_3, Fe_2O_3, and ZnS.

In photocatalysis, activation of the photocatalyst occurs under solar irradiation. This process involves the utilization of semiconductor materials either alone or in conjunction with metallic, organic, or organometallic co-catalysts to enhance light absorption. Within the photocatalytic cycle, two pivotal reactions, oxidation and reduction, take place, both of which are greatly advantageous for CO_2 conversion. The semiconductor has to have the right band gap and valence and conduction band edge potentials for every desired reaction [19]. In comparison to the Natural Hydrogen Electrode (NHE), a good photocatalyst should have a conduction band with a more negative potential and a valence band with a higher positive potential [20]. However, a lot of photocatalysts can only absorb UV light, which reduces their overall efficacy during the photocatalytic process. Inefficient charge transfer, recombination, and unwanted side reactions inside the semiconductor matrix are the causes of material instability [21]. Additionally, photocatalysis plays a pivotal role in solar energy harvesting [22].

The photocatalytic conversion of carbon dioxide (CO_2) stands as a promising approach for addressing global environmental challenges and ensuring sustainable energy resources for the future. Despite considerable research efforts thus far, significant hurdles remain in achieving efficient, selective, and stable CO_2 conversion. This chapter describes current developments in CO_2 photoreduction, highlighting some important challenges including optimizing light absorption, enabling charge separation, and activating CO_2 molecules. Furthermore, it elucidates methodologies demonstrating potential for enhancing photocatalytic efficacy, a pivotal aspect in accurately assessing performance and elucidating fundamental mechanisms underlying CO_2 photoreduction. A comprehensive examination is undertaken encompassing conventional photocatalysts, including metal oxides, alongside other emerging materials such as single atom nanomaterials and Two-Dimensional (2D) materials, is given in this chapter. To further our comprehension of the reaction mechanism, it also explores the source of carbon in CO_2 photoreduction. The chapter examines the photocatalytic viability of CO_2 reduction and ends with recommendations for future research paths.

8.1.1 Fundamentals of Photocatalytic CO_2 Conversion

The photocatalytic conversion of CO_2 into value-added compounds is a viable approach to tackle energy and environmental concerns because it uses water as a reactive component and abundant solar energy as its primary energy source. Additionally, it produces no waste or hazardous products. Despite advancements made in the recent past, photoconversion of CO_2 is still not up to the mark and it poses significant challenges. A few intrinsic obstacles [23] include

(a) The inert nature of the CO_2 molecule: Activation of CO_2 molecule requires –1.9 V potential which is formidable challenge that most semiconductors struggle to achieve [24].

(b) Multi-electron Kinetics: CO_2 reduction entails a complex process involving multiple proton and electron transfers, which ultimately necessitate reduction potentials similar to those required for proton reduction [25]. Consequently, reducing CO_2 is far more difficult than producing hydrogen using the easier two-electron proton reduction process.

(c) Unfavorable energy dynamics involved in water mediated CO_2 reduction [26].

(d) Challenging precise control over product selectivity towards desired compounds [27].

In addition to the aforementioned critical factors, the crux of CO_2 photoreduction centers on the choice of semiconductor photocatalysts. Li et al. [28] provided a comprehensive overview of the selectivity observed in the photocatalytic reduction of CO_2 with various cocatalysts. To summarize, Pd, Pt, or Au catalysts demonstrate a preference for CH_4 production, while Ag exhibits selectivity towards CO, CH_4, or CH_3OH. Cu cocatalysts favor the formation of hydrocarbons, and Cu_2O, RuO_2, or NiO_x are associated with enhanced CH_3OH production.

Photocatalysis via a variety of semiconductor photocatalysts uses the power of solar irradiation to transform CO_2 and water into useful solar fuels such as CH_4 (methane), CO (carbon monoxide), CH_3OH (methanol), HCOOH (formic acid), and HCHO (formaldehyde). Three main steps [29] usually make up this process: (i) production of electron-hole pairs in response to light; (ii) separation of these pairs and their migration to the photocatalyst's surface; and (iii) surface reactions in which electrons reduce CO_2 and holes oxidize H_2O.

The particular reduction product that is produced depends on the complex interactions of multistage CO_2 reduction, which are dictated by the movement of electrons across the system and the interaction of CO_2 and surface of the photocatalyst [30]. Both the thermodynamic and the kinetic aspects of these courses are necessary for the overall effectiveness of photocatalytic CO_2 reduction [31].

Among the different photocatalysts employed for CO_2 reduction, TiO_2 [32], ZnO [33], $SrTiO_3$ [34], and ZnS [35] stand out as the most commonly utilized materials. However, traditional photocatalysts suffer from shortcomings that hinder their efficiency in meeting modern requirements. These include:

(a) Wide band gaps that cause limited sensitivity to light within a narrow spectrum. (b) Short electron lifetimes which cause quick recombination of photogenerated charge carriers. (c) Insufficient electron transport and poor CO_2 adsorption at the reaction sites on photocatalysts.

Numerous studies have suggested improving photocatalyst efficiency in order to overcome these limitations. These techniques include alloying with different metals and non-metals, changing the structure to create unique forms, modifying the surface with noble metals, and creating heterojunctions and nanocomposites using different semiconductors [36]. These approaches enhance the effective segregation and propagation of photoinduced charges while simultaneously broadening the spectral range of sensitivity to light.

8.1.2 Nanomaterials for Photocatalytic CO_2 Conversion

Nanostructured materials exhibit enhanced physical and chemical properties as compared to bulk materials. These nanostructures, which include porous nanospheres, nanorods, nanowires, nanotubes, nanoparticles, and nanowires, have shown surprising properties that can greatly improve material selection and photocatalytic applications. For example, photocatalysts characterized by nanoscale dimensions offer a greater specific surface area, thereby furnishing an increased number of surface-active sites for electrochemical processes. Furthermore, some characteristics depend on the size and shape, as in the case of metallic nanostructures, plasmonic absorption and the quantum confinement effects seen in semiconductors with nanostructures. Photocatalysis also plays a pivotal role in harnessing solar energy [37].

8.1.3 Single Atom Nanoparticles for CO_2 Utilization

Single-Atom Catalysts (SACs) have garnered significant interest due to their remarkable atomic accessibility and active electronic states, often manifesting in exceptional catalytic efficacy and selectivity. Several combinations have demonstrated notable catalytic process in CO_2 reduction: Ni SA/ZrO_2, La SAs on CN, and Au SA/CN. Nonetheless, a primary drawback lies in the inherent instability of Single Atoms (SAs), leading to SAC aggregation into nanoparticles during synthesis and catalytic operations, consequently diminishing catalytic [38–40].

In order to Co-load Cu SAs and Au-Cu alloy NPs onto TiO_2 for the photocatalytic production of solar fuels from CO_2 and H_2O, a unique photodeposition technique has been devised. This technique produced an improved photocatalyst that performs remarkably well, producing 3578.9 μmol g^{-1} h^{-1} for CH_4 and 369.8 μmol g^{-1} h^{-1} for C_2H_4. This greatly increases the feasibility of synthesizing value-added solar fuels using sunlight.

It is believed that Cu SAs and Au-Cu alloy NPs cooperate to lower the total activation energy barrier for the ethylene and methane synthesis process, which includes the rate-determining phase, while increasing the activation of CO_2 and H_2O by adsorption. As a result, these constituents yield exceptionally stable and effective solar fuels, specifically methane and ethylene. The concept of metal alloy cocatalysts and synergistic SAs exhibits potential for application to several different systems, ultimately leading to the creation of more effective cocatalysts [41].

8.1.4 Metal-Oxide-Based Photocatalysts for CO_2 Utilization

Additionally, metal oxide composites can improve the stability of photocatalysts and make it possible to create sophisticated photocatalysts with unique topologies like microporous, hollow shells or hierarchical structures. The use of metal oxide nanocomposites in the photocatalytic reduction of CO_2 emphasizes the semiconductor nature of these materials as well as the structural and property aspects that influence their photocatalytic response [42, 43]. In addition to having good photocatalytic activity, transition metal oxides and their composites also have positive properties including stability and non-toxicity that make them valuable in a variety of applications. This paper covers a few properties of metal oxides and provides a quick overview of their numerous uses. It concentrates on the most prevalent metal oxides, namely TiO_2, ZnO, WO_3, CuO, and Cu_2O, which are known to be economical, stable, effective, and most importantly, ecologically benign when used in a sustainable manner for environmental restoration [44, 45].

Global warming is mostly caused by the uncontrolled combustion of fossil fuels to meet energy needs, which raises CO_2 emissions on a daily basis. Photocatalysis is the most cost-effective and ecologically friendly method for sustainable development. TiO_2 Z-scheme heterojunction composites and the usage of transition metals, which have suitable electronic and optical properties, low toxicity, high photoactivity, and low cost, are among the semiconductors that are being extensively investigated as photocatalysts for CO_2 reduction. Reducing CO_2 is facilitated by semiconductors. TiO_2 is one of the semiconductors that is most often studied for use as a photocatalyst for CO_2 reduction because to its great photoactivity, low cost, low toxicity, and attractive electrical and optical properties [46].

8.1.5 Mesoporous Oxide Nanomaterials for Photocatalytic CO_2 Conversion

Arranged mesoporous metal oxides as the active core or catalytic carrier. In the field of catalysis, the huge specific surface area and organized mesoporous structure have raised serious concerns. The ordered mesoporous materials in the catalytic reaction exhibited excellent mechanical properties, and there was substantial interaction between the active centers. These ordered mesoporous materials were clearly superior to bulk or standard nanomaterials in terms of their catalytic conversion of CO, NO_x, and other harmful gases, as well as volatile organic compounds [47, 48].

Iron oxide nanoparticles were encapsulated in a mesoporous silica shell to create core-shell structures using the wet impregnation and sol-gel techniques. The catalytic activity of these nanoparticles was investigated in a continuous packed bed laboratory flow reactor operating at 8.1 h^{-1} WHSV (Weight Hour Space Velocity) and between 1000 and 1173 K in temperature. Iron oxide nanoparticles immobilized in a mesoporous silica shell (FSCS) exhibited the highest activity of all the catalysts produced; at 1123 K, they converted approximately 82% and demonstrated over 100 hours of activity. Furthermore, 8.1 h^{-1} WHSV was used to assess the long term thermal stability up to 100 hours of time-on-stream at 1173. This catalyst's estimated activation energy was determined to be 127.39 kJ/mol. With less than 4% metal loss and agglomeration, the FSCS catalyst's noteworthy performance data showed that it was appropriate for the high temperature endothermic reaction in the continuous industrial scale hydrogen generating system [49–51].

8.1.6 Nanostructured Inorganic-Organic Hybrid Systems for CO_2 Photoreduction

Photocatalysis is an economical, rapid, and ecologically safe method for degrading organic contaminants and accelerating reactions such as CO_2 photoreduction. However, factors such as synthesis methodology, particle dimensions, surface morphology, and crystalline structure exert substantial influence over the photocatalytic behavior of nanocomposites. Among materials exhibiting desirable photoactivity characteristics at a low cost, C_3N_4 emerges as a notable candidate. In addition, this material exhibits chemical and thermal stability. Zinc oxide is another readily available, affordable, and non-toxic photocatalyst. However, certain restrictions, such as a large band gap, caused zinc oxide and carbon nitride materials, photoactivity to decrease in visible light. When inorganic zinc oxide and organic material like C_3N_4 are combined, a heterojunction material with intriguing properties is created that can degrade a variety of organic contaminants, including insecticides and organic dyes. Compared to ZnO and C_3N_4 nanoparticles, the ZnO–C_3N_4 nanocomposite exhibits enhanced photoactivity. Additionally, the photoactivity of the ZnO–C_3N_4 nanocomposite is greatly enhanced by doping it with other metals as Ag, Mg, Ni, Cu, and Al [52]·

The use of halide perovskite $CsPbBr_3$ in solar cells, optoelectronics, and artificial photosynthesis has recently attracted a lot of attention. However, additional work is required to create synthesis methods that are more environmentally friendly and scalable as well as to use the material in humid conditions. Here, we describe a quick and simple mechanochemical method for creating $CsPbBr_3$ perovskite nanocrystals, allowing for precise control over the crystals' size and structure (Figure 8.1).

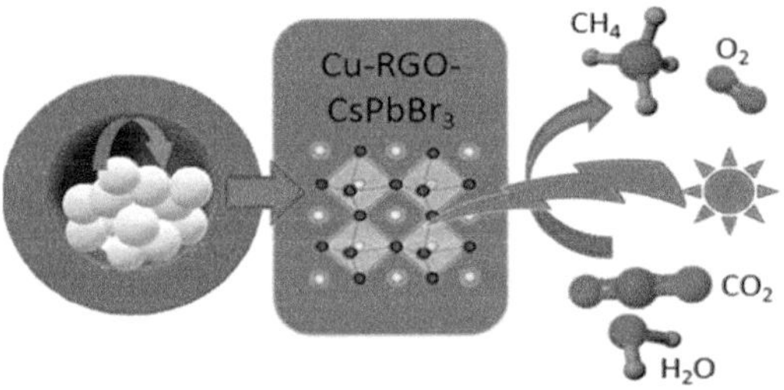

Figure 8.1 $CsPbBr_3$ nanocrystals in mechanochemical synthesis and selective photocatalytic CO_2 reduction to methane. Reproduced from Ref [53].

Through the manipulation of mechanochemical reaction parameters, such as ball size, ball milling duration, and Cs precursor, these perovskite nanocrystals can be shaped into a variety of shapes, including nanorods, nanospheres, and nanosheets. We investigate their potential as a proton source in gas-phase

Table 8.1 Some current reports involving the different nanomaterials for the conversion of CO_2 into value added products

S. No.	Nanoparticle/ Photocatalyst	Reaction Conditions	Remarks	Yield	Ref.
1.	Pt SA/ZrO_2	Visible light irradiation	Pt SA/ZrO_2 promoting photocatalytic reduction of CO_2 into CO	16.61 μmol/g/h	[38]
2.	Pd_{SA}/TiO_2	Visible light irradiation	Pd_{SA}/TiO_2 promoting CO_2 photoreduction to CH_4	271.6 $\mu mol \cdot g^{-1} \cdot h^{-1}$	[39]
3.	Pd/CN-SA, PdO_x/CN, Pd/CN-NP	Visible light irradiation	Pd/CN-SA, PdO_x/CN, Pd/CN-NP promoting CO_2 photoreduction to CH_4	2.25 μmol g^{-1}/1.08 μmol g^{-1}/ 0.44 μmol g^{-1}	[40]
4.	Cu SAs and Au–Cu alloy NPs on TiO_2	Visible light irradiation/8h	Cu SAs and Au–Cu alloy NPs on TiO_2 promoting CO_2 conversion to CH_4 and C_2H_4	3578.9 μmol g^{-1} h^{-1}/369.8 μmol g^{-1} h^{-1}	[41]
5.	Co_3O_4@CoO	300 W Xe lamp	Co_3O_4@CoO promoting CO_2 conversion to acetic acid and ethanol	73.44 μmol g^{-1} h^{-1} acetic acid/14.81 μmol g^{-1} h^{-1}/73.81 μmol g^{-1} h^{-1} ethanol	[48]
6.	ZrO_2–Mn_3O_4	UV light	ZrO_2–Mn_3O_4 exhibited photocatalytic reduction of CO_2 into CH_4	1.46 μmol g^{-1} h^{-1}	[50]
7.	Halide perovskite $CsPbBr_3$	Visible light absorption	Halide perovskite $CsPbBr_3$ promoting CO_2 conversion to CH_4, CO and H_2	0.43 (±0.03) μmol CH_4 g^{-1} h^{-1}, 2.25 (±0.09) μmol CO g^{-1} h^{-1}, 0.08 (±0.02) μmol H_2 g^{-1} h^{-1}	[53]
8.	Halide perovskite $CuPbBr_3$	Visible light absorption	Halide perovskite $CuPbBr_3$ promoting CO_2 conversion to CH_4, CO and H_2	12.7 (±0.95) μmol CH_4 g^{-1} h^{-1}, 0.46 (±0.11) μmol CO g^{-1} h^{-1}, and 0.27 (±0.02) μmol H_2 g^{-1} h^{-1}	[53]

photocatalytic CO_2 reduction utilizing water vapor, observed across three consecutive cycles under simulated sunlight, with $CsPbBr_3$ nanosheets retaining 30% of this activity. The photocatalytic activity of these $CsPbBr_3$ nanosheets is significantly increased upon mechanochemical preparation with Cu-loaded reduced graphene oxide (Cu-RGO), yielding 12.7 (±0.95) μmol CH_4 g^{-1} h^{-1}, 0.46 (±0.11) μmol CO g^{-1} h^{-1}, and 0.27 (±0.02) μmol H_2 g^{-1} h^{-1}, while maintaining 90% of this activity over three consecutive cycles. Significantly, at 523 nm, an apparent quantum efficiency of 1.10(±0.15)% is attained, and on an electron basis, CH_4 selectivity rises to 98.5(±0.93)%. Better charge separation, visible light absorption, CO_2 adsorption and activation, and the hydrophobic character of the resultant composites are all thought to be responsible for the increased activity, selectivity, and stability (Table 8.1) [53–55].

8.2 MECHANISTIC PATHWAYS

Photocatalytic CO_2 reduction has garnered significant attention for its unique potential advantages and applications in the realms of energy and the environment. Notably, the primary reduction products reported in photocatalytic CO_2 reduction, regarded as value added commodities, predominantly fall within the category of mono-carbon (C_1) products, such as CH_4, CO, CH_3OH, HCOOH, among others [36, 56–61]. Additionally, bicarbonate (C_2) products encompassing C_2H_4, C_2H_6, CH_3CHO, CH_3CH_2OH, and CH_3COOH, as well as multi-carbon (C^{2+}) products like C_3H_8, $(CH_3O)_2CO$, etc., have been identified [62–64].

Wang et al. [23] presented a comprehensive depiction of pathways (Figure 8.2) based on previous reports [65, 66]. The paths that these reactions follow are determined by whether they proceed via coordinated electron-proton transfer, electron-transfer, or proton-transfer (sometimes hydroxyl-transfer). If an intermediate molecule in one process quickly desorbs from the catalyst surface before undergoing additional transformations, it may function as the final product in another pathway. This overview described the several pathways—formaldehyde, carbene, formyl, and glyoxal—that could be involved in the photocatalytic production of C_1 and a number of C_2 compounds from CO_2.

8.2.1 Fuels Generated from CO_2 Conversion

Over the previous 60 years, the global CO_2 concentration in the atmosphere has increased by around 30% due to the massive use of fossil fuels during the industrialization process. At this point, the concentration has swiftly reached more than 400 parts per million. The ensuing ecological issues, such the ongoing rise in sea level and the average world temperature, have placed several limitations on both human existence and economic growth. Therefore, even if reducing or eliminating the greenhouse effect—which is primarily brought on by excessive CO_2 emissions—is a tough and long term task, it is also critical. Recognizing the potential benefits of reducing energy use and environmental degradation at the

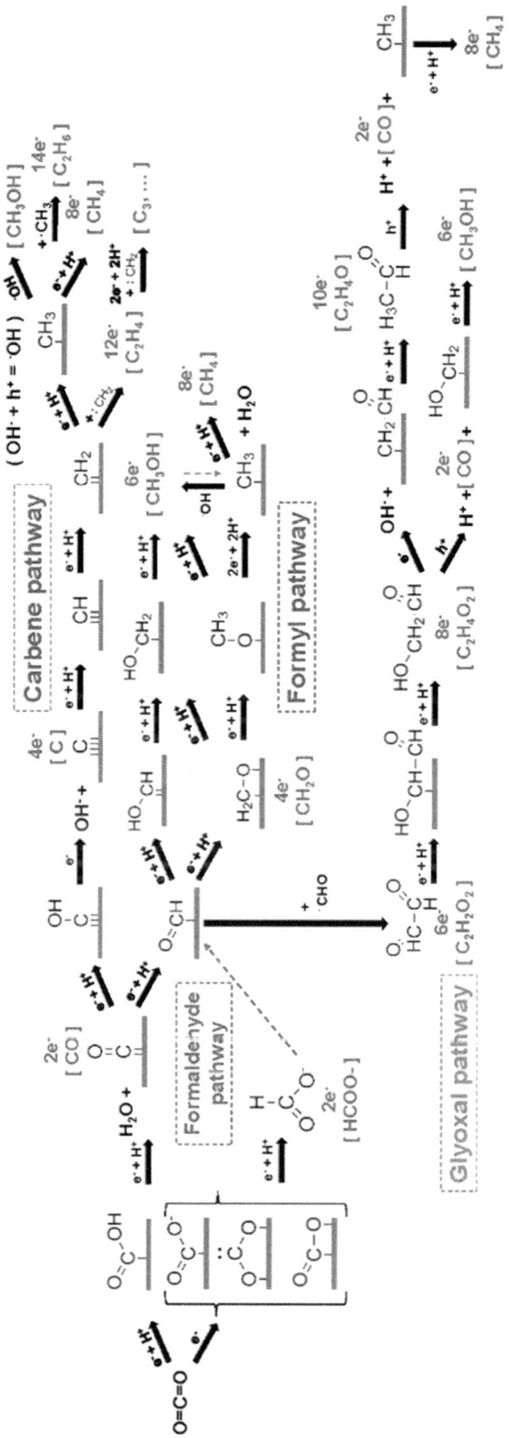

Figure 8.2 Mechanistic insights of the pathways of CO_2 reduction to usually observed C_1 and C_2 products [23].

same time, turning CO_2 into a variety of fuels such as formic acid (HCOOH), carbon monoxide (CO) methanol (CH_3OH), ethanol (CH_3CH_2OH), methane (CH_4), formaldehyde (HCHO), hydrogen (H_2), olefins and gasoline among other fuels that are now made from fossil resources (Figure 8.2) [67].

8.2.1.1 Formic Acid

One significant basic chemical that is being considered as a potential hydrogen storage option is Formic Acid (FA), which has the potential to be upgraded to greater added value CO_2. The initial byproduct of CO_2 hydrogenation is formic acid (FA), which can be used as a vector for storing hydrogen or as a precursor to products with higher added values [68].

Zhang et al. [69] created a titanium oxide nanocatalyst that uses H-TiO_2 as a hydride donor in order to achieve UV-light-induced selective photocatalytic reduction. According to the suggested approach, TiO_2 can function as a hydride reagent in photocatalysis by producing hydride-like sites by hydrogenation. The process of hydride transfer to the adsorbed CO_2 is the recommended way to produce formic acid. It is interesting to note that hydrogenated TiO_2 has the ability to function as a hydride reagent in reduction reactions other than CO_2.

Tian et al. successfully used photoreduction to convert CO_2 to formic acid by combining an inorganic photocatalyst-enzyme system. The catalytic efficacy of this approach is significantly limited by inadequate electron transfer and enzyme deactivation caused by reactive oxygen species during the photo-excitation phase. Thus, the conversion process toward the synthesis of formic acid has been hastened by the enzyme's combined protective action and improved electron transfer for NADH regeneration, which are made possible by the suggested construction of an inorganic photocatalyst enzyme system (Figure 8.3) [70].

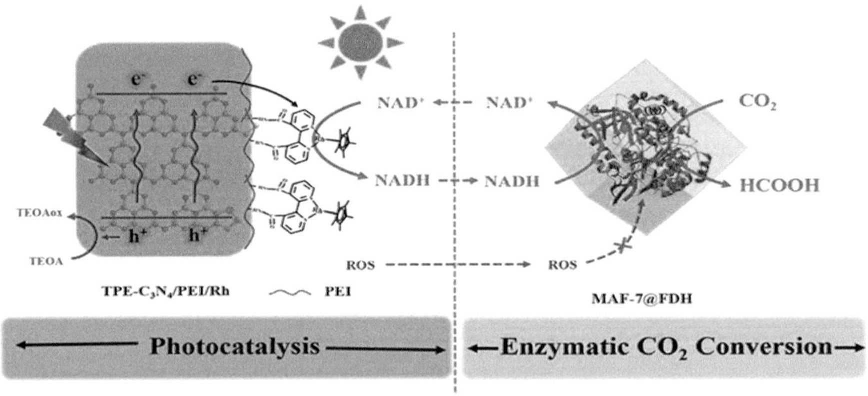

Figure 8.3 Photocatalytic TEOA oxidation (Utilizing TEOA oxidation, TPE-C_3N_4 uses light to excite electrons, which is then used to quench holes. In order to convert NAD^+ to NADH, the photoexcited electrons are subsequently quickly transported to the Rh complex. Both electrons and hydrides are carried by NADH and later transfer into FDH@ MAF-7, where FDH uses NADH as a reducing equivalent to convert CO_2 to formic acid). Reproduced from Ref [70]. Copyright © 2020 American Chemical Society.

For the photocatalytic conversion of CO_2 to formic acid, Kamada et al. created a strong, multifunctional PNNP type iridium complex photocatalyst. By reducing CO_2 to HCO_2H through outer sphere catalysis and CO_2 to CO through inner sphere catalysis, the produced photocatalyst functions as both a catalyst and a photosensitizer. Nevertheless, the robustness and reactivity of the catalyst are significantly increased by the addition of a PNNP-type ligand. This research provides fresh perspectives on creating an effective catalyst for the CO_2 reduction process [71].

Formic acid was synthesized by Chen et al. from biomass and the photocatalytic conversion of CO_2. The present step of this established method's CO_2 to formic acid conversion is still difficult and may require a costly catalytic device or energy intensive conditions. Therefore, solar-driven conversion and inexpensive catalysts are preferred. However, the process of turning biomass into formic acid is more advanced and has real-world uses. Finally, because HPAs can achieve high yields, high selectivities, and simple product separations from sugars or raw biomass resources at very low temperatures and pressures, catalytic oxidation of biomass by HPAs is preferred to hydrolysis and wet oxidation.Therefore, the production of formic acid from CO_2 and the conversion of biomass will support the low carbon, sustainable growth of contemporary enterprises [72].

8.2.1.2 Carbon-Monoxide

Since CO is a crucial industrial raw resource, research on the conversion of CO_2 to CO is very significant [73]. Each npAu leaf layer is 100 nm thick, and as their thickness increases—from 65 to 75 and 80% for layers 1, 5, and 10—so does their CO Faraday efficiency. This translates to an extraordinarily high CO current density of 955 A/g gold for the one npAu leaf layer configuration, indicating possible industrial use [74].

A pure inorganic photocatalyst known as quasi-keggin polyoxometalate (Co_4Mo_{24}) was successfully synthesized by Liu et al. It plays a great contribution to the CO production rate of 1848.3 $\mu mol^{-1} h^{-1}$ and outstanding activity selectivity of 97.0% for the photocatalytic reduction of CO_2 to CO. This approach confirms that Co_4Mo_{24} may remain stable during the photocatalytic reaction process using thermogravimetric and x-ray diffraction catalysis [75]. Light-driven photocatalyst was created by Le et al. to facilitate the photoreduction of CO_2 into sustainable and value-added fuels. This strategy might jointly address the energy crisis and the environmental problem. Using this technique, CO_2 is photocatalytically converted to produce C_1 and C_2 compounds. CO, HCOOH, HCHO, CH_3OH, and CH_4 are a few examples of potential C_1 products. C_2 products include C_2H_4, C_2H_6, C_2H_5OH, C_3H_6, and C_3H_8. In comparison to C_1 product, C_2 product has better profit and energy density [76].

Koci et al. created a Pt/C_3N_4 based photocatalyst for the photoreduction of CO_2 under solar light irradiations into solar fuels or valuable compounds such as carbon monoxide, methanol, formaldehyde, and formic acid. This modified Pt functioned as an effective co-photocatalyst, influencing the photocatalyst's photoactivity as well as the product formation's selectivity. C_3N_4 is another crucial element in the photoactivity of CO_2 reduction. These elements support the photocatalytic reaction's growing efficiency [77].

In order to study the photocatalytic conversion of CO_2 into carbon monoxide, methane, and hydrogen, Morawski et al. created a UV active TiO_2/rGO photocatalyst. The two stages used to manufacture this TiO_2/rGO photocatalyst were the solvothermal technique and the calcination at 500, 800, and 1000°C in an argon environment. In the photocatalytic reduction of CO_2, the photocatalyst's activity was assessed. The TiO_2/rGO–10 that was not calcined exhibited the greatest activity in terms of CO_2 conversion. After rGO modification, it was discovered that every sample showed good activity toward H_2 production with high selectivity. The presence of graphene was primarily responsible for the improved photocatalytic performance because of its superior electron transport and collection capabilities [78].

8.2.1.3 Methanol

Several methods, including enzymatic catalysis, photocatalysis, electrocatalysis, and heterogeneous and homogeneous catalytic hydrogenation are reported, for reducing CO_2 to CH_3OH. These days, heterogeneous hydrogenation catalysis and electrocatalysis are the most promising techniques for large-scale CO_2 reduction to CH_3OH. It seems that affordable and readily available sustainable electricity are necessary preconditions for effective CH_3OH synthesis [79].

A photocatalytic conversion of CO_2 into methanol and value added products was created by Shinde and others. Environmental changes and the global energy problem are the result of growing levels of hydrocarbon fuel consumption coupled with excessive carbon dioxide emissions. Photocatalysts include carbon, TiO_2, mixed metal oxide, MOFs, and plasmonic based materials. The process of photocatalytic reduction of CO_2 yields a number of compounds, including CH_3OH, OH, EtOH, and others, but one of the most significant and adaptable molecules is methanol, which is used extensively in both business and daily life [80].

A membrane-integrated green method was developed by Chakrabortty et al. for the photocatalytic conversion of CO_2 into MeOH. The modern world has a critical level of greenhouse gas abundance because of the industrialization, automobilization, and ever-increasing need for electricity. Use of atmospheric CO_2, a plentiful and renewable carbon feedstock, guarantees low cost involvement and thus reduces greenhouse gas emissions and global warming. Hence, there is a substantial anticipation for the development of a membrane-integrated photocatalytic reaction system capable of converting atmospheric CO_2 into pure methanol will have a pathway for both pollutant abatement and environmentally benign, inexpensive, and energy-intensive wealth development [81].

A $ZnFe_2O_4/TiO_2$ photocatalyst was created by Manzoor et al. to facilitate the photocatalytic reduction of CO_2 into methanol when exposed to UV light. This $ZnFe_2O_4/TiO_2$ photocatalyst was combined with titanium oxide (TiO_2) using the sol-gel technique of synthesis. The primary source of global warming is CO_2 emissions from the burning of fossil fuels. In order to reduce global warming, this photocatalyst converts CO_2 under UV light into beneficial chemicals as methanol, formic acid, formaldehyde, etc. This approach avoids converting CO_2 into several products while producing the highest yield of methanol as a single product by coupling TiO_2 with zinc ferrite ($ZnFe_2O_4$) under visible light [82].

A unique Fe_3O_4@N–C/Cu_2O nanostructure photocatalyst was created by Movahed et al. to improve photocatalysis-induced CO_2 to methanol conversion. This complex catalyst was produced by synthesizing copper oxide (I) nanoparticles and bonding them to nitrogen-doped carbon (N–C) rods in the form of a core shell nanostructure. The Fe_3O_4@N–C rod-shaped core shell nanostructure was first subjected to ascorbic acid reduction of Cu^{2+} ions, and then aerobic copper nanoparticle oxidation was carried out. In selective CO_2 reduction photocatalytic tests, methanol was the most common product compared to Fe_3O_4@N–C/Cu_2O. Interestingly, compared to Fe_3O_4@Cu_2O, Fe_3O_4@N–C/Cu_2O showed four times more photocatalytic activity for CO_2 reduction. This highlights Fe_3O_4@N–C/Cu_2O's exceptional performance as a catalyst for converting CO_2 to methanol, indicating its potential for use in sustainable energy applications. This difference can be explained by the higher capacity to efficiently distinguish electron-hole pairs produced by photolysis as well as the increased absorption of visible light.[83]

8.2.1.4 Ethanol

By reducing $LaCo_{0.5}Ga_{0.5}O_3$, a novel catalyst of $Co/La_4Ga_2O_9$ for CO_2 hydrogenation to create ethanol was developed. Under moderate reaction conditions, which are 270°C, 3.5 MPa, and 3000 mL g^{-1} h^{-1}, this catalyst demonstrated high selectivity to ethanol (≈35C-mol%). In addition to being a widely used chemical raw material, the solvent in many different industries like disinfection reagent, medicinal, chemical color, etc., ethanol is also a great fuel additive for gasoline [4, 5]. As a result, it is expected that CO_2 will be converted to ethanol; however, producing ethanol from CO_2 continues to be difficult, mostly because of CO_2's low conversion and poor selectivity to C_2H_5OH [84].

A Trimodal Porous Silica (TPS)/g-C_3N_4 nanotube photocatalyst was created by Wang and colleagues to aid in the photocatalytic conversion of CO_2 into ethanol. This two step hydrothermal process produced the photocatalyst. To drastically cut the price of producing photocatalysts, the TPS was isolated from rice husk, an agricultural waste. There were more catalytic reaction sites in the TPS/g-C_3N_4 due to its wide surface area, hollow tubular structure, and capacity to absorb CO_2. Ethanol was only detected by the photocatalytic activity measurement instrument when composite TGNT and g-C_3N_4 NT were utilized as photocatalysts. It was demonstrated that when exposed to visible light, TGNT is a stable, effective, and reasonably priced photocatalyst for turning CO_2 into ethanol [85].

For the photocatalytic conversion of CO_2 into ethanol, Lertthanaphol et al. synthesized metal-doped titanium dioxide-graphene oxide (TiO_2/GO) composites by a one step hydrothermal method. Using an autoclave hydrothermal reactor, they produced a variety of metal-doped TiO_2/GO composites, such as palladium-TiO_2/GO (Pd-TiO_2/GO), copper-TiO_2/GO (Cu-TiO_2/GO), and silver-TiO_2/GO (Ag-TiO_2/GO). To construct the synthesis, ethanol and water were mixed with titanium butoxide, graphene oxide, and the metal precursor. The photoreduction of CO_2 to ethanol was then used to evaluate the photocatalytic activity of these composites. Ag-TiO_2/GO, Pd-TiO_2/GO, and Cu-TiO_2/GO were found to generate ethanol at rates of 109, 125, and 233 μmol/gcat, respectively, after measurements were made. These results demonstrate Cu-TiO_2/GO's intriguing potential as an effective photocatalyst for the conversion of CO_2 to ethanol. The amalgamated influence of crucial attributes such crystalline dimensions, optical band gap, and BET surface area accounts for the exceptional performance of Ca-TiO_2/GO (Figure 8.4) [86].

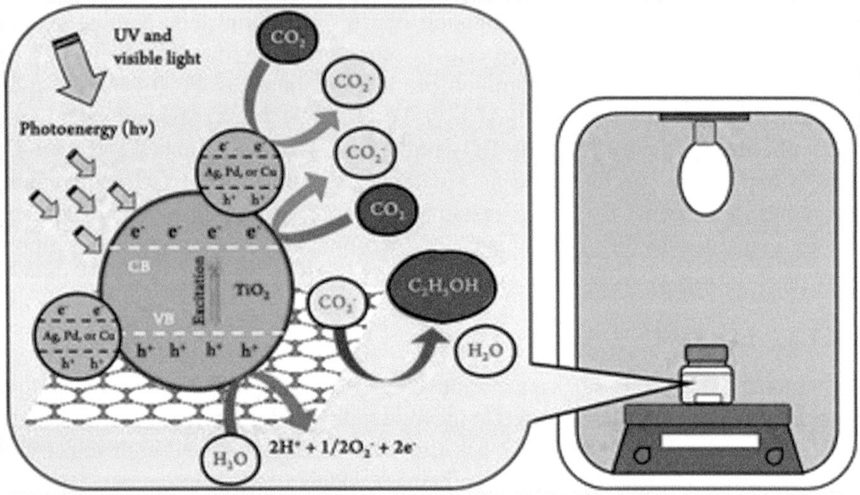

Figure 8.4 Proposed reaction mechanism of photoreduction of CO_2. Reproduced from Ref [86]. Copyright © 2021 American Chemical Society.

Using bicarbonate-assisted nanoparticles, Zhao et al. described a hierarchical Bi@Bi_2MoO_6 composite for photocatalytic reduction of CO_2 into ethanol. One inexpensive and environmentally benign solution to the energy problem and the greenhouse effect is photocatalytic CO_2 removal. A straightforward solvothermal method was used to create the high performance Bi@Bi_2MoO_6 photocatalyst, which is a Bi nanoparticle formed on the surface of Bi_2MoO_6 nanosheets containing oxygen vacancies. The Bi@Bi_2MoO_6 sample has good ethanol selectivity (92%), along with exceptional photocatalytic activity for CO_2 reduction. The synergistic interaction between thermodynamically accelerated bicarbonate routes and localized high electron density in active locations is responsible for this performance. Additionally, by encouraging C-C coupling between carbene intermediates and other carbon containing species, the addition of $NaHCO_3$ increases the overall catalytic efficiency, which increases the generation of C_2H_5OH and the formation of long chain products [87].

In order to construct a series of Pd_xIn_y@N_3–COF composites that may be employed for the photocatalytic reduction of CO_2 into alcohol (methanol/ethanol) with H_2O oxidation, Huang et al. created PdIn bimetallic nanoclusters, which then combined to photosensitizing N_3–COF. With H_2O oxidation serving as the H^+ and electron source, the resultant Pd_xIn_y@N_3–COF exhibits outstanding activity for CO_2 photoreduction, displaying a record high total yield of 798 μmol g^{-1} for producing methanol (74%) and ethanol (26%) among the framework materials. The non-copper catalyst has good activity for converting CO_2 into C_1/C_2 alcohols because the bimetallic nanocluster and cluster @N_3–COF cooperate to greatly increase C–C coupling and interfacial charge transfer. This work demonstrates the remarkable potential of bimetallic nanoclusters as photosensitizing supports for the production of alcohols through H_2O oxidation in artificial photosynthesis [88].

8.2.1.5 Methane

Given the challenges associated with storing and transporting hydrogen obtained via water electrolysis using renewable energy sources (commonly referred to as power-to-gas processes), the conversion of CO_2 to hydrogen-rich fuels such as methane (CH_4), characterized by a high hydrogen-to-carbon ratio, emerges as a significant avenue for hydrogen storage. Methane can be moved easily and safely utilizing the natural gas pipes that are currently in place, unlike hydrogen. In this work, Ni/SiO_2 catalysts supported by iron (Fe), cobalt (Co), and zinc (Zn) were fabricated using a simple wet impregnation technique. These catalysts were then assessed for their effectiveness in the process of converting CO_2 to methane [89].

Kulandaivalu et al. devised a photocatalytic reformation of CO_2 into value added fuels and methane. There is a significant chance that two significant greenhouse gases can be reformatted into fuels with additional value. Utilizing photon energy from endless solar radiation, photocatalysis has the enormous potential to carry out this reaction under benign conditions in order to meet the reaction's high temperature requirement, which gives rise to a number of significant challenges. Nevertheless, photoenergy can be used to transform extremely stable gases like CO_2 and CH_4 [90]·

Ge et al. used limited Cu clusters in TiO_2 nanotubes with a high Cu^+ ratio to photo catalytically convert CO_2 and create light olefin. Using trimethylaluminum (TMA) and (Bis/h-exafluoroacetylacetonato) copper (II) (Cu $(hfac)_2$) as precursors, template assisted Atomic Layer Deposition (ALD) is used here to create TiO_2 nanotubes with restricted Cu cocatalyst. The Cu species are widely distributed in multi atom clusters, according to numerous analyses. At 1500°C, this Cu/TiO_2 catalyst exhibits 60.4% selectivity to light olefins, demonstrating exceptional catalytic efficacy in the photocatalytic reduction of CO_2 with H_2O. The C–C coupling process active sites are supplied by the Cu^+ species. A high degree of olefin selectivity can be achieved by carefully regulating the reaction conditions, such as low light intensity, medium temperature, and the ratio of CO_2 to active hydrogen [91]·

A TiO_2-based photocatalyst was created by Jitan et al. to facilitate the photocatalytic conversion of CO_2 into valuable fuels. A process known as photocatalysis converts CO_2 into a variety of high value fuels, such as methanol, methane, formaldehyde, and formic acid, by using a semiconductor material and sustainable solar energy [92].

Renones et al. synthesized Ag/In_2O_3/TiO_2 to increase CO_2 photocatalytic conversion heterojunctions, improving methane generation. This well-proven technique uses water as the electron donor and UV irradiation to increase the photocatalytic activity for gas phase CO_2 reduction. The role of In_2O_3 in a heterojunction with TiO_2 is explored. The type of In_2O_3 used determines the behavior that it manifests. Consequently, commercial In_2O_3 high crystalline size is enhancing its activity in comparison to TiO_2; however, methane selectivity is only slightly improved. Even though the overall conversion is less than that of TiO_2, there is still a change in selectivity towards CH_4 when In_2O_3 produced at all laboratory conditions with low crystalline size is used. These heterojunctions with silver nanoparticles result in ternary systems that exhibit a significant shift in selectivity towards methane, indicating that silver serves as an electron collector to facilitate charge transfer to the reaction medium. Consequently, the Ag/In_2O_3 system enhanced selectivity and CH_4 production [93].

8.2.1.6 Formaldehyde

Formaldehyde is mostly used in the manufacturing of resins for the wood panel industry, but because of its high reactivity and versatility as a chemical building block, it is also used in the synthesis of a wide range of polymers, adhesives, and chemical intermediates. As a result, formaldehyde is needed in a large number of industries, including building, aviation, automotive, medicines, and cosmetics. The industrial manufacturing of formaldehyde uses two distinct process systems. The mixed metal oxide (iron, molybdenum, and/or vanadium) mediated process advances under excess air conditions and at lower temperatures, typically in the range of 250–400°C. In contrast, the silver catalyzed procedure occurs utilizing electrolytically generated silver particles under air lean conditions and higher temperatures ranging from 600 to 700°C [94]. Kumar et al. created a heterolepticrutherium metal complex-TiO_2 hybrid in an aqueous media to facilitate the reduction of CO_2 into formaldehyde with the aid of visible light. A hybrid photocatalyst has been fabricated that is active in visible light by using covalent interaction to apply the heterolepticrutherium metal complex over the TiO_2 nanoparticles. Formaldehyde was found to be the main reaction product; traces of CO and CH_4 were also found when triethanolamine (TEOA) was present as a sacrificial donor. After five hours of exposure to visible light, the maximum turnover number (720) for HCHO was determined using the metal complex that was applied to the surface. Additionally, under visible light, HCHO (*in-situ*) was used to react with the main amine (aniline, 4-aminobenzoic acid) to produce the equivalent imine. The work advances the green method of using CO_2 under visible light irradiation to produce useful compounds that will be expanded for several future uses [95]. Ranjbakhsh et al. reported on P-doped g-C_3N_4 based efficient photocatalysts for the photocatalytic conversion of CO_2 into value added chemicals. Phosphorus doping changed the photocatalytic yield of g-C_3N_4 for CO_2 reduction. This is the first study to use Density Function Theory (DFT) simulations to explore potential chemical pathways for CO_2 reduction on the P-doped g-C_3N_4 (PCN) surface. The experimental findings demonstrated that p-doping lengthens the carrier lifetime, which enhances CH_4 generation by raising the electrons driving force. Methane, methanol, formaldehyde, and formic acid are the most likely products, according to mechanistic studies [96]. Using graphite carbon nitride, Samanta et al. engineered a combined photocatalytic/thermocatalytic/electrocatalytic conversion of CO_2 into useful fuels. An appropriate energy source, such as thermal, photochemical, or electrochemical, in conjunction with an efficient heterogeneous catalyst was used to overcome the thermodynamic and kinetic barriers associated in CO_2 activation and conversion. A low cost, ecologically benign catalyst that is effective in converting CO_2 between carbon-based molecules is graphite carbon nitride (g-C_3N_4). For the use of visible light in the solar spectrum, g-C_3N_4 is the only carbon catalyst with a bandgap energy appropriate for valence and conduction band edge potential. It can effectively take part in the CO_2 photoreduction reaction by reducing CO_2 and oxidizing water at the same time [97].

He et al. and colleagues developed a defect engineering method for photocatalyzing the conversion of CO_2 into useful fuels. The photoreduction of CO_2 into useful fuel is a sustainable and environmentally benign method to mitigate the effects of the energy crisis and other environmental problems. The slow activation/reaction of CO_2 on the photocatalyst surface, narrow-band light adsorption, insufficient separation and transport of charge carriers, and these elements combined restrict the effectiveness of CO_2 photoreduction. In order to boost photocatalyst efficiency in the photocatalytic conversion of CO_2 into useable fuels, defect engineering is gaining popularity. This technique describes how structural flaws can be utilized to alter the photocatalyst's electrical structure and activate the inert CO_2 molecules [98].

8.2.1.7 Hydrogen

The process of producing radical species through light induced redox reactions on surrounding molecules is known as photocatalysis. These species are then used in various pollutant degradation processes, water splitting to produce H_2 and O_2, converting CO_2 into hydrocarbon fuels, disinfecting microorganisms, and other processes. For large scale real-time applications, a process like photocatalysis is excellent since it can take on several forms [99]. Hayat et al. developed a Polymeric Carbon Nitride (PCN) for the photocatalytic conversion of CO_2 into H_2. By using nanostructuring and copolymerization techniques, they were able to integrate the co-monomer 2,4-dibromophenyl (DBP) into PCN's triazine skeleton. This integration accelerates the evolution of hydrogen from water splitting and greatly improves CO_2 photoreduction. The addition of the electron-withdrawing co-monomer DBP accelerates electron charge transfer processes, changes the optical absorption characteristics, and affects electron-hole recombination in the CNU-DBP framework, despite the photocatalyst's exceptional performance as a heterogeneous catalyst. After photooxidation under solar light for about 4 hours, CNU-DBP shows maximal photocatalytic activity with respect to CO_2 reduction, producing 17.3 μmole of hydrogen and 65.7 μmole of CO in the reaction system. In the end, this work opens the door for the development of DBP-based PCN as a viable photocatalyst for practical uses, providing efficient light control for the evolution of hydrogen and CO_2 reduction using a synthetic methodology that makes sense [100].

A graphite-carbon nitride (g-C_3N_4) photocatalyst was created by Yuan et al. to facilitate the photocatalytic evolution of CO_2 and H_2. This photocatalyst (g-C_3N_4) exhibits quick electron hole pair recombination, tiny specific surface area, insolubility, and limited light absorption. This newly found method creates hydroxyl-grafted oxygen linked tri-s-triazine-based polymer (HGONTP) by polycondensing hydrothermally produced dicyandiamide (DCDA). By replacing the pendent NH_2 groups in HGONTP with OH groups, the degree of cyclization and hydrolysis of DCDA can regulate the number of C–O–C linkers and terminal OH groups. The hydrophilic surface, large specific surface area of 96.1 m^2g^{-1}, small band gap of 2.18 eV, and reduced charge recombination are all characteristics of the HGONTP photocatalyst. These properties enable remarkable absorption of light from UV to near-IR. Consequently, HGONTP shows a 27.7-fold greater rate of hydrogen evolution (654 versus 0.236 mmol g^{-1} h^{-1}) than pristine g-C_3N_4 [101].

Using water splitting, Shankar et al. created heterostructures and a 2D material for the photocatalytic conversion of CO_2 into useful chemicals and fuels. The long-term viability of plant and animal life on our planet has been severely strained by the rapidly rising global energy consumption that is fueled by conventional polluting fuels. The world needs switch to clean, renewable, carbon-neutral, and environmentally friendly energy sources more quickly. Simultaneously, there is an emerging need to manage CO_2 emissions by utilizing sunlight to convert them into valuable fuels or chemicals. Owing to the highly distinctive properties of the surface states, high surface area, and in depth surface chemical engineering, 2D materials hold great potential for enabling sustainable, commercially feasible clean energy harvesting solutions, as well as for effective CO_2 conversion and hydrogen production [102].

Ni (II) complex with N_4S_2 coordination was used by Lee et al. to create a visible light photocatalytic carbon dioxide conversion process that produced formate and hydrogen. Using this approach, CO_2 is photocatalytically converted to $HCOO^-$ by new mononuclear Ni (II) complexes. In an aqueous/organic solvent mixture, two Ni complexes of 2-(2-pyridyl)-benzimidazole and 2-(2-pyridyl)-benzothiazole demonstrated selective $HCOO^-$production under CO_2, but they were active towards H_2 photoreduction under argon. Under high pH conditions, the synthesis of $HCOO^-$ occurred with remarkable selectivity over proton reduction. Additonally, this study demonstrates that the Ni (II) catalyst significantly outperforms. H_2 evolution in CO_2 photoconversion, offering a controlled path to the generation of $HCOO^-$ with high energy efficiency utilizing components that are readily available on earth in a process that transforms an environmental contaminant into more useful organic compound [103].

8.2.1.8 Olefins and Gasoline

Light olefins (C_2–C_4), such as butylene, propylene, and ethylene, are necessary building blocks for the synthesis of polymer materials, which are typically made by dehydrogenating light alkanes and steam cracking naphtha. Moreover, light olefins can be generated from CO_2 hydrogenation by an altered process known as Fischer-Tropsch Synthesis (FTS). Another method for producing long chain hydrocarbons in the gasoline range (C_5–C_{11}) is direct CO_2 hydrogenation. With a methane content of only 4% and a selectivity of up to 78% towards hydrocarbons, the Na–Fe_3O_4/HZSM-5 catalyst can directly convert CO_2 into gasoline under industrially relevant conditions, all the while maintaining a 22% CO_2 conversion rate [104].

A multipurpose photocatalyst for the photocatalytic conversion of CO_2 into olefins and gasoline isoparaffins was described by Dokania et al. One way to significantly contribute to the reduction of global emissions is through the development of carbon neutral fuels from CO_2. In view of this, it is ideal to produce fuels high in isoparaffins. Here, we demonstrate the ability to synthesize a primarily isoparaffinic hydrocarbon mixture from CO_2 (up to ~85% isoparaffins selectivity among hydrocarbons) at a CO_2 conversion of >15% by combining a zinc-modified zeolite beta with a methanol producer (InCo). A wide range of methods were used to fully characterize the catalyst combination [105].

A novel method for converting CO_2 into valuable chemicals and fuels was developed by Saravanan et al. [106] Garba et al. reported on the photocatalytic conversion of CO_2 into hydrocarbon fuels. Fossil fuel consumption has a detrimental impact on the environment and adds to global warming. The atmospheric build-up of CO_2 is one of the main contributors to global warming. Creating hydrocarbon (HC) fuels from excess CO_2 is a key mitigation tactic. Using RWGS or dry methane reforming (DRM) to convert CO_2 to higher hydrocarbons and then hydrogenating the resultant CO to produce olefins and paraffin is a more practical and economical approach (FT process). While Fe-based catalysts have shown promising results in this field, they also have limitations, including strong selectivity for CH_4 molecules and oxygenates such as light olefins, CO, and formic acid, and low selectivity for C_5^+ compounds [107].

For the hydrogenation of carbon dioxide to produce gasoline-range hydrocarbons and olefins, ethylenediaminetetraacetic acid disodium salt (EDTA-2Na) was added to Fe-Co/Al_2O_3 catalysts, according to Dai et al. In this work, a range of highly dispersed iron-cobalt (Fe-Co) catalysts were produced by modifying EDTA-2Na, indicating an efficient way to convert CO_2 into valuable compounds (C_2-C_4 and C_5-C_{11}). The resistance of Fe species to sintering was significantly enhanced by the addition of trace amounts of Co and EDTA-2Na, according to X-Ray Diffraction (XRD) examination. EDTA-2Na breaks down into reducible tiny organic molecules during the calcination process, which helps to reduce iron species. Cobalt activates and adsorbs intermediate CO on the NaxFe-Co catalyst in CO Temperature-Programmed Desorption (CO-TPD) research, increasing the catalytic conversion of CO_2. Through EDTA complexation, the chemical adsorption of CO_2 was improved and the synergistic effect of Fe-Co bimetals was increased. Furthermore, the addition of sodium to EDTA-2Na increased the surface basicity of the catalysts, which enhanced the hydrocarbon product's distribution. The EDTA-2Na modified Fe-Co/Al_2O_3 catalyst showed notable selectivity towards high value added compounds and significant CO_2 conversion (32.8%), particularly at lower H_2/CO_2 (1/1) ratios [108].

Zhao and associates used a micro liquid film reactor-assisted coprecipitation technique to create a range of cobalt-doped Fe-based catalysts supported on ZrO_2.

It has been shown that effective interactions between Fe-containing species and the ZrO_2 support can promote the formation of faulty FeO_x components during reduction treatment and prevent metallic iron from carburizing during the entire process in the catalyst precursors. The catalyst that had the lowest CO selectivity (4.7%) and the highest Fe time yield to light olefins (9.2 μmol $CO_2.gFe^{-1}.S^{-1}$) out of all the Fe-based catalysts tested also had the best selectivity towards light olefins, at about 38 at 49% CO_2 conversion under reaction conditions (320°C, 2.0 MPa, and 4800 $mL.gcat^{-1}.h^{-1}$) (Table 8.2) [109].

8.2.2 Photocatalytic Electrochemical Reduction of CO_2

Photoelectrochemical (PEC) reduction of carbon dioxide (CO_2) is one of the most efficient methods to transform CO_2 into fuels and/or high value added chemicals, which has the potential to mitigate the climate and energy crises. Because of their vast sources, flexible energy band topologies, and environmental friendliness, zinc-based materials make excellent PEC catalysts. The many zinc-based PEC catalysts for CO_2 reduction are outlined, including zinc alloys, zinc porphyrin complex, Cu_2ZnSnS_4, Cu_2ZnGeS_4, ZnO, and ZnSe [110].

Using CO_2 from a point source or from an appropriate atmospheric or oceanic capture strategy, photoelectrochemical (PEC) reduction of CO_2 aims to use solar energy to convert CO_2 into usable molecules (fuel or chemicals such as carbon monoxide (CO), methanol, methane, ethanol, ethylene, etc.). It offers long term solar energy storage, supports ongoing initiatives to lower anthropogenic CO_2 emissions, and clears the way for the chemical industry to move away from fossil fuels. In recent years, traditional PEC—which featured direct contacts between semiconductors and liquid electrolytes—has become less prevalent. Alternatively, integrated devices, sometimes referred to as buried junctions, that directly connect an electrochemical (EC) cell with a photovoltaic (PV) cell have been investigated and show promising solar-to-fuel energy conversion efficiency (up to 20%) [111].

Ding et al. constructed a photocathode intended for effective photoelectrochemical reduction of CO_2. Photoelectrochemical (PEC) CO_2 reduction combines photocatalysis and electrolysis to produce chemical fuels from CO_2 reduction that can be produced solar-powered. There are two methods that show promise for turning solar energy into chemical energy, (i) photocatalysis and (ii) photo electrocatalysis. Because it facilitates charge separation and carries charges in a directed manner, PEC may be more efficient than the former. PEC CO_2 reduction yields a variety of compounds, some of which have relatively high activity and selectivity (often approaching 100%) for C_1 products like formate and CO, whereas multi carbon products typically have extremely low levels of both [112].

Cortes et al. fabricated a titania nanotube (TiNT) for photoelectrochemical reduction of CO_2. To reduce CO_2, a unique two-compartment photoelectrochemical cell (PEC) was created. This novel reactor consists of two phases, (i) a gas phase that contains CO fed to the cathode compartment and (ii) an aqueous phase inside the anode compartment. The anode of the assembly was constructed using perforated foil to allow aligned titania nanotubes to be exposed to the electrolyte, which is bonded to nafion, and to make a link to the Pt-deposited

Table 8.2 Some latest references reported the conversion of CO_2 into different fuels

S. No	Fuels	Photocatalyst	Reaction Conditions	Remarks	Yield	Ref.
1	Formic acid	TPE-C_3N_4/PEI/Rh	9 h illumination	Inorganic photocatalyst-enzyme system promoting CO_2 conversion to formic acid	16.75 mM	[70]
2	Formic acid	PNNP-type Ir	Visible light/electron source 1,3-dimethyl-2,3-dihydro-1H-benzo[d] imidazole (BIH)	PNNP type IR complex photocatalyst promoting CO_2 conversion to HCO_2H and CO	49%	[71]
3	Carbon-monoxide	Co_4Mo_{24}	8 h visible-light irradiation	Co_4Mo_{24} as heterogeneous catalyst promoting CO_2 conversion to CO	38.19%	[75]
4	Carbon-monoxide	Pt/C_3N_4	UV 8 W Hg lamp	Pt/C_3N_4 promoting CO_2 conversion to CH_4, CO and H_2	–	[77]
5	Carbon-monoxide	TiO_2/rGO	UV irradiation	TiO_2/rGO without calcination promotion CO_2 conversion to CH_4, CO and H_2	15.87, 2.98 and 113 $\mu mol/g_{photocatal}$	[78]
6	Methanol	Graphene based TiO_2/$CuSO_4$	UV irradiation	Graphene based, TiO_2/$CuSO_4$ promoting CO_2 conversion to CH_3OH	40 mg/g	[81]
7	Methanol	$ZnFe_2O_4$/TiO_2	Visible light irradiation	$ZnFe_2O_4$/TiO_2 promoting CO_2 conversion to CH_3OH	141.22 $\mu mol/g_{cat.}h$	[82]
8	Methanol	Fe_3O_4@N-C/Cu_2O	3 h/5 W Xenon HID	Fe_3O_4@N-C/Cu_2O promoting selective CO2 reduction to CH_3OH	146.7 $\mu molg^{-1}$	[83]
9	Ethanol	TPS/g-C_3N_4	Visible light	TPS/g-C_3N_4 promoting CO_2 photoreduction to ethanol	196 μm/g	[85]
10	Ethanol	Ag-TiO_2/GO, Pd-TiO_2/GO, Cu-TiO_2/GO	6 h/UV lamp	Ag-TiO_2/GO, Pd-TiO_2/GO, Cu-TiO_2/GO promoting CO_2 photoreduction to ethanol	109, 125, 233 μmol/ g_{cat}. h	[86]

(*Contd.*)

S. No	Fuels	Photocatalyst	Reaction Conditions	Remarks	Yield	Ref.
11	Ethanol	Bi@Bi_2MoO_6	Visible light irradiation	Bi@Bi_2MoO_6 promoting CO_2 conversion to ethanol	17.93 $\mu mol^{-1}g^{-1}\ h^{-1}$	[87]
12	Ethanol	Pd_xIn_y@N_3-COF	300 W Xe lamp	Pd_xIn_y@N_3-COF promoting CO_2 photoreduction to methanol and ethanol	798 $\mu molg^{-1}$	[88]
13	Methane	Cu/TiO_2	$150^{\circ}C$	Cu/TiO_2 promoting CO_2 photoreduction to light olefins	1.52 $\mu mol^{-1}\cdot g_{cat.}^{-1}\ .\ h^{-1}$	[91]
14	Methane	$Ag/In_2O_3/TiO_2$	UV irradiation	$Ag/In_2O_3/TiO_2$ promoting CO_2 photoreduction to methane	46%	[93]
15	Formaldehyde	TiO_2/RuL1OH and TiO_2/ RuL2NO	Visible light irradiation	TiO_2/RuL1OH and TiO_2/RuL2NO promoting CO_2 photoreduction to formaldehyde including CO and CH_4 as minor product	L1 and L2 gives 86% and 93% yield	[95]
16	Hydrogen	PCN and PCN-DBP	Visible light illumination	PCN and PCN-DBP promoting CO_2 conversion to H_2	PCN/11.9 μmol/h and PCN-DBP/804.2 μmol/h	[100]
17	Hydrogen	g-C_3N_4 and HGONTP	Visible light irradiation	g-C_3N_4 and HGONTP promoting photoreduction of CO_2 to CO and H_2	CO/3.3 $\mu mol^{-1}g^{-1}\ h^{-1}$ and H_2/12.6% at 420 nm and 4.1% at 500 nm	[101]
18	Hydrogen	Ni(pbt) $(pyS)_2$ and Ni(pbi) $(pyS)_2$	Visible light irradiation	Ni(pbt) $(pyS)_2$ and Ni(pbi) $(pyS)_2$ promoting CO_2 reduction to formate and hydrogen	–	[103]
19	Olefins and gasoline	Fe-Co/Al_2O_3		Fe-Co/Al_2O_3 promoting CO_2 conversion to olefin and gasoline	–	[108]
20	Olefins and gasoline	ZrO_2-supported cobalt-doped Fe-based	320°C	ZrO_2-supported cobalt-doped Fe-based photocatalyst promoting CO_2 conversion to light olefins	9.2 μmol	[109]

carbon cloth cathode electrode situated in the gas phase compartment. Products of CO_2 reduction was found using a GC attached to the reactor. Comparing the PEC reactor to our earlier research with photocatalytic reactors, we were able to increase both the formal quantum efficiency and yield [113].

Zhang et al. synthesized a sandwich-like organic-inorganic hybrid perovskite based photocathode, incorporating carbon encapsulation, for the purpose of photoelectrochemical reduction of CO_2. In addition to providing protection for the perovskite, the carbon encapsulation enables effective photogenerated electron conductance. The prepared photocathode functionalized with CoPc molecular catalyst displayed the state-of-the-art in PEC CO_2 reduction under AM 1.5G illumination (100 mW cm^{-2}), with an onset potential of 0.58 V versus RHE and a photocurrent density of −15.5 mA cm^{-2} at −0.11 V vs RHE. In addition, the photocathode demonstrated remarkable stability throughout a continuous 25 h response to PEC CO_2 decrease. This perovskite photocathode was integrated with an IrO_x/a-Si photoanode to create the established tandem cell, which accomplished unbiased PEC CO_2 removal at around −3 mA cm^{-2} [114].

8.2.3 Photocatalytic Heterogeneous Thermal Catalysis for the Hydrogenation of CO_2

Reactions involving hydrogenation are crucial in the chemical industry of today. In hydrogenation processes, an unsaturated link in olefins or other polyenes is usually removed by thermally catalyzing the reaction with hydrogen. Since hydrogenation reactions frequently occur at temperatures of several hundred degrees, they need a large amount of energy, much of which is provided by burning fossil fuels. For hydrogenation processes, researchers are currently examining photothermal catalysts, which use concentrated sunlight to reach the necessary reaction temperatures or incorporate sunlight into thermally driven reaction systems to lower reaction temperatures. These initiatives seek to reduce CO_2 emissions and preserve fossil fuels. Consequently, photothermal catalysts must exhibit high activity and selectivity to drive the hydrogenation process effectively while also efficiently absorbing sunlight [115].

Specifically, Jia et al. used Ni nanoparticle/N-doped CeO_2 nanocomposite to develop catalysts for the photothermal reduction of CO_2 into CO. In this work, three steps were taken to create Ni nanoparticle-loaded N-doped CeO_2 (Ni/N–CeO_2) composite catalysts. The optimal Ni/$N_{5.0}$–CeO_2 demonstrated a CO yield rate of 20.9 mmol•$gcat^{-1}$•h^{-1}, with photothermal CO_2 reduction providing almost 100% CO selectivity, under light irradiation. On the other hand, Ni/CeO_2 that was not nitrogen doped showed better CH_4 selectivity. The increased oxygen vacancies in $N_{5.0}$–CeO_2 from N-doping, which improved CO_2 molecule adsorption/activation, and the significant light absorption of Ni/$N_{5.0}$–CeO_2, which was mostly caused by loading Ni nanoparticles, were credited with greater catalytic performances. Nitrogen doping was necessary to create N-H bonds, which might change the chemical route to promote CO_2 reduction toward CO while blocking the unfavorable CO_2 methanation process [116].

A Ru catalyst supported by Mg $(OH)_2$ was created by Kong et al. for the photothermal hydrogenation of CO_2. The ion-exchange pathway serves as the foundation for this technique, which supports Ru nanoparticles with enhanced activity and comparatively good stability in photothermal CO_2 catalysis. There are significant loadings and dispersions of these nanoparticles. Ru^{3+} and $Mg(OH)_2$ chemically bond and evenly distribute Ru precursors on $Mg(OH)_2$ supports via an ion-exchange process. Ru nanoparticles are produced by the subsequent low temperature reduction by H_2, which are widely distributed and hardly alter in size as loading rises. Our findings pave the way for the effective production of widely dispersed, potently light-absorptive metal catalysts that can be used to convert carbon dioxide into solar fuels [117].

Ma et al. fabricated a N-doped carbon-coated cobalt nanoparticle as a photothermal catalyst for photothermal reduction of CO_2. It is reported that this photothermal catalyst was created by pyrolyzing a ZIF-67 precursor based on cobalt in two steps. Under full spectrum light irradiation, the catalyst shows outstanding activity and stability for photothermal hydrogenation to CO, with a 0.75 mol $gcat^{-1}$ h^{-1} CO production rate. The combination of the N doping species, the thickness of the carbon layers, and the precisely calibrated size of the Co NPs is primarily responsible for the Co NP's exceptional activity and durability. Extensive theoretical calculations and realistic characterizations show how such a simple N-doped carbon coating method can greatly aid in the activation of reactants and the desorption of generated CO due to its high photothermal action [118].

8.3 CONCLUSION AND FUTURE PERSPECTIVES

Climate change, driven historically by solar cycle fluctuations, has seen a profound shift since the 1800s due to human activities, especially fossil fuel combustion, disrupting the atmosphere's energy balance with rising carbon dioxide levels. This imbalance has triggered global temperature increases, accelerating polar ice cap melting and turning fertile lands into barren wastelands, compounding sea-level rise. With carbon dioxide emissions now surpassing levels of the past 2 million years, urgent action is needed to curb greenhouse gases, mainly from hydrocarbon fuel burning. Beyond just increasing temperatures, climate change has disastrous effects for ecosystems that are closely related to each other. Efforts in catalytic conversion of carbon dioxide for value added products hold promise but require efficient catalysts to mitigate energy barriers.

Looking ahead, addressing climate change demands a multifaceted approach that goes beyond carbon reduction efforts. Future perspectives emphasize innovative technologies like carbon capture, utilization, and storage (CCUS) to curb greenhouse gas emissions effectively. Developing efficient catalysts for carbon dioxide conversion into valuable products remains pivotal, requiring advancements in photocatalytic and other conversion processes. Additionally, nanomaterial research could revolutionize catalysis by enhancing surface effects and quantum size properties, optimizing energy efficiency in carbon conversion. Embracing renewable energy sources and fostering international cooperation are essential for sustainable solutions. Ultimately, a collective global commitment to innovation and action is imperative to mitigate climate change and secure a habitable future for generations to come.

REFERENCES

[1] (WMO), W.M.O. 2023. The Global Climate 2011–2020: A decade of accelerating climate change. WMO-No. 1338.

[2] Baloch, M.A., Mahmood, N. and Zhang, J.W. 2019. Effect of natural resources, renewable energy and economic development on CO_2 emissions in BRICS countries. Science of the Total Environment. 678: 632–638.

[3] Hiragond, C., Ali, S., Sorcar, S. and In, S.I. 2019. Hierarchical nanostructured photocatalysts for CO_2 photoreduction. Catalysts. 9(4): 370.

[4] Hiragond, C.B., Powar, N.S. and In, S.I. 2020. Recent developments in lead and lead-free halide perovskite nanostructures towards photocatalytic CO_2 reduction. Nanomaterials. 10(12): 2569.

[5] Kuk, S.K., Ham, Y., Gopinath, K., Boonmongkolras, P., Lee, Y., Lee, Y.W., et al. 2019. Continuous 3D titanium nitride nanoshell structure for solar-driven unbiased biocatalytic CO_2 reduction. Advanced Energy Materials. 9(25): 1900029.

[6] Ye, J., Yu, J., Zhang, Y., Chen, M., Liu, X., Zhou, S., et al. 2019. Light-driven carbon dioxide reduction to methane by Methanosarcina barkeri-CdS biohybrid. Applied Catalysis B: Environmental. 257: 117916.

[7] Lu, Q. and Jiao, F. 2016. Electrochemical CO_2 reduction: Electrocatalyst, reaction mechanism, and process engineering. Nano Energy. 29: 439–456.

[8] Lee, M.Y., Park, K.T., Lee, W., Lim, H., Kwon, Y. and Kang, S. 2020. Current achievements and the future direction of electrochemical CO_2 reduction: A short review. Critical Reviews in Environmental Science and Technology. 50(8): 769–815.

[9] Pullar, R.C., Novais, R.M., Caetano, A.P., Barreiros, M.A., Abanades, S. and Oliveira, F.A.C. 2019. A review of solar thermochemical CO_2 splitting using ceria-based ceramics with designed morphologies and microstructures. Frontiers in Chemistry. 7: 601.

[10] Wu, X.Y. and Ghoniem, A.F. 2019. Mixed ionic-electronic conducting (MIEC) membranes for thermochemical reduction of CO_2: A review. Progress in Energy and Combustion Science. 74: 1–30.

[11] Meng, X., Wang, T., Liu, L., Ouyang, S., Li, P., Hu, H., et al. 2014. Photothermal conversion of CO_2 into CH_4 with H_2 over group VIII nanocatalysts: an alternative approach for solar fuel production. Angewandte Chemie International Edition. 53(43): 11478–11482.

[12] Jia, J., O'Brien, P.G., He, L., Qiao, Q., Fei, T., Reyes, L.M., et al. 2016. Visible and near-infrared photothermal catalyzed hydrogenation of gaseous CO_2 over nanostructured Pd@ Nb_2O_5. Advanced Science. 3(10): 1600189.

[13] Bui, M., Adjiman, C.S., Bardow, A., Anthony, E.J., Boston, A., Brown, S., et al. 2018. Carbon capture and storage (CCS): The way forward. Energy and Environmental Science. 11(5): 1062–1176.

[14] Bai, X.J., Lu, X.Y., Ju, R., Chen, H., Shao, L., Zhai, X., et al. 2021. Preparation of MOF film/aerogel composite catalysts via substrate-seeding secondary-growth

for the oxygen evolution reaction and CO_2 cycloaddition. Angewandte Chemie International Edition. 60(2): 701–705.

[15] Roduner, E. 2006. Size matters: why nanomaterials are different. Chemical Society Reviews. 35(7): 583–592.

[16] Yang, C.C. and Mai, Y.W. 2014. Thermodynamics at the nanoscale: A new approach to the investigation of unique physicochemical properties of nanomaterials. Materials Science and Engineering: R: Reports. 79: 1–40.

[17] Namazu, T., Isono, Y. and Tanaka, T. 2000. Evaluation of size effect on mechanical properties of single crystal silicon by nanoscale bending test using AFM. Journal of Microelectromechanical Systems. 9(4): 450–459.

[18] Li, Y. and Ding, Y. 2010. Porous AgCl/Ag nanocomposites with enhanced visible light photocatalytic properties. The Journal of Physical Chemistry C. 114(7): 3175–3179.

[19] Akhtar, J., Tahir, M.B., Sagir, M. and Bamufleh, H.S. 2020. Improved photocatalytic performance of Gd and Nd co-doped ZnO nanorods for the degradation of methylene blue. Ceramics International. 46(8): 11955–11961.

[20] Oelgemöller, M. and Hoffmann, N. 2016. Studies in organic and physical photochemistry–an interdisciplinary approach. Organic and Biomolecular Chemistry. 14(31): 7392–7442.

[21] Tahir, M.B., Farman, S., Rafique, M., Shakil, M., Khan, M.I., Ijaz, M., et al. 2021. Photocatalytic performance of hybrid WO_3/TiO_2 nanomaterials for the degradation of methylene blue under visible light irradiation. International Journal of Environmental Analytical Chemistry. 101(10): 1448–1460.

[22] Hunge, Y.M. and Yadav, A.A. 2018. Basics and advanced developments in photocatalysis—A review (Mini review). International Journal of Hydrology. 2(4): 539–540.

[23] Wang, Y., Chen, E. and Tang, J. 2022. Insight on reaction pathways of photocatalytic CO_2 conversion. ACS Catalysis. 12(12): 7300–7316.

[24] Koppenol, W.H. and Rush, J.D. 1987. Reduction potential of the carbon dioxide/carbon dioxide radical anion: a comparison with other C1 radicals. Journal of Physical Chemistry. 91(16): 4429–4430.

[25] Albero, J., Peng, Y. and García, H. 2020. Photocatalytic CO_2 reduction to C^{2+} products. ACS Catalysis. 10(10): 5734–5749.

[26] Chang, X., Wang, T. and Gong, J. 2016. CO_2 photo-reduction: insights into CO_2 activation and reaction on surfaces of photocatalysts. Energy and Environmental Science. 9(7): 2177–2196.

[27] Nitopi, S., Bertheussen, E., Scott, S.B., Liu, X., Engstfeld, A.K., Horch, S., et al. 2019. Progress and perspectives of electrochemical CO_2 reduction on copper in aqueous electrolyte. Chemical Reviews. 119(12): 7610–7672.

[28] Li, X., Yu, J., Jaroniec, M. and Chen, X. 2019. Cocatalysts for selective photoreduction of CO_2 into solar fuels. Chemical Reviews. 119(6): 3962–4179.

[29] Wang, Y., Liu, X., Han, X., Godin, R., Chen, J., Zhou, W., et al. 2020. Unique hole-accepting carbon-dots promoting selective carbon dioxide reduction nearly 100% to methanol by pure water. Nature Communications. 11(1): 2531.

[30] Wang, J., Lin, S., Tian, N., Ma, T., Zhang, Y. and Huang, H. 2021. Nanostructured metal sulfides: classification, modification strategy, and solar-driven CO_2 reduction application. Advanced Functional Materials. 31(9): 2008008.

[31] Li, H., Chen, K., Fu, J. and Liu, M. 2022. Photocatalytic CO_2 Reduction Reactions. Materials Today. 32: 222–243.

[32] Li, Y., Wei, Y., He, W., Tang, Z., Xiong, J. and Zhao, Z. 2023. Ordered macroporous structured TiO_2-based photocatalysts for CO_2 reduction: a review. Chinese Chemical Letters. 34(12): 108417.

[33] Patial, S., Kumar, R., Raizada, P., Singh, P., Van Le, Q., Lichtfouse, E., et al. 2021. Boosting light-driven CO_2 reduction into solar fuels: Mainstream avenues for engineering ZnO-based photocatalysts. Environmental Research. 197: 111134.

[34] Kuspanov, Z., Umirzakov, A., Serik, A., Baimenov, A., Yeleuov, M. and Daulbayev, C. 2023. Multifunctional strontium titanate perovskite-based composite photocatalysts for energy conversion and other applications. International Journal of Hydrogen Energy. 48(98): 38634–38654.

[35] Meng, Y., Liu, G., Zuo, G., Meng, X., Wang, T. and Ye, J. 2022. A review on ZnS-based photocatalysts for CO_2 reduction in all-inorganic aqueous medium. Nanoscale. 14(39): 14455–14465.

[36] Wang, Y., Shang, X., Shen, J., Zhang, Z., Wang, D., Lin, J., et al. 2020. Direct and indirect Z-scheme heterostructure-coupled photosystem enabling cooperation of CO_2 reduction and H_2O oxidation. Nature Communications. 11(1): 3043.

[37] Tahir, M.B. and Sagir, M. 2019. RETRACTED: Carbon nanodots and rare metals (RM = La, Gd, Er) doped tungsten oxide nanostructures for photocatalytic dyes degradation and hydrogen production.

[38] Dong, S., Liu, W., Liu, S., Li, F., Hou, J., Hao, R., et al. 2022. Single atomic Pt on amorphous ZrO_2 nanowires for advanced photocatalytic CO_2 reduction. Materials Today Nano. 17: 100157.

[39] Tang, R., Dong, X.A., Sheng, J., Xi, S., Zhang, L. and Dong, F. 2022. Single-metal catalytic sites via high-throughput mechanochemistry enable selective and efficient CO_2 photoreduction. Applied Catalysis B: Environmental. 316: 121661.

[40] Li, Q., Tang, Q., Xiong, P., Chen, D., Chen, J., Wu, Z., et al. 2023. Effect of palladium chemical states on CO_2 photocatalytic reduction over g-C_3N_4: distinct role of single-atomic state in boosting CH_4 production. Chinese Journal of Catalysis. 46: 177–190.

[41] Yu, Y., Dong, X.A., Chen, P., Geng, Q., Wang, H., Li, J., et al. 2021. Synergistic effect of Cu single atoms and Au–Cu alloy nanoparticles on TiO_2 for efficient CO_2 photoreduction. ACS Nano. 15(9): 14453–14464.

[42] Gusain, R., Kumar, N. and Ray, S.S. 2022. Metal Oxide-Based Nanocomposites for Photocatalytic Reduction of CO_2. In Advanced Materials for a Sustainable Environment. 293–315. CRC Press.

[43] Wu, M., Ouyang, C., Ye, Z., Li, S., Hong, Z. and Zhi, M. 2022. Ag–CeO_2 composite aerogels as photocatalysts for CO_2 reduction. ACS Applied Energy Materials. 5(6): 7335–7345.

[44] Danish, M.S.S., Estrella, L.L., Alemaida, I.M.A., Lisin, A., Moiseev, N., Ahmadi, M. et al. 2021. Photocatalytic applications of metal oxides for sustainable environmental remediation. Metals. 11(1): 80.

[45] Nunes, D., Pimentel, A., Branquinho, R., Fortunato, E. and Martins, R. 2021. Metal oxide-based photocatalytic paper: A green alternative for environmental remediation. Catalysts. 11(4): 504.

[46] Rehman, Z.U., Bilal, M., Hou, J., Butt, F.K., Ahmad, J., Ali, S., et al. 2022. Photocatalytic CO_2 reduction using TiO_2-based photocatalysts and TiO_2 Z-scheme heterojunction composites: A review. Molecules. 27(7): 2069.

[47] Fu, Z., Zhang, G., Tang, Z. and Zhang, H. 2020. Preparation and application of ordered mesoporous metal oxide catalytic materials. Catalysis Surveys from Asia. 24(1): 38–58.

[48] Xiong, Y., Zhao, W., Gu, D., Tie, Z., Zhang, W. and Jin, Z. 2023. Tunable C_2 Products via Photothermal Steam Reforming of CO_2 over Surface-Modulated Mesoporous Cobalt Oxides. Nano Letters. 23(11): 4876–4884.

[49] Pathak, S. and Upadhyayula, S. 2023. Insights into the stability of the iron oxide immobilized into mesoporous silica catalysts in iodine–sulfur cycle for hydrogen production. Reaction Kinetics, Mechanisms and Catalysis. 136(6): 2977–2996.

[50] Farid, M.A., Ijaz, S., Ashiq, M.N., Ehsan, M.F., Gul, F., Batool, S.R., et al. 2022. Synthesis of mesoporous zirconium manganese mixed metal oxide nanowires for photocatalytic reduction of CO_2. Journal of Materials Research. 1–11.

[51] Reli, M., Nadrah, P., Edelmannová, M.F., Ricka, R., Škapin, A.S., Štangar, U.L., et al. 2024. Photocatalytic CO_2 reduction over mesoporous TiO_2 photocatalysts. Materials Science in Semiconductor Processing. 169: 107927.

[52] Mohammadzadeh Kakhki, R. 2023. Polymeric organic–inorganic C_3N_4/ZnO high-performance material for visible light photodegradation of organic pollutants. Polymer Bulletin. 80(10): 10697–10717.

[53] Kumar, S., Regue, M., Isaacs, M.A., Freeman, E. and Eslava, S. 2020. All-inorganic $CsPbBr_3$ nanocrystals: gram-scale mechanochemical synthesis and selective photocatalytic CO_2 reduction to methane. ACS Applied Energy Materials. 3(5): 4509–4522.

[54] Liao, J.F., Cai, Y.T., Li, J.Y., Jiang, Y., Wang, X.D., Chen, H.Y., et al. 2021. Plasmonic $CsPbBr_3$–Au nanocomposite for excitation wavelength dependent photocatalytic CO_2 reduction. Journal of Energy Chemistry. 53: 309–315.

[55] Yang, H., Zhang, J. and Dai, K. 2022. Organic amine surface modified one-dimensional CdSe0. 8S0. 2-diethylenetriamine/two-dimensional $SnNb_2O_6$ S-scheme heterojunction with promoted visible-light-driven photocatalytic CO_2 reduction. Chinese Journal of Catalysis. 43(2): 255–264.

[56] Chen, L., Tang, Q., Wu, S., Zhang, L., Feng, L., Wang, Y., et al. 2023. Covalent coupling promoting charge transport of CdSeTe/UiO-66 for boosting photocatalytic CO_2 reduction. Chinese Chemical Letters. 34(6): 107903.

[57] Yu, M., Wang, J., Li, G., Zhang, S. and Zhong, Q. 2023. Construction of 3D/2D indium vanadate/graphite carbon nitride with nitrogen defects Z-scheme heterojunction for improving photocatalytic carbon dioxide reduction. Journal of Materials Science and Technology. 154: 129–139.

[58] Zuo, Y., Wang, Y., Dai, G., Ge, F., Fang, L., Zhou, X., et al. 2023. Dual interfacial build-in electric field effect induced by sandwich-type heterojunction for propelling photocatalytic fuel extraction from CO_2 in water. Separation and Purification Technology. 308: 122971.

[59] Mao, Y., Zhang, M., Si, S., Zhai, G., Bao, X., Song, K., et al. 2023. Electronic structure manipulation via site-selective atomically dispersed Ni for efficient photocatalytic CO_2 reduction. ACS Catalysis. 13(13): 8362–8371.

[60] Lei, Q., Yuan, H., Du, J., Ming, M., Yang, S., Chen, Y., et al. 2023. Photocatalytic CO_2 reduction with aminoanthraquinone organic dyes. Nature Communications. 14(1): 1087.

[61] Li, C., Zhang, P., Gu, F., Tong, L., Jiang, J., Zuo, Y., et al. 2023. Atomically dispersed Au confined by oxygen vacancies in Au-θ-Al_2O_3/Au/PCN hybrid for boosting photocatalytic CO_2 reduction driven by multiple built-in electric fields. Chemical Engineering Journal. 476: 146514.

[62] Velty, A. and Corma, A. 2023. Advanced zeolite and ordered mesoporous silica-based catalysts for the conversion of CO_2 to chemicals and fuels. Chemical Society Reviews. 52(5): 1773–1946.

[63] Verma, R., Belgamwar, R. and Polshettiwar, V. 2021. Plasmonic photocatalysis for CO_2 conversion to chemicals and fuels. ACS Materials Letters. 3(5): 574–598.

[64] Park, J.H. and Kim, J.H. 2020. Selective CO_2 photoconversion on low-coordinate TiO_2. Chem. 6(10): 2435–2436.

[65] Habisreutinger, S.N., Schmidt-Mende, L. and Stolarczyk, J.K. 2013. Photocatalytic reduction of CO_2 on TiO_2 and other semiconductors. Angewandte Chemie International Edition. 52(29): 7372–7408.

[66] Tasbihi, M., Kočí, K., Troppová, I., Edelmannová, M., Reli, M., Čapek, L., et al. 2018. Photocatalytic reduction of carbon dioxide over Cu/TiO_2 photocatalysts. Environmental Science and Pollution Research. 25: 34903–34911.

[67] Sun, M., Zhao, B., Chen, F., Liu, C., Lu, S., Yu, Y., et al. 2021. Thermally-assisted photocatalytic CO_2 reduction to fuels. Chemical Engineering Journal. 408: 127280.

[68] Weilhard, A., Argent, S.P. and Sans, V. 2021. Efficient carbon dioxide hydrogenation to formic acid with buffering ionic liquids. Nature Communications. 12(1): 231.

[69] Zhang, H., Li, Y., Wang, J., Wu, N., Sheng, H., Chen, C., et al. 2021. An unprecedent hydride transfer pathway for selective photocatalytic reduction of CO_2 to formic acid on TiO_2. Applied Catalysis B: Environmental. 284: 119692.

[70] Tian, Y., Zhou, Y., Zong, Y., Li, J., Yang, N., Zhang, M., et al. 2020. Construction of functionally compartmental inorganic photocatalyst–enzyme system via imitating chloroplast for efficient photoreduction of CO_2 to formic acid. ACS Applied Materials and Interfaces. 12(31): 34795–34805.

[71] Kamada, K., Jung, J., Wakabayashi, T., Sekizawa, K., Sato, S., Morikawa, T., et al. 2020. Photocatalytic CO_2 reduction using a robust multifunctional iridium complex toward the selective formation of formic acid. Journal of the American Chemical Society. 142(23): 10261–10266.

[72] Chen, X., Liu, Y. and Wu, J. 2020. Sustainable production of formic acid from biomass and carbon dioxide. Molecular Catalysis. 483: 110716.

[73] Dong, WANG., ZHONG, DZ., HAO, G.Y., LI, J.P. and Qiang, ZHAO. 2021. ZnOHF nanorods for efficient electrocatalytic reduction of carbon dioxide to carbon monoxide. Journal of Fuel Chemistry and Technology. 49(9): 1379–1388.

[74] Qi, Z., Biener, M.M., Kashi, A.R., Hunegnaw, S., Leung, A., Ma, S., et al. 2021. Electrochemical CO_2 to CO reduction at high current densities using a nanoporous gold catalyst. Materials Research Letters. 9(2): 99–104.

[75] Liu, X.M., Kang, R.K., Wang, J.L., Li, J.N., Chen, Q.L. and Xu, Y. 2021. A Purely Inorganic Quasi-Keggin Polyoxometalate for Photocatalytic Conversion of Carbon Dioxide to Carbon Monoxide. ChemPlusChem. 86(7): 1014–1020.

[76] Van Le, Q., Nguyen, V.H., Nguyen, T.D., Sharma, A., Rahman, G. and Nguyen, D.L.T. 2021. Light-driven reduction of carbon dioxide: altering the reaction

pathways and designing photocatalysts toward value-added and renewable fuels. Chemical Engineering Science. 237: 116547.

[77] Kočí, K., Van, H.D., Edelmannová, M., Reli, M. and Wu, J.C. 2020. Photocatalytic reduction of CO_2 using Pt/C_3N4 photocatalyts. Applied Surface Science. 503: 144426.

[78] Morawski, A.W., Kusiak-Nejman, E., Wanag, A., Narkiewicz, U., Edelmannová, M., Reli, M., et al. 2021. Influence of the calcination of TiO_2-reduced graphite hybrid for the photocatalytic reduction of carbon dioxide. Catalysis Today. 380: 32–40.

[79] Navarro-Jaén, S., Virginie, M., Bonin, J., Robert, M., Wojcieszak, R. and Khodakov, A.Y. 2021. Highlights and challenges in the selective reduction of carbon dioxide to methanol. Nature Reviews Chemistry. 5(8): 564–579.

[80] Shinde, G.Y., Mote, A.S. and Gawande, M.B. 2022. Recent advances of photocatalytic hydrogenation of CO_2 to methanol. Catalysts. 12(1): 94.

[81] Chakrabortty, S., Nayak, J., Ruj, B., Pal, P., Kumar, R., Banerjee, S., et al. 2020. Photocatalytic conversion of CO_2 to methanol using membrane-integrated green approach: a review on capture, conversion and purification. Journal of Environmental Chemical Engineering. 8(4): 103935.

[82] Manzoor, N., Sadiq, M., Naqvi, M., Sikandar, U. and Naqvi, S.R. 2020. Retracted: Experimental study of CO_2 conversion into methanol by synthesized photocatalyst ($ZnFe_2O_4/TiO_2$) using visible light as an energy source. Catalysts. 10(2): 163.

[83] Movahed, S.K., Najinasab, A., Nikbakht, R. and Dabiri, M. 2020. Visible light assisted photocatalytic reduction of CO_2 to methanol using Fe_3O_4@NC/Cu_2O nanostructure photocatalyst. Journal of Photochemistry and Photobiology A: Chemistry. 401: 112763.

[84] An, K., Zhang, S., Wang, J., Liu, Q., Zhang, Z. and Liu, Y. 2021. A highly selective catalyst of Co/$La_4Ga_2O_9$ for CO_2 hydrogenation to ethanol. Journal of Energy Chemistry. 56: 486–495.

[85] Wang, Y., Jia, H., Gong, H., Zhou, L., Qiu, Z., Fang, X., et al. 2021. Fabrication of trimodal porous silica/g-C_3N_4 nanotubes for efficient visible light photocatalytic reduction of CO_2 to ethanol. Chemical Engineering Journal. 426: 130877.

[86] Lertthanaphol, N., Pienutsa, N., Chusri, K., Sornsuchat, T., Chanthara, P., Seeharaj, P., et al. 2021. One-step hydrothermal synthesis of precious metal-doped titanium dioxide–graphene oxide composites for photocatalytic conversion of CO_2 to ethanol. ACS Omega. 6(51): 35769–35779.

[87] Zhao, D., Xuan, Y., Zhang, K. and Liu, X. 2021. Highly selective production of ethanol over hierarchical Bi@ Bi_2MoO_6 composite via bicarbonate-assisted photocatalytic CO_2 reduction. ChemSusChem. 14(16): 3293–3302.

[88] Huang, Y., Du, P., Shi, W.X., Wang, Y., Yao, S., Zhang, Z.M., et al. 2021. Filling COFs with bimetallic nanoclusters for CO_2-to-alcohols conversion with H_2O oxidation. Applied Catalysis B: Environmental. 288: 120001.

[89] Dias, Y.R. and Perez-Lopez, O.W. 2021. CO_2 conversion to methane using Ni/SiO_2 catalysts promoted by Fe, Co and Zn. Journal of Environmental Chemical Engineering. 9(1): 104629.

[90] Kulandaivalu, T., Mohamed, A.R., Ali, K.A. and Mohammadi, M. 2020. Photocatalytic carbon dioxide reforming of methane as an alternative approach for solar fuel production-a review. Renewable and Sustainable Energy Reviews. 134: 110363.

[91] Ge, H., Zhang, B., Liang, H., Zhang, M., Fang, K., Chen, Y., et al. 2020. Photocatalytic conversion of CO_2 into light olefins over TiO_2 nanotube confined Cu clusters with high ratio of Cu^+. Applied Catalysis B: Environmental. 263: 118133.

[92] Al Jitan, S., Palmisano, G. and Garlisi, C. 2020. Synthesis and surface modification of TiO_2-based photocatalysts for the conversion of CO_2. Catalysts. 10(2): 227.

[93] Reñones, P., Fresno, F., Oropeza, F.E. and de la Peña O'Shea, V.A. 2022. Improved methane production by photocatalytic CO_2 conversion over $Ag/In_2O_3/TiO_2$ heterojunctions. Materials. 15(3): 843.

[94] Lervold, S., Lødeng, R., Yang, J., Skjelstad, J., Bingen, K. and Venvik, H.J. 2021. Partial oxidation of methanol to formaldehyde in an annular reactor. Chemical Engineering Journal. 423: 130141.

[95] Kumar, A. and Ananthakrishnan, R. 2020. Visible light-assisted reduction of CO_2 into formaldehyde by heteroleptic ruthenium metal complex–TiO_2 hybrids in an aqueous medium. Green Chemistry. 22(5): 1650–1661.

[96] Ranjbakhsh, E., Izadyar, M., Nakhaeipour, A. and Habibi-Yangjeh, A. 2020. P-doped g-C_3N_4 as an efficient photocatalyst for CO_2 conversion into value-added materials: a joint experimental and theoretical study. International Journal of Quantum Chemistry. 120(23): e26388.

[97] Samanta, S. and Srivastava, R. 2020. Catalytic conversion of CO_2 to chemicals and fuels: the collective thermocatalytic/photocatalytic/electrocatalytic approach with graphitic carbon nitride. Materials Advances. 1(6): 1506–1545.

[98] He, Y., Lei, Q., Li, C., Han, Y., Shi, Z. and Feng, S. 2021. Defect engineering of photocatalysts for solar-driven conversion of CO_2 into valuable fuels. Materials Today. 50: 358–384.

[99] Shwetharani, R., Chandan, H.R., Sakar, M., Balakrishna, G.R., Reddy, K.R. and Raghu, A.V. 2020. Photocatalytic semiconductor thin films for hydrogen production and environmental applications. International Journal of Hydrogen Energy. 45(36): 18289–18308.

[100] Hayat, A., Taha, T.A.M., Alenad, A.M., Yingjin, L., Mane, S.K.B., Hayat, A., et al. 2021. Organic conjugation of polymeric carbon nitride for improved photocatalytic CO_2 conversion and H_2 fixation. Energy Technology. 9(10): 2100091.

[101] Yuan, J., Yi, X., Tang, Y., Liu, C. and Luo, S. 2020. Efficient photocatalytic hydrogen evolution and CO_2 reduction: enhanced light absorption, charge separation, and hydrophilicity by tailoring terminal and linker units in g-C_3N_4. ACS Applied Materials and Interfaces. 12(17): 19607–19615.

[102] Shanker, G.S., Biswas, A. and Ogale, S. 2021. 2D materials and their heterostructures for photocatalytic water splitting and conversion of CO_2 to value chemicals and fuels. Journal of Physics: Energy. 3(2): 022003.

[103] Lee, S.E., Nasirian, A., Kim, Y.E., Fard, P.T., Kim, Y., Jeong, B., et al. 2020. Visible-light photocatalytic conversion of carbon dioxide by Ni (II) complexes with N_4S_2 coordination: Highly efficient and selective production of formate. Journal of the American Chemical Society. 142(45): 19142–19149.

[104] Gao, D., Li, W., Wang, H., Wang, G. and Cai, R. 2022. Heterogeneous catalysis for CO_2 conversion into chemicals and fuels. Transactions of Tianjin University. 28(4): 245–264.

[105] Dokania, A., Ould-Chikh, S., Ramirez, A., Cerrillo, J.L., Aguilar, A., Russkikh, A., et al. 2021. Designing a multifunctional catalyst for the direct production of gasoline-range isoparaffins from CO_2. JACS Au. 1(11): 1961–1974.

[106] Saravanan, A., Vo, D.V.N., Jeevanantham, S., Bhuvaneswari, V., Narayanan, V.A., Yaashikaa, P.R., et al. 2021. A comprehensive review on different approaches for CO_2 utilization and conversion pathways. Chemical Engineering Science. 236: 116515.

[107] Garba, M.D., Usman, M., Khan, S., Shehzad, F., Galadima, A., Ehsan, M.F., et al. 2021. CO_2 towards fuels: A review of catalytic conversion of carbon dioxide to hydrocarbons. Journal of Environmental Chemical Engineering. 9(2): 104756.

[108] Dai, C., Zhao, X., Hu, B., Zhang, X., Luo, Q., Guo, X., et al. 2021. Effect of EDTA-2Na modification on Fe-Co/Al_2O_3 for hydrogenation of carbon dioxide to lower olefins and gasoline. Journal of CO_2 Utilization. 43: 101369.

[109] Zhao, Q., Xu, X., Fan, G. and Li, F. 2023. Crucial Role of Surface FeO_x Components on Supported Fe-Based Nanocatalysts for CO_2 Hydrogenation to Light Olefins. Industrial and Engineering Chemistry Research. 62(24): 9420–9432.

[110] Wang, Z., Wang, Y., Ning, S. and Kang, Q. 2022. Zinc-based materials for photoelectrochemical reduction of carbon dioxide. Energy and Fuels. 36(19): 11380–11393.

[111] Boutin, E., Patel, M., Kecsenovity, E., Suter, S., Janáky, C. and Haussener, S. 2022. Photo-electrochemical conversion of CO_2 under concentrated sunlight enables combination of high reaction rate and efficiency. Advanced Energy Materials. 12(30): 2200585.

[112] Ding, P., Jiang, T., Han, N. and Li, Y. 2020. Photocathode engineering for efficient photoelectrochemical CO_2 reduction. Materials Today Nano. 10: 100077.

[113] Cortes, M.A.L.R.M., McMichael, S., Hamilton, J.W.J., Sharma, P.K., Brown, A. and Byrne, J.A. 2020. Photoelectrochemical reduction of CO_2 with TiNT. Materials Science in Semiconductor Processing. 108: 104900.

[114] Zhang, H., Chen, Y., Wang, H., Wang, H., Ma, W., Zong, X., et al. 2020. Carbon encapsulation of organic–inorganic hybrid perovskite toward efficient and stable photo-electrochemical carbon dioxide reduction. Advanced Energy Materials. 10(44): 2002105.

[115] Liu, H., Shi, L., Zhang, Q., Qi, P., Zhao, Y., Meng, Q., et al. 2021. Photothermal catalysts for hydrogenation reactions. Chemical Communications. 57(11): 1279–1294.

[116] Jia, Z., Ning, S., Tong, Y., Chen, X., Hu, H., Liu, L., et al. 2021. Selective photothermal reduction of CO_2 to CO over Ni-nanoparticle/N-doped CeO_2 nanocomposite catalysts. ACS Applied Nano Materials. 4(10): 10485–10494.

[117] Kong, N., Han, B., Li, Z., Fang, Y., Feng, K., Wu, Z., et al. 2020. Ruthenium nanoparticles supported on $Mg(OH)_2$ microflowers as catalysts for photothermal carbon dioxide hydrogenation. ACS Applied Nano Materials. 3(3): 3028–3033.

[118] Ma, J., Yu, J., Chen, G., Bai, Y., Liu, S., Hu, Y., et al. 2023. Rational design of N-doped carbon-coated cobalt nanoparticles for highly efficient and durable photothermal CO_2 conversion. Advanced Materials. 35(42): 2302537.

Chapter **9**

Nanomaterials for Air Pollution Control and Remediation

Sushma Yadav*[1], Neeraj Kumari[2] and Babita Singh[3]
[1]Department of Environmental Sciences, J.C. Bose University of Science and Technology, Faridabad
[2]Department of Chemistry, SBAS, K.R. Mangalam University, Gurugram
[3]Department of Environmental Studies, Satyawati College Evening, University of Delhi

9.1 INTRODUCTION

Air pollution has become a reason for global concern and challenge worldwide in recent years. With the beginning of newer technologies, speedy industrialisation, and exponential growth of the world population, the problem of air pollution is going to increase to a larger extent. Globally, air pollutants are responsible for a higher mortality rate of about 7 million people and about 4.2 million premature deaths annually [1]. It is also estimated that about 99% of the world's population breathes unhealthy air exceeding WHO guidelines limits, of which low and middle-income groups constitute the majority and face the highest exposures [2]. Toxic air pollutants from diverse sources have detrimental effects on human health and substantial impacts on the environment and economy [3, 4], which need to be addressed using appropriate scientifically sound more reliable, and economical greener technologies [5, 6]

In the past few decades, many physical, chemical, and biological techniques have been developed to trace, monitor, and remediate air pollutants but these conventional

*For Correspondence: sushma@jcboseust.ac.in; sushma.yadav4@gmail.com

DOI: 10.1201/9781003492184-9

methods are more complex, time-consuming, and have monetary constraints [7]. However, in recent years nanotechnology has emerged as a promising solution for improving the conditions of polluted components of environments [8–10] since these particles possess different atomic structures and electronic configurations and have distinctive properties of their lofty ratio of surface area to volume as compared to conventionally used counterpart bulk materials. Increased surface area-volume enhances their reactivity, adsorption, and catalytic capacity which can be used as novel applications in different fields [11, 12]. Nanomaterials have magnetic, optical, electrical, and high structural properties that make them efficient sensors for monitoring and remediating air pollution and for understanding the potential toxicity and fate of atmospheric pollutants [13, 14].

Nanomaterials possess different structural forms such as nanowires, nanorods, nano branches, Nano aerogels, etc which can be designed for control of air pollution [15, 16]. Apart from these structural forms various methods based on nanotechnology for air pollution remediation are nanosensors, nanophotocatalysts, membrane technology, nanostructured filters, nano- adsorbents, photocatalytic anti-bacterial/viral coatings, integrated methods, recyclable SERS substrates, etc. [17].

9.2 NANOTECHNOLOGICAL STRATEGIES FOR AIR POLLUTION REMEDIATION

Advanced materials, particularly nanomaterials like carbon-based and metal-based nanoparticles, have been developed for efficient removal of air pollutants. These materials offer high surface area and reactivity, enhancing their ability to adsorb or catalyse the decomposition of pollutants [18]. Nanofibres and biofiltration techniques utilising microbial approaches have also shown promise in air pollutant removal. Additionally, functional materials designed for air quality management employ physicochemical and reactive adsorption mechanisms to interact with and remove pollutants [19].

Currently, the major problem is poor air quality which can be effectively addressed using nanoscale adsorbents known as nanoadsorbents. These minute particles possess the ability to adsorb multiple airborne contaminants. There are various nanoadsorbents which have been classified to investigate and removal of pollutants like microbes, Volatile Organic Compounds (VOCs), and metal oxides [20].

Various nanomaterials used for air pollution remediations are as follows:

9.2.1 Nano Adsorbents

Recently nanoscale adsorbents have gained popularity in air pollution remediation than their conventional counterparts due to advantages in terms of higher efficiency, greater energy binding active sites, and porosity. These properties of nanoadsorbents are explored by researchers for successful air pollution remediation worldwide. Some of the commonly used nanoadsorbents are discussed below:

9.2.1.1 Carbon Nanotubes

Carbon nanotubes possess exclusive properties like good porosity, large surface area ratio, and high electrical conductivity with outstanding physio-chemical and mechanical properties that make them exhibit higher durability and adsorption capacity which can be applied for air pollution detection and remediation. Structurally CNTs can be Single-Wall CNTs (SWCNTs) and Multi-Wall CNTs (MWCNTs) based on the number of cylindrical shells [21, 22]. Carbon nanotubes show greater selectivity, adsorption capacity, and affinity for many air pollutants. Also, the structures of CNTs can be modified to increase the number of active sites and surface properties using additional functional groups. Hence CNTs can be employed for separation as well as degradation of several pollutants in the environment [23]. g-C_3N_4 NSs, r-GO, and CNTs, modified carbon nanostructured materials were synthesised by Attia et al. (2024) and used for the adsorption of particulate materials present in the air [24].

9.2.1.2 Zinc Oxide

Due to the photocatalytic degradation characteristics of ZnO nanoparticles properties, they have been used for air purification on a large scale [25]. These nanoparticles have been integrated to some advanced techniques to improve air quality especially in urban areas. Some green methods have been incorporated for the synthesis of ZnO nanoparticles in a sustainable way, highlighting the promising role of ZnO nanoparticles in reducing air pollution [26]. Recently, ZnO nanoparticles synthesised through ultrasonic-assisted precipitation methods have been used for removal of hydrogen sulphide (H2S) from air through adsorption process [27]. Ghosh et al. 2018 used wire explosion method to synthesise ZnO nanoparticles and used further to adsorb CO_2 at low temperature and high pressure [28]. Consequently, it can be concluded that carbon nanotube and ZnO-based nanoparticles can also be used to remove harmful gases like CO, CH_3OH, and H_2S from air.

9.2.1.3 Nanocatalysts

Nanocatalysts are substances formed of nanoparticles or some other nanostructured materials having high resistant against bacteria, viruses, pathogens, and organic contaminants. Nanocatalysts can be considered as alternate to address the global challenges such as air and water pollution. Due to to the reduction in catalyst size, enhancement in the surface area and active sites of nanocatalysts, they show higher efficiency to remove the toxic gases from the atmosphere. On exposure to light, the nanocatalyst generates electron hole pairs resulting in initiation of the breakdown of pollutants and convert them into innocuous gases [29]. The nanocatalyst can be homogeneous and heterogenous. Recently, various nanocatalysts have been developed for the remediation of air pollutants. Some of the important nanocrystals are discussed below:

9.2.1.4 Titanium Oxide

Titanium dioxide has garnered significant research attention as a semiconductor photocatalyst in recent years due to its cost-effectiveness, abundance, stability, and non-toxic nature, making it highly suitable for air purification. The presence of surface hydroxyl groups plays a pivotal role in contaminant adsorption and hydroxyl radical generation, making the fabrication method crucial for regulating their population on the titanium dioxide surface, thereby influencing air decontamination [30]. Surface-fluorinated titanium dioxide nanosheets, synthesised via the hydrothermal method using titanium tetra butoxide, have demonstrated enhanced activity in gaseous ammonia separation, nearly doubling the efficiency compared to other titanium dioxide variants [31].

Titanium dioxide nanoparticles (TiO_2NPs) were used for the oxidation and reduction of a large number of organic and inorganic air and water contaminants in air through the mechanism of photocatalytic degradation. Due to their 3.2 eV bandgap constraining TiO_2 performance should be enhanced using photons from visible light [32]. It was concluded that TiO_2-based building and construction material is a promising technique for the remediation of indoor and outdoor environmental pollutants [33]. Although TiO_2 nanoparticles exhibit high photoactivity, their lower adsorption affinity towards air contaminants, such as VOCs, coupled with smaller surface areas, diminishes reaction rates and separation efficiency. Hence, combining TiO_2 with materials that enhance contaminant adsorption is preferred.

Currently, carbon-based nanostructures such as graphene nanosheets and carbon nanotubes play a significant role in enhancing the photocatalytic efficiency of TiO_2. In TiO_2-CNT composites, electrons can efficiently transfer through the carbon nanotubes, inhibiting electron-hole recombination. Giannakopoulou et al. synthesised graphitic carbon nitride/titania composite photocatalysts which exhibited superior photocatalytic activity in NO oxidation compared to pristine semiconductors [34]

Formaldehyde (HCHO) is a hazardous VOC, and its degradation rate under visible light was evaluated over titanium dioxide-reduced-graphene oxide (TiO_2-rGO). TiO_2-rGO demonstrated significantly higher removal efficiency than pure titanium dioxide or reduced graphene oxide alone, attributed to the prolonged lifespan of charge carriers [35]. Jampaiah et al. fabricated titanium dioxide nanorods-supported mixed metal oxide catalysts, showing high Hg separation efficiency due to their crystallisation and large surface area. These catalysts, especially MnOx-FeOx-CrOx supported on titanium dioxide nanorods, demonstrated superior performance, aided by nitric oxide addition [36]. Titanium dioxide nanoparticles are utilised in self-cleaning coatings, harnessing photocatalytic properties to decompose atmospheric pollutants under UV radiation [37].

Therefore, it is concluded from the discussion, that TiO_2 serves as an efficient photocatalyst for separating gaseous ammonia and formaldehyde. Combining TiO_2 with different carbon-based nanostructures like TiO_2-CNT and TiO_2-graphitic carbon nitride/titania can enhance TiO_2's photocatalytic efficiency for hazardous gas removal [38, 39]. Furthermore, there are various other nanocatalysts like ZnO, CuO, Al_2O_3 etc. which can be used effectively for the degradation of various toxic gases present in the atmosphere.

9.2.2 Nanofilters and Nanostructured Membranes

Another alternate for the reduction of air pollution is nanostructured membranes. The presence of small pores in the membrane mainly help for filtration of contaminants from exhaust gases. Recently, various studies have been conducted for the developing nanostructured membrane and further used for capturing of toxic gases from the atmosphere. Zhao et al. (2024) developed 3D curly nanofibre membrane and investigated its efficiency for removing air pollutants [40]. The nanostructured membranes can also be used for the removal of volatile organic vapours. About a decade ago, it was very difficult to remove the formaldehyde from indoor air using conventional techniques, but researchers developed its alternate. They developed electrospun polyacrylonitrile-based carbon nanofibre membranes which showed superior efficiency against the toxic gases due to having microporosity and presence of nitrogen-containing functional groups [41]. Bioaerosol can also be removed using copper and silver-based nanoparticles [20].

9.2.3 Metal-Organic Frameworks

Due to diverse functionalities and large surface to volume ratio, Metal-Organic Frameworks (MOFs) are used on a large scale to address air pollution especially for formaldehyde. The efficiency to remove the air pollutant can be increased by incorporating MOFs into polymer matrices or by modifying their structure. Furthermore, the metal ions and defects present in MOFs also enhance their adsorption efficiencies toward particulate matter having size below 2.5 µm through electrostatic interaction [42, 43]. Bian et al. 2018 developed an advanced MOF-based nanofibre filter and investigated its efficiency for the removal of formaldehyde and particulate matter [44]. Nanocrystal-based nanofibrous filters were developed by Zhang et al. (2016) and used further for the adsorption of particulate matter and sulphfur dioxide over an extended period of time [45]. The study focuses on the potential use of MOFs and nanocrystal-based filters in removing air polluting contaminants and reducing the risk of health generated from air pollutants.

9.2.4 Carbon Nanotube-Based Filters

Carbon Nanotube-based Filters are the part of novel nanostructured materials which are used for controlling air pollution due to their unique physical and chemical characteristics like high surface area. Recently, carbon nanotube-based filters have been developed successfully and used further to remove the biodegradable pollutants present in water and to remove bioaerosols and particulate matters from the atmosphere. The interaction of carbon-based nanomaterials with the pollutants can be enhanced through functionalisation which enable the fabricated carbon-based nanotube filters to adsorb targeted pollutants. The functionalisation can be possible using small surfactant molecule and large polymer through fluorination, hydrogenation, and many other processes. A porous material, graphitic carbon nitride with high specific surface area was synthesised and utilised further to remove particulate matters. Romero-Guzman et al. synthesised carbon-based nanotube filters packed in a stainless steel filter using different prototypes and used efficiently to eliminate NO_x, CO, CO_2, and hydrocarbon from the air [46]. Li et al. discussed the application of high-efficiency particulate (HEPA) air filters based on CNTs for pollution remediation and detection [47].

9.3 DETECTION OF AIR POLLUTANTS USING NANOTECHNOLOGY

It is well known that the respiratory problems generally occur in human beings due to poor air quality as the toxic and harmful gases present in the air has deleterious effects on human health. Detection of air pollutants and their remediation using an integrated approach of modern day engineering techniques coupled with nanotechnological approaches can be done to achieve the target of a clean sustainable environment.

9.3.1 Nanosensors-Based Air Pollution Detection Approaches

Nanosensors based on nanotechnological advancements are more advantageous in their pollution sensing and remediation abilities owing to their greater efficiency than conventional methods in terms of power consumption, greater sensitivity ranges, low sample requirements of analysts, and lower interaction distance unique properties of nanomaterials that also differentiate them from current materials used for sensing air pollutants, these properties can be tailored to improve the performance of nanomaterials in terms of reactivity and electrical conductivity. Since the sensing ability of these sensors falls into the nano range to sub-picomolar level these can be applied for the estimation of toxic nano-range particles in the different environmental components. The structural components of nanosensors comprise (1) Nanomaterial, (2) a Detector element with specificity (3) a Signal transducer that offers a means of reliance on analyte existence. Nanosensors are nowadays gaining popularity due to their use to detect and monitor heavy metals, hydrocarbons, and other toxic contaminants present in the air [48]. Some of the most frequently used nanomaterials used for sensor applications include graphene, CNT, metal, and metal-oxide-based nanoparticles. Recently researchers have developed gas-sensing nanomaterials for the detection of air pollutants like hydrogen sulphide, sulphur oxide, and nitrogen oxides, sulphur oxide gases. Some of the air pollutants and application of nanosensors are discussed below:

9.3.1.1 Detection of NO_2

9.3.1.1.1 Use of Graphene

Gas sensors have important applications in pollution monitoring and mitigation one of the commonest and widely used nanomaterials used for the sensing and detection of toxic gases is graphene and its chemically designed composites (Graphene Oxide (GO), and reduced Graphene Oxides (rGO).The unique characteristics of graphene such as its high thermal stability, good electrical conductivity, low electrical noise and excellent mechanical properties make it an efficient material that can be used for the detection and removal of pollutants present in the air. Several researchers have used Graphene Oxide (GO), and reduced graphene oxides (rGO) for NO_2 sensing and detection [49, 50]. Novikov et al. 2016 fabricated an ultrasensitive gas sensor based on epitaxial graphene. The sensor showed good performance for measurements of NO_2 concentration in the range of 5 ppb–50 ppb for environmental pollution and was able to reduce it to 1 ppb [51]. In another study, NO_2 detection using epitaxial graphene sensors was observed in a complex environment and it was seen that single-layer graphene is highly effective in terms of carrier-concentration response upon NO_2 exposure as compared to bilayer graphene [52]. Doped reduced graphene oxide (rGO) showed high sensitivity and selectivity towards NO_2. Moreover, the N-rGO sensor showed highest efficiency, sensitivity, and low limit of detection of NO_2 among doped N-rG,O B-rGO and NB-rGO sensors [53]. Hussain et al. 2024 studied zigzag graphene nanoribbon (ZGNR)-based sensor device for detection of the NO_x family (NO, NO_2, and N_2O) at low concentrations and observed good performance, high selectivity, and better stability [54].

9.3.1.1.2 Use of Metal-Oxide-Based Nanostructures

Fabrication of Semiconducting Metal Oxides (SMOs) at the nanoscale provides a unique class of sensor technology wherein metallic species are functionalised with different dopants to modify their crystalline scaffolding together with chemical resistance properties [55]. For increasing the sensing performance, several nanostructured forms such as nanosheets, nanocomposites, nanoflowers, nanotubes, nanoparticles, and nanowires, are used. The metal oxide nanomaterial sensors are advantageous due to ease of handling, portability, and detection ability of environmental gas pollutants at low temperatures. Ding et al. prepared the NiO-SnO_2 nanostructure using the solvothermal then the nanomaterials were calcined using Ni/Sn bimetallic organic frameworks (Ni/Sn-BTC MOFs, BTC: benzene-1,3,5-tricarboxylic) as sacrificial templates [56]. NiO-SnO_2 has high porosity and a large specific surface area, increased active sites which enhanced the performance to capture NO_2.

To effectively detect NO_2 at room temperature Liang et al. used the hydrothermal method and synthesised MoS_2/WS_2 composites by a facile one-pot. The MoS_2/WS_2 heterostructure composite showed high performance and good sensitivity to NO_2 at room temperature. Zheng et al. also used hydrothermal to synthesise Black Phosphorus–Tungsten Oxide sandwich-like nanostructures and observed good selectivity towards NO_2 sensing and it was found that fabricated nanostructures were largely insensitive to interfering gases, especially H_2S and had excellent resolving power and repeatability [57].

9.3.1.2 SO_2 Detection

9.3.1.2.1 Carbon Nanotubes (CNTs)

Recently, carbon nanotubes (CNTs) have garnered significant attention as nanoadsorbents for various metal ions, benefitting from their exceptional physical properties including pore volume, diameter, and surface area. Based on these characteristics of carbon nanotube, they can be effectively used for the removal of air pollutants like carbon dioxide, methane and many other greenhouse gases emitted from vehicles, industries, coal mines and powerplants [58]. Furthermore, the adsorption capacity of CNTs can also be enhanced through functionalisation. The functionalisation with a specific group provides more active sites to CNTs results in increment of the binding sites of CNTs [59]. The enhancement of adsorption of CO_2 was obtained through functionalisation of CNTs with (3-aminopropyl triethoxysilane (APTES). Similarly, favourable adsorption of CO_2 was achieved when multi-walled carbon nanotube was modified using APTS [59]. A lot of research has been done for the detection and mitigation of SO_2 and H_2S using CNTs [60–62]. Kuganathan et al. confirmed the detection and mitigation of SO_2 and H_2S using Ru-doped single-walled carbon nanotubes [63].

9.3.1.2.2 SO_2 Detection by Nanostructure of Metal and Metal-Oxide-Based

The gas sensors made up of metal and metal oxide nanomaterials are gaining a lot of attention in the last few years due to their ease of preparation, minimal cost, high specific surface area, more active sites, high response against air pollutants and less recovery time. Since 1960, different p-type and n-type metal oxides like SnO_2, WO_3, Nb_2O_5, ZnO etc. have been synthesised successfully for their effective use as a sensor for the detection of toxic gases like H_2, H_2S, CO, CH_4, and other volatile organic compounds. The semiconducting metal oxide was used at nanoscale for the detection of SO_2 [52, 64]. Recently, SnSe/SnO_2 nanoparticles (NPs) were used as gas-sensing materials by Pan et al. [65]. For the detection of SO_2, pristine and Al-doped ZnO were used at a low concentration by Chaitra et al. [66]. Cu-doped flower like SnO_2 architecture were fabricated by Li et al. (2023) and used as sensing materials for sensing SO_2 [67].

9.3.1.3 H_2S Detection

9.3.1.3.1 Carbon-Based Nano-Structures

From the past few decades, several researchers have used carbon-based nanostructured materials like graphene oxide, carbon nanotube for the sensing of H_2S. H_2S is known for its toxic nature if a human being is exposed to H_2S for a longer time, it may cause nausea, headache, coma, and death. The gas penetrates into blood streams after be present in the alveolar membrane resulting in death due to the failure of respiratory system. The efficiency of CuO-based single walled carbon nanotube has been compared for the detection of H_2S gas in the range of 100 ppb–50 ppm and it was found that CuO- based single walled carbon nanotube showed higher efficiency (70%) toward H_2S gas at 150 ppm concentration while only single walled carbon nanotube showed only 50% response toward H_2S at 50 ppm concentration. Cu_2O/CuO-based multi walled carbon nanotube was fabricated using the sputtering technique and it was confirmed that the sensor response time and recovery time was tremendous as compared to multi walled carbon nanotube [68]. The efficiency of Cu-based multi walled carbon nanotube was investigated by Musaveya et al. who developed sensing of H_2S gas and it was confirmed that the sensitivity and selectivity was optimum at room temperature [69].

9.3.1.3.2 H_2S Detection by Nanomaterials Based on Metal and Metal-Oxide

In a recent study, CuO/ZnO multifaceted nanostructures (NS) prepared by Cu^{2+}/ZIF-8 impregnation precursors using the heat treating method, and their microstructure was examined for their gas-sensing ability. It was observed that hollow CuO/ZnO multifunctional nanostructures had exceptionally high gas response values with quick response and recovery time and low detection limit for H_2S [70]. The detection of H_2S at 100ppm concentration was investigated by Shi et al. 2022 using porous multifaceted CO_3O_4/ZnO nanomaterials synthesised through the pyrolysis process. Hassanzadeh-Tabrizi et al. studied H_2S gas sensing using iron oxide nano powder and found high H_2S detection of 100 ppb [71].

9.4 FACTORS AFFECTING THE REMEDIATION OF AIR

Nanotechnology is playing a large role to address the challenges of air pollution due to their significant characteristics like high specific surface area, high reactivity, and enhanced functionality. However, there are several factors like surface area, size, shape, functionalisation of surface etc. which effect the efficiency of nanomaterials due to alteration of their chemical composition. Therefore, it is very necessary to understand these factors for the optimisation of synthesis and the implementation process of nanomaterial-based air remediation technologies.

9.4.1 Surface Area and its Impact on Air Remediation with Nanomaterials

Surface area of the nanomaterials has a great impact on the efficiency of nanomaterials when used in air remediation processes. Due to their high surface area-to-volume ratios at nanoscale dimensions, the nanomaterials have a large number of active sites which make them an effective adsorbent for the adsorption of pollutants. Various studies have confirmed that nanomaterials having high specific surface area showed excellent efficiency toward the air pollutants remediation when compared to their bulk counterparts or larger sized particles [72, 73]. The large number of active sites generated due to high surface area accelerate the degradation of the air pollutant through the catalytic reaction [74]. Furthermore, the high surface area enhances the rate of mass transfer and diffusion of pollutants through the porous structure of nanomaterials as hierarchical pore networks of nanostructured materials providing them a large surface area which allows rapid diffusion and transportation of pollutants toward the binding sites of nanostructured materials results in ore adsorption of air pollutants [75].

9.4.2 Chemical Composition

The chemical composition of nanomaterials generally decides their role of interaction for a specific pollutant and their catalytic activity. There are different nanomaterials which shows their interaction towards specific pollutants due to their unique physicochemical properties and specific chemical composition. The functional groups present on the surface and their chemical structure generally decides the interaction towards a specific pollutant. Khan et al. (2019) confirmed the degradation of organic pollutants using titanium dioxide (TiO_2) and zinc oxide ZnO nanoparticles which are known for their photocatalytic activity [76]. TiO_2 and ZnO nanoparticles initiate the degradation process of pollutants in the presence of light through oxidation process [77]. On the other hand, activated carbon and zeolites are known for adsorption of volatile organic compounds [78]. The nanomaterials specifically metal-based nanoparticles like palladium (Pd), platinum (Pt), cobalt (Co) and iron (Fe) are better known for their catalytic activity which convert the toxic and harmful gases like carbon monoxide (CO) and nitrogen oxides (NOx) into less harmful products. Based on the chemical composition, the stability and durability of nanomaterials can also be decided. The nanomaterial with robust chemical structure is found to be stable and resist against high temperature, humidity by ensuring their better performance for air pollutants. The performance of nanomaterials can also be enhanced by surface modification or the doping method [72].

9.4.3 Size and Shape

The size and shape of nanomaterials is one most of the most important parameters. The surface area, reactivity, mass transfer rate, diffusion rate and transport properties depend on the size and shape of nanomaterials which directly affect their performance in air remediation. The smaller size of nanoparticles exhibits higher surface area-to-volume ratios results in formation of more active site for the interaction of pollutants and catalytic reactions. Furthermore, the shape of nanoparticles generally decides the rate of dispersal and transport of the pollutant for interaction with them. The nanoparticles with certain uniform shapes like spherical or needle shape enhanced their interaction rate [79–81]. The better dispersion of nanoparticles in air allowed them a better coverage of pollutants due to better interaction. Furthermore, the aerodynamic characteristics of nanoparticles can also be considered as an important parameter to design air remediation process as these characteristics are responsible for the deposition pattern and residence time of pollutants in air. By altering the shape and size of nanoparticles, the photocatalytic performance can be enhanced for specific air pollutant. The nanoparticles with specific size and shape generate Reactive Oxygen Species (ROS) due to adsorption of light in specific region results in enhanced degradation of pollutants [82].

9.4.4 Surface Functionalisation and its Impact on Air Remediation with Nanomaterials

The surface functionalisation of nanomaterials through different methods shows a significant impact in air remediation processes. The specific functional group present on the surface of nanomaterials enhance their selectivity, stability, reactivity, and recyclability towards a specific pollutant. The functional groups like amino, carboxyl, or hydroxyl also facilitate the compatibility of nanomaterials with the pollutant through chemical interaction like hydrogen bonding or π-π interactions during the remediation process. Graphene or carbon-based nanostructured materials showed their affinity towards volatile organic compounds (VOCs) due to the presence of oxygen containing a functional group on their surface results in π-π interactions or hydrogen bonding with the air pollutant. Similarly, the surface of titanium oxide (TiO_2) modified with nitrogen containing the functional group, showed higher affinity toward nitrogen oxide (NO_x) through Lewis's acid-base interactions. The surface functionalisation also enhances stability of nanomaterials by preventing them from aggregation, oxidation or leaching results in better performance in air remediation process. The hydrophobic or hydrophilic group present on the surface of nanomaterials make them preferable for remediation of humidity by enhancing their dispersibility. When magnetic nanoparticles were modified with organosilane, the stability and recyclability were enhanced [83]. By modifying the surface chemistry of nanomaterials, the catalytic performances can also be modified for a specific air pollutant as the functional group or metal act as a catalyst and immobilised various active sites for the oxidation, reduction, or transformation of pollutants. The photocatalytic characteristics of nanomaterials provides a sustainable process for the remediation of air pollutants. When metal oxide nanoparticles decorated with some noble metals like platinum or palladium, their photocatalytic performances enhanced by imitating the chemical reaction by utilising the light energy where the harmful organic pollutants degraded into less harmful products and contribute to a clean and healthier environment. The photocatalytic properties of carbon based nanomaterials can be tailored using metal/metal oxide nanoparticles for the removal of NOx and Sox from the atmosphere [84]. The generation of electron-hole pairs in nanomaterials in the presence of light initiate the redox reaction which further enhances the adsorption of pollutants [85, 86]. Various nanomaterials like titanium dioxide (TiO_2), zinc oxide (ZnO), and tungsten trioxide (WO_3) can be used as photocatalysts for removal of toxic gases, particulate matter, and volatile organic substances from the air by generating ROS which oxidise organic pollutants leading to their degradation [87, 88]. There are several other factors like surface area, band gap energy, crystallinity and defects which directly or indirectly impact the photocatalytic performances of nanomaterials. Nitrogen doped TiO_2 nanoparticles having reduced gap energy can be used for the photocatalytic degradation of volatile organic matter [89]. The doping of metal ions, surface functionalisation or immobilisation of photocatalysts onto substrates are also helpful for improving the recyclability and remediation processes [72, 90].

9.4.5 Environmental Conditions

Environmental conditions like temperature, humidity, ambient light, and air flow rate can affect the efficiency of nanomaterials in air remediation processes by impacting their adsorption, catalytic activity, and stability of nanomaterials [72, 91]. Higher temperature accelerates the volatile organic compounds desorption form activated carbon nanomaterials results in reduction of adsorption efficiency of nanomaterials. The rate of chemical reaction and desorption of pollutants is found to be higher at high temperatures [92]. Komaraiah et al. investigated the role of temperature on photocatalytic degradation of NOx where they confirmed better photocatalytic degradation of NOx using TiO_2 nanoparticles due to their increased rate of reaction [93]. However, excess high temperatures can also reduce the stability and performances of nanomaterials responsible by their thermal degradation. High humidity could promote the condensation of water vapour on nanomaterial surfaces, affecting their adsorption capacity and catalytic activity. Additionally, moisture can lead to the formation of surface hydroxyl groups on nanomaterials, which may enhance their interaction with certain pollutants or affect their surface properties. Liu et al. confirmed the reduction of adsorption capacity of zeolite nanoparticles for volatile organic compounds (VOCs) due to increased humidity levels [94]. Sun and his co-workers reported the enhancement of removal of pollutants such as Nitrogen Oxides (NOx) using metal oxide nanoparticles when humidity level was high as high humidity levels promoting the formation of hydroxyl radicals (•OH) results in the enhancement of pollutant removal [95].

The flow rate of air passing through air remediation systems can affect the contact time between nanomaterials and pollutants, as well as the distribution of pollutants within the system. The rate of mass transfer of pollutants toward the surface of nanomaterials improves the removal efficiency of nanomaterials when air flow rate is high. Under excess high low rate, the contact time of nanomaterials with air pollutants decreases thereby reducing their efficiencies [96]. Activated carbon nanomaterials adsorb volatile organic matter effectively under high air flow rate [97]. The photocatalytic activities of nanomaterials can also be impacted under ambient light conditions including UV and visible light [98].

9.5 NEGATIVE IMPACT OF NANOMATERIALS USED FOR AIR POLLUTANT REMEDIATION ON ENVIRONMENT

While nanomaterials have been used on a large scale for removing toxic and hazardous gases and organic pollutants from the atmosphere, they have a negative impact on the environment. It is still unknown about the impact of engineered nanomaterials (NMs), especially once they are released into the air. At each stage of their life cycle—from production and processing to transportation, handling, application, and eventual disposal—nanomaterials could potentially be emitted into the air, but the specific impacts remain uncertain [99]. Nanoparticles found in airborne pollution may play a significant role in human health effects [100]. It is worth noting that in atmospheric and combustion emission research, particles smaller than 100 nm are often termed ultrafine particles. These nanomaterials can directly contact with human organs due to their nanoscale dimension and biochemical reactivity. Despite contributing minimally to the mass concentrations of particulate matter, nanoparticles make up most airborne particle numbers (currently unregulated) [101]. The adverse effect of metal-based nanoparticles and carbon-based nanomaterials will be discussed in the following section:

9.5.1 Ecotoxicity

Nanomaterials once released into the atmosphere leads to ecotoxic for human beings, aquatic life, and terrestrial ecosystems. Metal oxide nanoparticles have been used on a large scale for photocatalytic degradation of hazardous chemical compounds to less harmful compounds under UV/ visible light, some secondary pollutants may also be generated during the degradation reaction. For example, when benzene or toluene undergoes photodegradation, ozone (O_3) may also be generated as a secondary pollutant which has an adverse effect on the health of human beings and are also responsible for smog formation [102]. Certain studies have confirmed the toxic nature of metal oxide nanoparticles for the aquatic environment when present in high concentrations [103]. Several studies have been conducted to investigate the effect of nanomaterials on the fertility of soil, nutrient cycle and functioning of soil ecosystem. ZnO and Fe_2O_3 nanoparticles when present in soil reducing the enzymatic activities in the soil results in reduction of soil fertility. Additionally, these nanoparticles have been consumed by plants which may show the risk to human health [104]. Nanomaterials also affect the life of human beings through bioaccumulation and biomagnification. In bioaccumulation, the nanoparticles accumulate in human organs through inhalation or ingestion. Once accumulated, these nanoparticles can be transferred to other organisms through the food chain leading to potential toxic effects. The nanomaterials present in the atmosphere interact with plants affecting their growth and physiology. While some nanomaterials are beneficial for plants, help them to grow, others may have phytotoxic effect inhibiting the growth of development of the plant. Therefore, it is very necessary to understand the mechanism of interaction between plants and nanomaterials to assess their ecotoxicity [105, 106].

9.5.2 Air Quality Degradation

Hence, it is true that the nanostructured based materials have been designed for the degradation of toxic and harmful air pollutants through various processes, like photocatalysis itself produces various harmful byproducts such as ozone (O_3) and volatile organic compounds (VOCs) results in degradation of air quality thereby posing the risk on human health and the environment. When metal oxide nanoparticles are present in the atmosphere for a longer time, the particulate matter and smog are generally generated causing various respiratory and cardiovascular problems in human beings [107]. Some synthesis processes like combustion, chemical precipitation used for the synthesis of nanomaterials are mainly responsible for the emission of particulate matter, volatile organic compounds, and many more pollutants, contributing to the increment of air pollution and compromising with air quality. Spraying and grinding of nanomaterials can also produce dust and aerosol which disperse into the atmosphere and affect the air quality [108].

9.5.3 Potential Health Risks

Nanomaterials specifically metal oxide nanoparticles once released into the air during air remediation activities can pose health risks to workers who are involved in the installation, operation, and maintenance of air purification systems. The metal oxide nanoparticles are generally penetrated into the respiratory system leading to inflammation, oxidative stress, and pulmonary disorders. Furthermore, nanoparticles are also present in houses which may further foster the indoor air quality particularly in poorly ventilated or confined spaces. Therefore, it is very necessary to mitigate all the measures which are essential to reduce the health hazard. Toxicity of Nanoparticles: Some nanomaterials used for air remediation, such as metal oxides (e.g., titanium dioxide, zinc oxide) or carbon-based nanomaterials (e.g., carbon nanotubes, graphene), have been associated with adverse health effects due to their inherent toxicity. Studies have shown that certain nanoparticles can induce oxidative stress, inflammation, genotoxicity, and cytotoxicity in exposed cells and tissues. Prolonged or high-level exposure to toxic nanoparticles may increase the risk of developing respiratory diseases, cardiovascular disorders, or even cancer. Nanomaterials used for air remediation, particularly those employed in photocatalytic processes, may generate secondary pollutants as by products of chemical reactions. For example, the degradation of air pollutants by photocatalytic nanoparticles (e.g., titanium dioxide) can produce reactive oxygen species (ROS) and ozone (O_3), both of which can have harmful effects on respiratory health. If a human being is in contact with ozone for a longer duration, respiratory problems may occur by increasing the cardiovascular diseases. Children, the elderly, and individuals with compromised immune systems or respiratory conditions are more vulnerable to the adverse effects of nanoparticle exposure. Additionally, long-term exposure to nanoparticles or cumulative exposure to multiple environmental stressors may exacerbate health risks over time [109, 110].

9.6 CONCLUSION

In this chapter, the potential use of nanomaterials for removal of air pollutants has been described. Nanomaterials offer high surface area, enhanced reactivity, and tailored functionality, making them effective material for air remediation applications. The most popular NMs used in various air pollution remediation methods for are nanoadsorbents, zinc oxide, nano-catalysts, titanium dioxide, nano-filters, metal-organic frameworks etc. The efficiency of nanomaterial-based air remediation processes is influenced by various factors, such as surface area, catalytic activity, in the case on nanocatalysts, mass transfer and diffusion, chemical composition, selectivity towards pollutants, stability and durability, light absorption in the case of photocatalysis, environmental conditions like temperature, humidity, ambient light etc. The main concern of NMs is itself as an environmental pollutant due to their small size and high stability and durability which can persist in the environment for a long time and cause eco-toxicity in aquatic and terrestrial ecosystems, air quality degradation due to deposition of nanoparticles and a generation of secondary pollutants and other potential health risks related to respiratory diseases in the lungs.

REFERENCES

[1] World Health Organization. 2021. WHO global air quality guidelines: Particulate matter ($PM_{2.5}$ and PM_{10}), ozone, nitrogen dioxide, sulfur dioxide and carbon monoxide. World Health Organization.

[2] World Health Organization. 2022. Billions of people still breathe unhealthy air: New WHO data. World Health Organization. 4.

[3] Pandey, A., Brauer, M., Cropper, M.L., Balakrishnan, K., Mathur, P., Dey, S., et al. 2021. Health and economic impact of air pollution in the states of India: The Global Burden of Disease Study 2019. The Lancet Planetary Health. 5(1): e25–e38.

[4] Orellano, P., Quaranta, N., Reynoso, J., Balbi, B. and Vasquez, J. 2020. Short-term exposure to particulate matter (PM_{10} and $PM_{2.5}$), nitrogen dioxide (NO_2), and ozone (O_3) and all-cause and cause-specific mortality: Systematic review and meta-analysis. Environment International. 142. 105876. https://doi.org/10.1016/j.envint.2020.105876

[5] Liu, Q., Xiong, B., Shi, J., Tao, M., He, Y. and Shi, Y. 2014. Enhanced tolerance to flue gas contaminants on carbon dioxide capture using amine-functionalized multiwalled carbon nanotubes. Energy and Fuels. 28(10): 6494–6501. https://doi.org/10.1021/ef501439z

[6] Bradu, P., Biswas, A., Nair, C., Sreevalsakumar, S., Patil, M., Kannampuzha, S., et al. 2023. Recent advances in green technology and Industrial Revolution 4.0 for a sustainable future. Environmental Science and Pollution Research. 30: 124488–124519. https://doi.org/10.1007/s11356-022-20024-4

[7] Ganie, A.S., Bano, S., Khan, N., Sultana, S., Rehman, Z., Rahman, M.M., et al. 2021. Nanoremediation technologies for sustainable remediation of contaminated environments: Recent advances and challenges. Chemosphere. 275: 130065. https://doi.org/10.1016/j.chemosphere.2021.130065

[8] Samuel, M.S., Selvarajan, E., Chidambaram, R., Patel, H. and Brindhadevi, K. 2021. Clean approach for chromium removal in aqueous environments and role of nanomaterials in bioremediation: Present research and future perspective. Chemosphere. 284: 131368.

[9] Yadav, N., Garg, V.K., Chhillar, A.K. and Rana, J.S. 2021. Detection and remediation of pollutants to maintain ecosustainability employing nanotechnology: A review. Chemosphere. 280: 130792.

[10] Sushma and Yadav, A. 2020. Nanomaterials for air remediation. *In*: Biological and Physicochemical Combination Processes. 361–372. Elsevier. ISBN: 9780128188217

[11] Saravanan, A., Kumar, P.S., Jeevanantham, S., Karishma, S., Yaashikaa, P.R. and Narayanan, M. 2021. A comprehensive review on sources, analysis and toxicity of environmental pollutants and its removal methods from water environment. Environmental Pollution. 268: 115937. https://doi.org/10.1016/j.envpol.2020.115937

[12] Thakur, S., Sharma, B., Verma, A., Chaudhary, J., Tamulevicius, S. and Thakur, V.K. 2022. Recent developments in recycling of polymeric waste for environmental sustainability. Bioresource Technology. 351: 126956. https://doi.org/10.1016/j.biortech.2022.126956

[13] Alebachew, N., Ananda Murthy, H.C., Abdisa, B. and Demissie, T.B. 2023. Important features of nanomaterials for environmental remediation. *In*: Green Nanoremediation: Sustainable Management of Environmental Pollution. 1–23. Springer International Publishing.

[14] Sheoran, A.R., Lakra, N., Luhach, A., Saharan, B.S., Debnath, N. and Sharma, P. 2024. Nano-engineered solutions for sustainable environmental cleanup. BioNanoScience. 1–25. https://doi.org/10.1007/s12668-024-01135-z

[15] Nasrollahzadeh, M., Issaabadi, Z., Sajjadi, M., Sajadi, S.M. and Atarod, M. 2019. Types of nanostructures. *In*: Interface Science and Technology. 28: 29–80. Elsevier. https://doi.org/10.1016/B978-0-12-813586-0.00002-1

[16] Gupta, D., Thakur, A. and Gupta, T.K. 2023. Green and sustainable synthesis of nanomaterials: Recent advancements and limitations. Environmental Research. 116316. https://doi.org/10.1016/j.envres.2023.116316

[17] Rumyantseva, M., Sun, S., Kuznetsov, A. and Prakash, J. 2023. Emerging nanomaterials in the detection and degradation of air pollutants. Current Opinion in Environmental Science and Health. 35: 100497.

[18] Asghar, N., Hussain, A. and Nguyen, D.A. 2024. Advancement in nanomaterials for environmental pollutants remediation: A systematic review on bibliometrics analysis, material types, synthesis pathways, and related mechanisms. Journal of Nanobiotechnology. 22(26). https://doi.org/10.1186/s12951-023-02151-3

[19] Mohamed, E.F. and Awad, G. 2021. Nanostructured ceramics for air pollution control: Removal of gaseous pollutants and pathogenic organisms. *In*: Advanced Ceramics for Energy and Environmental Applications. 299–312. CRC Press. https://doi.org/10.1201/9781003005155-14

[20] Saleem, H., Zaidi, S.J., Ismail, A.F. and Goh, P.S. 2022. Advances of nanomaterials for air pollution remediation and their impacts on the environment. Chemosphere. 287: 132083. https://doi.org/10.1016/j.chemosphere.2021.132083

[21] Thiruvengadam, M., Rajakumar, G., Swetha, V., Ansari, M.A., Alghamdi, S., Almehmadi, M., et al. 2021. Recent insights and multifactorial applications of carbon nanotubes. Micromachines. 12(12): 1502. https://doi.org/10.3390/mi12121502

[22] Yu, G., Wang, X., Liu, J., Jiang, P., You, S., Ding, N., et al. 2021. Applications of nanomaterials for heavy metal removal from water and soil: A review. Sustainability. 13(2): 713.

[23] Gulati, S., Kumar, S., Goyal, K., Arora, A. and Varma, R.S. 2022. Improving the air quality with functionalized carbon nanotubes: Sensing and remediation applications in the real world. Chemosphere. 299: 134468. https://doi.org/10.1016/j.chemosphere.2022.134468

[24] Attia, Y.A., Ezet, A.E., Saeed, S. and Galmed, A.H. 2024. Nano carbon-modified air purification filters for removal and detection of particulate matters from ambient air. Scientific Reports. 14(1): 621. https://doi.org/10.1038/s41598-023-50902-x

[25] Ameta, R., Solanki, M.S., Benjamin, S. and Ameta, S.C. 2018. Photocatalysis. *In*: Advanced Oxidation Processes for Waste Water Treatment. 135–175. https://doi.org/10.1016/b978-0-12-810499-6.00006-1

[26] Modi, S., Yadav, V.K., Gacem, A., Ali, I.H., Dave, D., Khan, S.H., et al. 2022. Recent and emerging trends in remediation of methylene blue dye from wastewater by using zinc oxide nanoparticles. Water. 14(11): 1749. https://doi.org/10.3390/w14111749

[27] Huy, N.N., Thuy, V.T.T., Thang, N.H., Thuy, N.T., Quynh, L.T., Khoi, T.T., et al. 2019. Facile one-step synthesis of zinc oxide nanoparticles by ultrasonic-assisted precipitation method and its application for H_2S adsorption in air. Journal of Physics and Chemistry of Solids. 132: 99–103. https://doi.org/10.1016/j.jpcs.2019.04.018

[28] Ghosh, S., Ranjan, P., Ramaprabhu, S. and Sarathi, R. 2018. Carbon dioxide adsorption of zinc oxide nanoparticles synthesized by wire explosion technique. INAE Letters. 3(4): 197–202. https://doi.org/10.1007/s41403-018-0049-9

[29] Le Pivert, M., Oliver, K., Zerelli, B. and Leprince-Wang, Y. 2021. ZnO nanostructures-based innovative photocatalytic road for air purification. Journal of Cleaner Production. 318: 128447. https://doi.org/10.1016/j.jclepro.2021.128447

[30] Theerthagiri, J., Duraimurugan, K., Kim, H. and Madhavan, J. 2019. Graphitic carbon nitride-based nanostructured materials for photocatalytic applications. *In*: Photocatalytic Functional Materials for Environmental Remediation. 291–307. https://doi.org/10.1002/9781119529941.ch10

[31] Salabat, A. and Keshavarz, A. 2018. Effect of HCl on the structure and catalytic activity of Pt/Al_2O_3 nanocatalyst prepared in microemulsion system. Scientia Iranica. 0(0): 0–0. https://doi.org/10.24200/sci.2018.5287.1184

[32] Rajendran, R., Vignesh, S., Suganthi, S., Raj, V., Kavitha, G., Palanivel, B., et al. 2022. g-C_3N_4/TiO_2/CuO S-scheme heterostructure photocatalysts for enhancing organic pollutant degradation. Journal of Physics and Chemistry of Solids. 161: 110391.

[33] Binas, V., Venieri, D., Kotzias, D. and Kiriakidis, G. 2017. Modified TiO_2 based photocatalysts for improved air and health quality. Journal of Materiomics. 3(1): 3–16. https://doi.org/10.1016/j.jmat.2016.11.002

[34] Giannakopoulou, T., Papailias, I., Todorova, N., Boukos, N., Liu, Y., Yu, J., et al. 2017. Tailoring the energy band gap and edges' potentials of g-C_3N_4/TiO_2 composite photocatalysts for NO_x removal. Chemical Engineering Journal. 310. 571–580. https://doi.org/10.1016/j.cej.2015.12.102

[35] Mamaghani, A.H., Haghighat, F. and Lee, C. 2018. Gas phase adsorption of volatile organic compounds onto titanium dioxide photocatalysts. Chemical Engineering Journal. 337: 60–73. https://doi.org/10.1016/j.cej.2017.12.082

[36] Wu, H., Ma, J., Li, Y., Zhang, C. and He, H. 2014. Photocatalytic oxidation of gaseous ammonia over fluorinated TiO_2 with exposed (001) facets. Applied Catalysis B: Environmental. 152–153, 82–87. https://doi.org/10.1016/j.apcatb.2014.01.021

[37] Portela, R., Jansson, I., Suárez, S., Villarroel, M., Sánchez, B. and Avila, P. 2017. Natural silicate–TiO_2 hybrids for photocatalytic oxidation of formaldehyde in gas phase. Chemical Engineering Journal. 310: 560–570. https://doi.org/10.1016/j.cej.2016.06.018

[38] Jampaiah, D., Chalkidis, A., Sabri, Y.M., Mayes, E.L., Reddy, B.M. and Bhargava, S.K. 2019. Low-temperature elemental mercury removal over TiO_2 nanorods-supported MnO_x–FeO_x–CrO_x. Catalysis Today. 324: 174–182. https://doi.org/10.1016/j.cattod.2018.11.049

[39] Yu, L., Wang, L., Sun, X. and Ye, D. 2018. Enhanced photocatalytic activity of rGO/TiO_2 for the decomposition of formaldehyde under visible light irradiation. Journal of Environmental Sciences. 73: 138–146. https://doi.org/10.1016/j.jes.2018.01.022

[40] Moroz, B., Pyrjaev, P., Zaikovskii, V. and Bukhtiyarov, V. 2009. Nanodispersed Au/Al_2O_3 catalysts for low-temperature CO oxidation: Results of research activity at the Boreskov Institute of Catalysis. Catalysis Today. 144(3–4): 292–305. https://doi.org/10.1016/j.cattod.2008.10.038

[41] Zhao, W., Tulugan, K., Zhang, X., Li, X. and Tian, P. 2024. Electrospun 3D curly electret nanofiber air filters for particulate pollutants. Sustainability. 16(7): 2808. https://doi.org/10.3390/su16072808

[42] Yunus, I.S., Harwin, Kurniawan, A., Adityawarman, D. and Indarto, A. 2012. Nanotechnologies in water and air pollution treatment. Environmental Technology Reviews. 1(1): 136–148. https://doi.org/10.1080/21622515.2012.733966

[43] Latif, S., Imran, M., Ahmad, M.N., Siddique, M.H. and Boczkaj, G. 2024. Metal-organic frameworks as versatile materials for air purification. *In*: Recent Advancements in Wastewater Management: Nano-based Remediation. 173–190. https://doi.org/10.1016/bs.apmp.2023.08.002

[44] Bian, Y., Wang, R., Wang, S., Yao, C., Ren, W., Chen, C., et al. 2018. Metal–organic framework-based nanofiber filters for effective indoor air quality control. Journal of Materials Chemistry A. 6(32): 15807–15814. https://doi.org/10.1039/c8ta04539a

[45] Zhang, Y., Yuan, S., Feng, X., Li, H., Zhou, J. and Wang, B. 2016. Preparation of nanofibrous metal–organic framework filters for efficient air pollution control. Journal of the American Chemical Society. 138(18): 5785–5788. https://doi.org/10.1021/jacs.6b02553

[46] Romero-Guzmán, L., Reyes-Gutiérrez, L.R., Romero-Guzmán, E.T. and Savedra-Labastida, E. 2018. Carbon nanotube filters for removal of air pollutants from mobile sources. Journal of Minerals and Materials Characterization and Engineering. 6(1): 105–118. https://doi.org/10.4236/jmmce.2018.61009

[47] Li, P., Wang, C., Zhang, Y. and Wei, F. 2014. Air filtration in the free molecular flow regime: A review of high-efficiency particulate air filters based on carbon nanotubes. Small. 10(22): 4543–4561. https://doi.org/10.1002/smll.201400191

[48] Javaid, M., Haleem, A., Singh, R.P., Rab, S. and Suman, R. 2021. Exploring the potential of nanosensors: A brief overview. Sensors International. 2: 100130. https://doi.org/10.1016/j.sintl.2021.100130

[49] Dunst, K.J., Scheibe, B., Nowaczyk, G., Jurga, S. and Jasiński, P. 2017. Graphene oxide, reduced graphene oxide and composite thin films NO_2 sensing properties.

Measurement Science and Technology. 28(5): 054005. https://doi.org/10.1088/1361-6501/aa62e7

[50] Novikov, S., Lebedeva, N., Satrapinski, A., Walden, J., Davydov, V. and Lebedev, A. 2016. Graphene based sensor for environmental monitoring of NO_2. Sensors and Actuators B: Chemical. 236: 1054–1060. https://doi.org/10.1016/j.snb.2016.06.066

[51] Melios, C., Panchal, V., Edmonds, K., Lartsev, A., Yakimova, R. and Kazakova, O. 2018. Detection of ultralow concentration NO_2 in complex environment using epitaxial graphene sensors. ACS Sensors. 3(9): 1666–1674.

[52] Kanan, S., Obeideen, K., Moyet, M., Abed, H., Khan, D., Shabnam, A., et al. 2024. Recent advances on metal oxide-based sensors for environmental gas pollutants detection. Critical Reviews in Analytical Chemistry. 1–34. https://doi.org/10.1080/10408347.2024.2300246

[53] Ding, Y., Du, B., Guo, X., Dong, Y., Zhang, M., Jin, W., et al. 2024. An ultrasensitive NO_2 gas sensor based on a NiO-SnO_2 composite with a sub-ppb detection limit at room temperature. Sensors and Actuators B: Chemical, 135916. https://doi.org/10.1016/j.snb.2024.135916

[54] Liang, Z., Wang, M., Liu, S., Hassan, M., Zhang, X., Lei, S., et al. 2024. One-pot hydrothermal synthesis of self-assembled MoS_2/WS_2 nanoflowers for chemiresistive room-temperature NO_2 sensors. Sensors and Actuators B: Chemical. 403: 135215. https://doi.org/10.1016/j.snb.2024.135215

[55] Zheng, C., Shi, Y., Tang, B. and Zhang, J. 2024. Black phosphorus–tungsten oxide sandwich-like nanostructures for highly selective NO_2 detection. Sensors. 24(5): 1376. https://doi.org/10.3390/s24051376

[56] Chu, J., Wang, X., Wang, D., Yang, A., Lv, P., Wu, Y., et al. 2018. Highly selective detection of sulfur hexafluoride decomposition components H_2S and SOF_2 employing sensors based on tin oxide modified reduced graphene oxide. Carbon. 135: 95–103. https://doi.org/10.1016/j.carbon.2018.04.037

[57] Gui, M.M., Yap, Y.X., Chai, S. and Mohamed, A.R. 2013. Multi-walled carbon nanotubes modified with (3-aminopropyl)triethoxysilane for effective carbon dioxide adsorption. International Journal of Greenhouse Gas Control. 14: 65–73. https://doi.org/10.1016/j.ijggc.2013.01.004

[58] Kuganathan, N. and Chroneos, A. 2021. Ru-doped single-walled carbon nanotubes as sensors for SO_2 and H_2S detection. Chemosensors. 9(6): 120. https://doi.org/10.3390/chemosensors9060120

[59] Pan, Q.N., Yang, Z.M., Wang, W.W. and Zhang, D.Z. 2021. Sulfur dioxide gas sensing at room temperature based on tin selenium/tin dioxide hybrid prepared via hydrothermal and surface oxidation treatment. Rare Metals. 40(6): 1588–1596. https://doi.org/10.1007/s12598-020-01555-3

[60] Li, R., Wang, S., Li, S., Zhao, F., Dong, T., He, P., et al. 2023. Cu-doped flower-like SnO_2 architecture toward promoting SO_2 detection: Fast equilibrium and low trace monitoring. Sensors and Actuators B: Chemical. 390: 133953. https://doi.org/10.1016/j.snb.2023.133953

[61] Bang, J.H., Mirzaei, A., Choi, M.S., Han, S., Lee, H.Y., Kim, S.S., et al. 2021. Decoration of multi-walled carbon nanotubes with CuO/Cu_2O nanoparticles for selective sensing of H_2S gas. Sensors and Actuators B: Chemical. 344: 130176. https://doi.org/10.1016/j.snb.2021.130176

[62] Musayeva, N., Khalilova, H., Izzatov, B., Trevisi, G., Ahmadova, S. and Alizada, M. 2023. Highly selective detection of hydrogen sulfide by simple Cu-CNTs nanocomposites. C. 9(1): 25. https://doi.org/10.3390/c9010025

[63] Hussain, S., Amu-Darko, J.N.O., Wang, M., Alothman, A.A., Ouladsmane, M., Aldossari, S.A., et al. 2023. CuO-decorated MOF-derived ZnO polyhedral nanostructures for exceptional H_2S gas detection. Chemosphere. 317: 137827. https://doi.org/10.1016/j.chemosphere.2023.137827

[64] Hassanzadeh-Tabrizi, S.A. 2024. Mild oxidizing synthesis of non-agglomerated Fe_2O_3 nanoparticles for H_2S gas sensing. Results in Chemistry. 6: 101395. https://doi.org/10.1016/j.rechem.2024.101395

[65] Yang, K., Sun, Q., Xue, F. and Lin, D. 2011. Adsorption of volatile organic compounds by metal–organic frameworks MIL-101: Influence of molecular size and shape. Journal of Hazardous Materials. 195: 124–131. https://doi.org/10.1016/j.jhazmat.2011.08.020

[66] Missaoui, N., Kahri, H. and Demirci, U.B. 2022. Rapid room-temperature synthesis and characterizations of high-surface-area nanoparticles of zeolitic imidazolate framework-8 (ZIF-8) for CO_2 and CH_4 adsorption. Journal of Materials Science. 57(34): 16245–16257. https://doi.org/10.1007/s10853-022-07676-w

[67] Yu, J., Zhao, X. and Zhao, Q. 2000. Effect of surface structure on photocatalytic activity of TiO_2 thin films prepared by sol-gel method. Thin Solid Films. 379(1–2): 7–14. https://doi.org/10.1016/s0040-6090(00)01542-x

[68] Ayad, M.M., Elmorsy, E., Amer, W.A. and Mahrous, A. 2022. Insight into the novel ZIF-8@N-CQDs/ZIF-67 nanocomposite for photocatalytic degradation of methylene blue under visible light irradiation. SSRN Electronic Journal. https://doi.org/10.2139/ssrn.4286781

[69] Khan, I., Saeed, K. and Khan, I. 2019. Nanoparticles: Properties, applications and toxicities. Arabian Journal of Chemistry. 12(7): 908–931. https://doi.org/10.1016/j.arabjc.2017.05.011

[70] Qin, W., Cao, W., Liu, H., Li, Z. and Li, Y. 2014. Metal–organic framework MIL-101 doped with palladium for toluene adsorption and hydrogen storage. RSC Advances. 4(5): 2414–2420. https://doi.org/10.1039/c3ra45983g

[71] Zhu, Z. and Xu, B. 2022. Purification technologies for NOx removal from flue gas: A review. Separations. 9(10): 307. https://doi.org/10.3390/separations9100307

[72] Liao, D. and Liao, B. 2007. Effects of surfactant composition and concentration on shape, size and photocatalytic activity of TiO_2 nanoparticles. International Journal of Chemical Reactor Engineering. 5(1). https://doi.org/10.2202/1542-6580.1390

[73] Sambaer, W., Zatloukal, M., Kimmer, D. and Zatloukal, M. 2011. 3D air filtration modeling for nanofiber-based filters in the ultrafine particle size range. AIP Conference Proceedings. https://doi.org/10.1063/1.3604492

[74] Fu, C., Hsieh, C., Juang, R., Yang, J., Gu, S. and Gandomi, Y.A. 2019. Highly efficient carbon quantum dot suspensions and membranes for sensitive/selective detection and adsorption/recovery of mercury ions from aqueous solutions. Journal of the Taiwan Institute of Chemical Engineers. 100: 127–136. https://doi.org/10.1016/j.jtice.2019.04.012

[75] Radhakrishnan, S., Naidu, D.J., Senbagarajan, S., Vanaraj, R., Kanthapazham, R. and Balachandran, S. 2023. Silver nanoparticles decorated cellulose fibrous silica hybrid materials for surface-enhanced Raman spectroscopy. ACS Applied Nano Materials. 6(24): 22846–22856. https://doi.org/10.1021/acsanm.3c04059.s001

[76] Ma, K. 2014. Synthesis, sulfonic acid functionalization and catalytic application of metal-organic frameworks [Master's thesis, Unknown University]. https://doi.org/10.14711/thesis-b1333477

[77] Van Cat, V., Dinh, N.X., Ngoc Phan, V., Le, A.T., Nam, M.H., Dinh Lam, V., et al. 2020. Realization of graphene oxide nanosheets as a potential mass-type gas sensor for detecting NO_2, SO_2, CO, and NH_3. Materials Today Communications. 25: 101682. https://doi.org/10.1016/j.mtcomm.2020.101682

[78] Khan, M., Rao, M. and Li, Q. 2019. Recent advances in electrochemical sensors for detecting toxic gases: NO_2, SO_2 and H_2S. Sensors. 19(4): 905. https://doi.org/10.3390/s19040905

[79] Zhang, J., Zhu, Z., Chen, C., Chen, Z., Cai, M., Qu, B., et al. 2018. ZnO-carbon nanofibers for stable, high response, and selective H_2S sensors. Nanotechnology. 29(27): 275501. https://doi.org/10.1088/1361-6528/aabd72

[80] Khan, M.M., Adil, S.F. and Al-Mayouf, A. 2015. Metal oxides as photocatalysts. Journal of Saudi Chemical Society. 19(5): 462–464. https://doi.org/10.1016/j.jscs.2015.04.003

[81] Yadav, A. and Sinha, N. 2022. Nanomaterial-based gas sensors: A review on experimental and theoretical studies. Materials Express. 12(1): 1–33. https://doi.org/10.1166/mex.2022.2121

[82] Sahoo, T.R. and Prelot, B. 2020. Adsorption processes for the removal of contaminants from wastewater. *In*: Nanomaterials for the Detection and Removal of Wastewater Pollutants. 161–222. Elsevier. https://doi.org/10.1016/B978-0-12-818489-9.00007-4

[83] Komaraiah, D., Madhukar, P., Vijayakumar, Y., Ramana Reddy, M. and Sayanna, R. 2016. Photocatalytic degradation study of methylene blue by brookite TiO_2 thin film under visible light irradiation. Materials Today: Proceedings. 3(10): 3770–3778. https://doi.org/10.1016/j.matpr.2016.11.026

[84] Liu, Y., Shen, J. and Zhu, X.D. 2010. Effect of temperature, relative humidity and ACH on the emission of volatile organic compounds from particleboard. Advanced Materials Research. 113–116: 1874–1877. https://doi.org/10.4028/www.scientific.net/AMR.113-116.1874

[85] Sun, S., Bao, J., Gao, C. and Ding, J. 2011. Photocatalytic degradation of gaseous o-xylene over M-TiO_2 (M = Ag, Fe, Cu, Co) in different humidity levels under visible-light irradiation: Activity and kinetic study. Rare Metals. 30(S1): 147–152. https://doi.org/10.1007/s12598-011-0258-9

[86] Laskar, I.I., Hashisho, Z., Phillips, J.H., Anderson, J.E. and Nichols, M. 2019. Modelling the effect of relative humidity on adsorption dynamics of volatile organic compound onto activated carbon. Environmental Science and Technology. 53(5): 2647–2659. https://doi.org/10.1021/acs.est.8b05664.s001

[87] Chaitra, U., Ali, A.M., Viegas, A.E., Kekuda, D. and Rao, K.M. 2019. Growth and characterization of undoped and aluminium doped zinc oxide thin films for SO_2 gas sensing below threshold value limit. Applied Surface Science. 496: 143724. https://doi.org/10.1016/j.apsusc.2019.143724

[88] Marugán, J., López-Muñoz, M.J., Fernández-Ibáñez, P. and Malato, S. 2016. Solar photocatalysis: Fundamentals, reactors and applications. *In*: Energy and Environment Series. 92–129. https://doi.org/10.1039/9781782627104-00092

[89] Cucurachi, S. and Blanco Rocha, C.F. 2019. Life-cycle assessment of engineered nanomaterials. *In*: Pacheco-Torgal, F., Diamanti, M.V., Nazari, A. and Leonelli, C. (Eds). Nanotechnology in Eco-efficient Construction. 815–846. Woodhead Publishing. https://doi.org/10.1016/B978-0-08-102641-0.00031-1

[90] Roberto, M. and Christofoletti, C. 2020. How to assess nanomaterial toxicity? An environmental and human health approach. *In*: Nanomaterials—Toxicity, Human Health and Environment. IntechOpen. https://doi.org/10.5772/intechopen.88970

[91] Rönkkö, T. and Timonen, H. 2019. Overview of sources and characteristics of nanoparticles in urban traffic-influenced areas. Journal of Alzheimer's Disease. 72(1): 15–28. https://doi.org/10.3233/jad-190170

[92] Roy, A., Sharma, A., Yadav, S., Jule, L.T. and Krishnaraj, R. 2021. Nanomaterials for remediation of environmental pollutants. Bioinorganic Chemistry and Applications. 2021: 1–16. https://doi.org/10.1155/2021/1764647

[93] Gambardella, C. and Pinsino, A. 2022. Nanomaterial ecotoxicology in the terrestrial and aquatic environment: A systematic review. Toxics. 10(7): 393. https://doi.org/10.3390/toxics10070393

[94] Loan, T.T. 2018. Synthesis and characterization of anatase TiO_2:Cu^{2+} powders prepared via a sol-gel technique. VNU Journal of Science: Mathematics—Physics. 34(3). https://doi.org/10.25073/2588-1124/vnumap.4272

[95] Ningthoujam, R., Singh, Y.D., Babu, P.J., Tirkey, A., Pradhan, S. and Sarma, M. 2022. Nanocatalyst in remediating environmental pollutants. Chemical Physics Impact. 4: 100064. https://doi.org/10.1016/j.chphi.2022.100064

[96] Smijs, T. and Pavel, S. 2011. Titanium dioxide and zinc oxide nanoparticles in sunscreens: Focus on their safety and effectiveness. Nanotechnology, Science and Applications. 4: 95–112. https://doi.org/10.2147/NSA.S19419

[97] Cormier, S.A., Lomnicki, S., Backes, W. and Dellinger, B. 2006. Origin and health impacts of emissions of toxic by-products and fine particles from combustion and thermal treatment of hazardous wastes and materials. Environmental Health Perspectives. 114(6): 810–817. https://doi.org/10.1289/ehp.8629

[98] Geldasa, F.T., Kebede, M.A., Shura, M.W. and Hone, F.G. 2023. Experimental and computational study of metal oxide nanoparticles for the photocatalytic degradation of organic pollutants: A review. RSC Advances. 13(27): 18404–18442. https://doi.org/10.1039/d3ra01505j

[99] Ashori, A., Sepahvand, S. and Jonoobi, M. 2024. Development of biodegradable nanofiber filters based on surface-modified cellulose nanofibers with graphene oxide for high removal of airborne particulate matter. International Journal of Biological Macromolecules. 261: 129687. https://doi.org/10.1016/j.ijbiomac.2024.129687

[100] El-Shamy, A.G. 2021. New nano-composite based on carbon dots (CDots) decorated magnesium oxide (MgO) nano-particles (CDots@MgO) sensor for high H_2S gas sensitivity performance. Sensors and Actuators B: Chemical. 329: 129154. https://doi.org/10.1016/j.snb.2020.129154

[101] Feng, S., Li, X., Zhao, S., Hu, Y., Zhong, Z., Xing, W., et al. 2018. Multifunctional metal-organic framework and carbon nanotube-modified filter for combined ultrafine dust capture and SO_2 dynamic adsorption. Environmental Science: Nano. 5(12): 3023–3031. https://doi.org/10.1039/C8EN00694E

[102] Fu, T. 2013. Sensing behavior of CdS nanoparticles to SO_2, H_2S and NH_3 at room temperature. Materials Research Bulletin. 48(5): 1784–1790. https://doi.org/10.1016/j.materresbull.2013.01.037

[103] Gordijo, J.S., Rodrigues, N.M. and Martins, J.B. 2023. CO_2 and CO capture on the ZnO surface: A GCMC and electronic structure study. ACS Omega. 8(49): 46830–46840. https://doi.org/10.1021/acsomega.3c06457

[104] Husain, M.M., Ansari, M.T. and Almohammedi, A. 2024. Analyzing the electronic and conductive characteristics of zigzag graphene nanoribbons upon NOx and N_2O adsorption: An ab-initio study. Materials Today Communications. 39: 108725. https://doi.org/10.1016/j.mtcomm.2024.108725

[105] Jha, R.K., Nanda, A. and Bhat, N. 2022. Sub-ppm sulfur dioxide detection using MoS_2 modified multi-wall carbon nanotubes at room temperature. Nano Select. 3(1): 98–107. https://doi.org/10.1002/nano.202100166

[106] Jung, S., Yusuf, M., Son, Y., Han, S., Lee, H., Mahadadalkar, M.A., et al. 2024. An efficient atmospheric pollution control using hierarchical porous nanofibers containing zeolitic-imidazolate-frameworks and hydroxyapatite nanoparticles. Journal of Environmental Chemical Engineering. 12(1): 111798. https://doi.org/10.1016/j.jece.2023.111798

[107] Kumar, A., Kumar, R., Singh, R., Prasad, B., Kumar, D. and Kumar, M. 2021. Biosynthesis of ZnO nanoparticles and effect of silver doping in gas sensing characteristics of volatile organic compounds. Journal of Coatings Technology and Research. 1–13. https://doi.org/10.1007/s11998-021-00445-9

[108] Lam, S.M., Sin, J.C., Zeng, H., Lin, H., Li, H., Chai, Y.Y., et al. 2021. Green synthesis of Fe-ZnO nanoparticles with improved sunlight photocatalytic performance for polyethylene film deterioration and bacterial inactivation. Materials Science in Semiconductor Processing. 123: 105574. https://doi.org/10.1016/j.mssp.2020.105574

[109] Laskar, I.I., Hashisho, Z., Phillips, J.H., Anderson, J.E. and Nichols, M. 2019. Modelling the effect of relative humidity on adsorption dynamics of volatile organic compound onto activated carbon. Environmental Science and Technology. 553(5): 2647–2659.

[110] Lee, M., Ojha, G.P., Oh, H.J., Kim, T. and Kim, H.Y. 2020. Copper//terbium dual metal-organic frameworks incorporated side-by-side electrospun nanofibrous membrane: A novel tactic for efficient adsorption of particulate matter and luminescence property. Journal of Colloid and Interface Science. 578: 155–163. https://doi.org/10.1016/j.jcis.2020.05.085

Chapter **10**

Nanomaterials for Water Purification and Treatment

Deepak Kumar and Meenakshi Suhag*
Institute of Environmental Studies,
Kurukshetra University, Kurukshetra, Haryana, India-136119

10.1 INTRODUCTION

Clean water is one of the most crucial components for the survival of all living beings. However, water resources have become contaminated worldwide due to massive population growth and quick industrialization [1]. All living things depend on water; thus, protecting its safety and purity is essential for public health. Probably the most important global concern is that water purification and treatment are crucial processes for ensuring access to clean and safe water [2]. These processes are particularly important in areas where clean water is scarce or contaminated [3]. Due to increasing populations, industrial activities, and human practices in agriculture, water pollution has become a crucial and devastating concern in terms of impairing human health and ecosystems. Some of the world's inhabitants are predicted to reside in water-stressed regions by 2025 [4]. Before 2015, only 20% of wastewater globally had undergone successful treatment. According to the UN (2016), in developing countries, almost 70% of industrial effluent is dumped without being properly destroyed [1]. Currently, 2.2 billion people (27% of the population) are at risk of not having safe drinking water [5]. Water purification remains a daunting task due to issues like the availability of water resources and over-exploitation, so some innovations are necessary to evade

*For Correspondence: msuhag@kuk.ac.in

DOI: 10.1201/9781003492184-10

the devastating consequences of climate change and conserve such resources. Nanotechnology is relevant in water purification. 'Nanotechnology' was first used by Norio Taniguchi in the 1970s during material processing [6]. He stated that several commendable functions occur when nanotechnology is applied to materials, such as separation, solidification, and deformation by atoms or molecules [7]. Precisely, saving water from contaminants like viruses, heavy metals, inorganic solutes, complex organic compounds, and natural organic matter can be effectively achieved using nanoparticles. Due to the significant mechanical strength, enlarged overall chemical reactions, altered size-dependent properties, and huge surface area, nanomaterials are preferred for water remediation [8]. While this is in effect, the expenditure towards nanomaterials is less expensive and simpler, compared to conventional flowing methods. Regeneration of the nanomaterial accounts for a reusable option towards less consumption of the novel and simplified water purification procedure. Interventions based on nanowater treatment methods will emerge to enhance processes such as desalinating brackish and seawater, safe reuse of wastewater, and fumigation and sanitization of water [9]. Specific virtual properties of nanomaterials could yield revolutionary interventions in the field of water purification.

Materials having dimensions on average between 1 and 100 nm are referred to as nanomaterials. Since they have a small size, immense surface area, and strong reactivity, nanomaterials are effective adsorbents for achieving their goals [10, 11]. At this scale, materials display unique physical, chemical, and biological characteristics in contrast to their bulk counterparts. Nanomaterials are very useful in the treatment and purification of water because of their special qualities. With its many uses, including polymer removal, heavy metal adsorption, dye adsorption, dye degradation, and antibacterial properties, nanomaterials have become an advanced improvement in water treatment. These uses help to purify and remediate water resources [12, 13]. Many nanomaterials, such as polymer nanomaterials, nanosorbents, nanocatalysts, metal-oxide nanoparticles, carbon nanotubes, zeolites, fullerenes, graphene, metallic nanoparticles, graphene quantum dots, and graphene oxides, have been developed for the purification of water and wastewater [14–17]. With the use of innovative nanoparticles made possible by nanotechnology, pathogens, metal oxides, hazardous metal ions, and inorganic and organic solutes can be removed from wastewater, external water, groundwater, and other natural resources. These procedures include ultrafiltration sheaths, carbon nanotubes, reverse osmosis, nanofiber filters, and nanofiltration [17, 18]. Revolutionizing water treatment systems has great potential for nanomaterials in the future. The integration of these advanced materials leads to a significant improvement in killing off contaminants and harmful microorganisms by water treatment technologies [19]. This improvement not only makes water treatment processes more effective but also enhances the sustainability, accessibility, and durability of water resources [3]. Nanotechnology provides new ways to deal with current challenges relating to water quality, with fresh techniques for purification and treatment. Some existing issues on water quality might be solved when these problems are addressed utilizing nanotechnology. Furthermore, as compared to conventional wastewater treatment methods which includes adsorption, filtration, centrifugation, catalysis, and biological treatment facing challenges like energy consumption and generation of secondary pollution, nanotechnology offers better efficiency due to their unique characteristics as having nanosize size, large surface area, highly reactive, thermal stability, biodegradability, renewability and durable in removing pollutants [17, 20–23].

10.2 CLASSIFICATION OF NANOMATERIALS

Nanomaterials have been considered a viable option for water treatment and purification because they are highly reactive, possess large specific surface areas, and behave differently in solution [24]. Classification of nanomaterials during synthesis mainly depends on their material's origin, structure, and dimensional properties (Figure 10.1) [25].

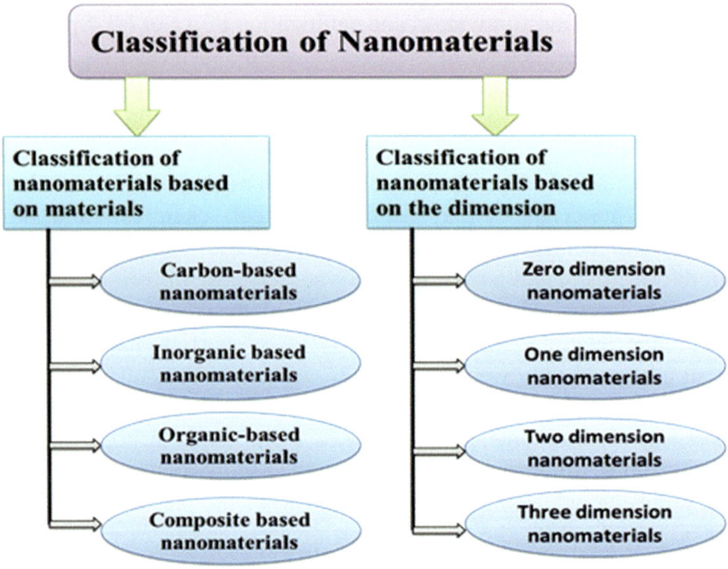

Figure 10.1 A schematic diagram shows the basic classification of nanomaterials.

Water treatment technologies can significantly increase the efficacy and efficiency of water purification and treatment procedures by creating sophisticated filtration materials and employing nanosorbents, nanocatalysts, bioactive nanoparticles, and nanostructured catalytic membranes. More water reuse, recycling, and desalination are made possible by nanotechnology, which can also help address the world water crisis. Four groups of nanoscale materials are highlighted in Figure 10.2, and are being considered as efficient materials for water decontamination: Zeolites, metal-containing nanomaterials, dendrimers, and carbon-based nanomaterials. These are particularly appealing as parting and volatile media for the decontamination of water because of their wide range of physicochemical features [17, 26].

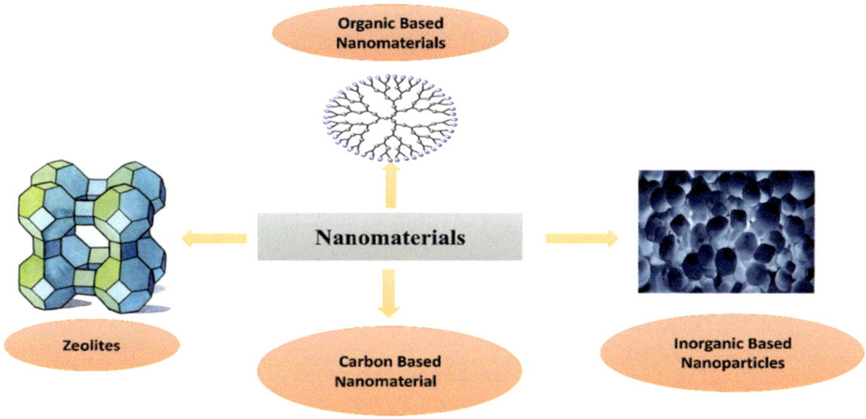

Figure 10.2 Certain nanoparticles are presently being assessed as useful materials for the filtration of water.

The present section is devoted to the application of nanomaterials in water treatment and the process of sanitization, covering the creation of innovative substances for saltwater extraction, filtration, and disinfection.

10.2.1 Zeolites

Pollution and water crises have currently become pressing issues globally as they relate to first-world and third-world countries. One of the most common methods in the purification of water is the use of zeolites [27]. Zeolites are naturally occurring minerals that are microporous, crystalline materials with a special architecture comprising silicon, oxygen, and aluminum atoms arranged in a framework with pores forming small cages that allow them to selectively absorb and remove different pollutants from water [28]. They have a high ability to bind with contaminants present in water because they have high surface areas and ion exchange capacities [29]. Phosphorus ions, nitrates, as well as ammonia, are also removed by zeolites from water, making it suitable for irrigation [30]. To overcome the global issue of a lack of clean water, zeolites are versatile and effective materials used for the purification of contaminated water [31]. That is why using zeolites for purifying water should be encouraged.

10.2.2 Carbon-Based Nanomaterials

Carbon-containing nanomaterials are growing more enticing in the field of science and technology [32, 33]. Carbon nanomaterials recently gained value due to the discovery of numerous alternative forms of carbon, from well-known allotropic phases like amorphously formed carbon, graphite, and diamonds to recently found beneficial carbon nanotubes, graphene, fullerenes, carbon nanofibers, and carbon quantum dots [34]. Because of their superior physicochemical qualities, carbon-based nanomaterials have garnered a lot of interest in the arena of pollution treatment because of their huge specific surface areas, excellent acid stability, and thermal resistance [35]. It has been discovered that carbon nanomaterials, with an adsorption proficiency of more than 80% or a sunlight degradation efficacy greater than 98%, can efficiently remove impurities from the environment, including nitric oxide, heavy metals, H_2S, colorants, and medicine components [36]. According to Liu et al. (2022), carbon-based nanomaterials have been widely used in the treatment and purification of wastewater in recent years [37]. Some of the carbon-based nanomaterials that are used in wastewater treatment are listed below:

10.2.2.1 Fullerenes

The tertiary allotropic form of carbon compound is fullerenes, which were discovered by Harold in 1985. Buckminsterfullerene (C_{60}) is a well-known type of fullerene having a morphology like a soccer ball. The fullerene has 12 pentagons and 20 hexagonal carbon rings, resulting in the formation of spherical molecules consisting of carbon atoms having sp^2 hybridization [25]. These are widely used in various applications, including solar cells, drug delivery systems, superconductors, and many more.

10.2.2.2 Graphene

For eliminating different types of contaminants from water, graphene and graphene-based nanomaterials have been widely used in recent years [38, 39]. Due to single layered atom thickness and an intriguing two dimensional carbon sheet, graphene has distinct physical as well as chemical features, having sp^2 hybridization [39–42]. Graphene and graphene-based composites are generally used as catalysts, adsorbents, photocatalysts, electrocatalysts, disinfectants, and desalination for water purification systems [43]. Graphene has excessive electrical conductivity, mechanical strength, and flexibility. Graphene is widely used in the fields of electronics and energy storage.

10.2.2.3 Carbon Nanotubes

Carbon nanotubes are graphene sheets rolled up into tubes with diameters as small as 1 nm. Due to their effective physicochemical characteristics, carbon nanotubes are widely used in various environmental applications. The properties include high surface area, capacity to absorb, mechanical strength, electrical and thermal conductivity, and flexibility [25, 44]. Single-walled nanotubes are composed of one layer of carbon atoms and have a diameter of 0.7 nm. Double-walled carbon nanotubes (DWCNTs) are formed of two overlapping spherical tubes embraced by carbon atoms. Multiple rolled sheets result in the formation of multi-walled carbon nanotubes (MWCNTs), which have a diameter of 100 nm.

10.2.2.4 Carbon Nanofibers

Carbon nanofibers are discontinuous strands that are different from carbon fibers. Carbon nanofibers have sp^2 hybridization [45]. These have a high surface-to-volume ratio, elasticity, absorbency, and low density [46, 47]. Due to the presence of fibers as well as small pores, they are highly effective in removing pollutants from wastewater [48–50].

10.2.2.5 Carbon Quantum Dots

In some recent cases, a type of nanomaterial called carbon quantum dots has been created. Owing to their inexpensiveness, safety in terms of being environmentally friendly, and high stability in solution, meaning they are inactive chemically among others, CQDs have been used extensively in several areas [51]. These particles are usually smaller than even 10 nm, having sp^2/sp^3 hybridization [52]. Carbon quantum dots are widely used in various applications such as chemical sensing, bioimaging, biosensing, drug delivery, photocatalysis, and electrodialysis [53]. Principally among the features of carbon quantum dots, their fluorescence properties have garnered a lot of attention recently [54].

10.2.3 Inorganic-Based Nanomaterials

Inorganic nanomaterials are defined as materials consisting of inorganic compounds or elements, such as metals, metal oxides, semiconductors, etc., and typically have diameters between 1 and 100 nm. Since they have a small size, large surface area-to-volume ratio, and quantum effects, these are more stable than organic based nanomaterials [55]. Metals like Silver, Gold, Platinum, and Iron can be used to generate them, and the metal oxides that result are Titanium oxide, Zinc oxide, and MnO_2 (Figure 10.3). Ceramic and silicon constituents can also be used to form semiconductor nanoparticles [25, 56].

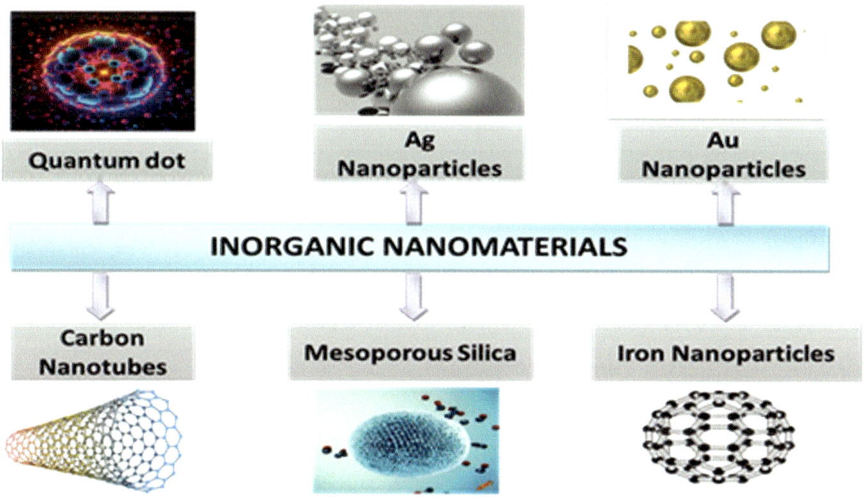

Figure 10.3 Metal and metal oxide nanoparticles are classified as inorganic nanomaterials.

10.2.3.1 Metal Nanomaterials

Metal-based nanoparticles are manufactured from alloys by any productive or destructive processes. The uncontaminated metal nanoparticles are formed by consuming the metal precursors. Due to their plasma-on resonance uniqueness, metallic nanoparticles have distinct optoelectrical features [57, 58]. Similar nanoparticles of the element include those of the metals iron, zinc, cobalt, aluminum, copper, silver, lead, gold, and cadmium. Lesser sizes (10–100 nm), surface characteristics (aperture size, surface charge, environmental influences, surface area-to-volume ratio), shapes (asymmetrical, rod, circular, tube shaped, tetragonal, and hexagonal), surface charge density, and color can all be used to categorize nanoparticles [55].

10.2.3.2 Metal-Oxide Nanomaterials

These nanoparticles are extremely small elements of metal oxides, usually with a size between 1 and 100 nm. Because of their compact size and high surface area-to-volume ratio, they have special qualities that make them appealing for a variety of applications in different sectors. To create these nanomaterials, various applications are widely used, including sol-gel synthesis, thermal dilapidation, chemical precipitation, and hydrothermal synthesis. There are various types of metal oxide nanoparticles, some of which are listed here: zinc oxide, iron oxide, titanium oxide, copper oxide, calcium oxide, and aluminum oxide [55, 59].

10.2.3.3 Ceramic Nanomaterials

Ceramics are substances that lie between metallic and nonmetallic elements; the most prevalent types include nitrides, oxides, carbides, and phosphates [60]. Some typical examples include cement, glass, firebricks, and China clay, which are frequently used in a variety of applications. As compared to polymers and metals, ceramic nanoparticles are extremely impervious to harsh environmental conditions as well, and are also insulating to heat and energy. At the nanoscale, these have extremely high heat capacities, toughness, erosion resistance, and electrical lining properties. By repeatedly heating and cooling, nanomaterials are formed with a variety of morphologies such as dense, hollow, polycrystalline, and amorphous [60, 61]. According to Thomas et al. ceramic nanomaterials are widely used in a variety of applications, including photodegradation of dyes, catalysis, imaging, and many more [62].

10.2.4 Organic Nanomaterials

Ferritin, micelles, dendrimers, liposomes, nanogels, polymeric NPs, and layered biopolymers are various organic nanomaterials as shown in Figure 10.4. Some organic nanoparticles have a hollow sphere (micelles and liposomes), and are nontoxic and recyclable, as well as they are also decomposed naturally. It is also known as nanocapsules, which are highly susceptible to light and heat [58, 63]. Due to these properties, organic nanomaterials are an excellent choice for drug delivery and are extensively used in target drug delivery. Polymeric nanoparticles are another term used for organic nanomaterials. A nanosphere or nanocapsule is the main form of organic or polymeric nanoparticles [64, 65]. The outermost border of the circular surface of the matrix particles, whose earlier prevalent mass is solid, adsorbs other molecules. In the latter sample, the solid mass is encased in particles [55, 63].

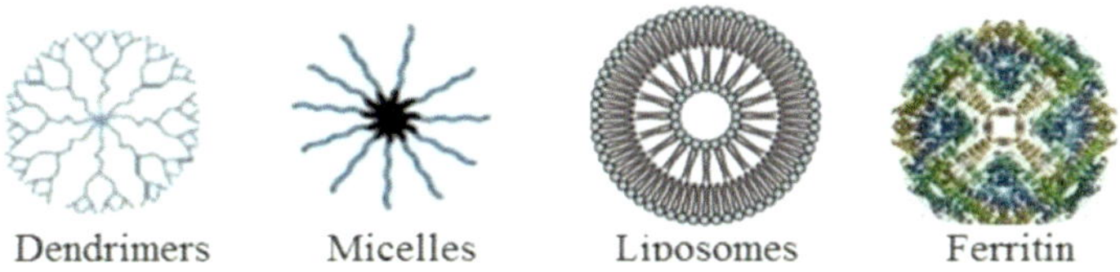

Figure 10.4 Various types of organic nanomaterials.

10.3 DESALINATION

Almost half of all desalination systems used worldwide to produce drinkable water are based on the filter membrane technique for desalinating salt seawater. It is a widely used method for treating water because it is simple to use, economically viable, requires fewer chemical additives, has a lower phase change and higher productivity, is readily scalable, and has an effective removal capacity. Significant progress has been achieved in the past in enhancing the effectiveness of TFC (Thin Film Composite membrane) by employing hydrophilic nanoparticles such as zeolites, graphene oxide, carbon nanotubes, titanium dioxide, silica, and alumina [66].

10.3.1 Solar Thermal Desalination

Lacking electricity and water infrastructure, solar thermal desalination provides a sustainable and affordable method of producing freshwater that meets high standards [67]. Reverse osmosis, membrane distillation, electrodialysis, vapor compression, multistage flash desalination, and other traditional desalination methods can all be combined with solar collectors to generate heat in an indirect form of solar desalination. Direct desalination uses solar energy [68]. Water is heated to the point of evaporation by solar radiation. Water vapor rises when the water evaporates and condenses to gather on the glass surface. Solar thermal desalination removes salts, heavy metals, and microbial organisms.

For purification and wastewater treatment, there has been a great interest in the use of nanomaterials in filtration techniques over the last few years. Compared to conventional methods, nanomaterials have several benefits, including great efficiency and selectivity in the removal of pollutants from contaminated water. Nanoparticles show great adsorption capacity in the removal of pollutants as they have a high surface area-to-volume ratio. Additionally, certain pollutants are targeted by these nanomaterials, so they are very useful in cleaning a variety of wastewater types [31]. Depending on the physicochemical properties of certain contaminants, a membrane acts as a selective barrier or permeable covering that permits certain substances to pass through the membrane while restricting others [69]. These nanomembranes act as a selective block to isolate the two different phases [70]. Based on the size of pores, these membranes are of four types: Microfiltration, Ultrafiltration, Nanofiltration, and Reverse Osmosis. All these different types of membranes are based on pressure [69].

10.4 TYPES OF MEMBRANES USED IN WATER PURIFICATION AND TREATMENT

10.4.1 Microfiltration

The membrane, which has a pore size of about 0.1 to 10 micrometers, is considered a microfiltration membrane that is used to remove elements and microbes from an aqueous solution. It is used in various sectors like food and beverage industries, wastewater treatment, and pharmaceuticals, as well as in biotechnology to eradicate pathogens, suspended solids particles, as well as other pollutants from liquids. Compared to the traditional raw filtration method, microfiltration is typically more effective at low differential pressures of 2-bar.

10.4.2 Ultrafiltration

Ultrafiltration is a separation process used to remove suspended solids and solutes with high molecular weight from a liquid solution. It is a pressure-driven separation unit operation that is almost always based on membrane filtration. Ultrafiltration is equipped with membrane pores in the range of 0.001 to 0.1 micrometers. Therefore, the membrane pores are large for water and low molecular weight which allows them to pass through. However, it is much smaller than the larger molecules that comprise proteins, colloids, and suspended solids and does not pass through the membrane persona. Ultrafiltration removes divalent ions from water, which is chiefly responsible for water rigidity in textile industries. The use of this high colored textile discharge is restricted in the fabric industry since the molecular masses of the dyes are significantly smaller than the molecular mass cut off of ultrafiltration membranes, restraining its application in the fabric industry [71].

10.4.3 Nanofiltration

A membrane filtration technique called nanofiltration uses a microscopic scale to extract different materials from water, such as salts, organic debris, and certain microbes. Nanofiltration membranes have an aperture size in the range of 0.001–0.01 micrometers. Nanofiltration has multiple uses in wastewater purification, such as removal of salts, organic matter, color and odor, pathogens, and ion removal. By creating a high quality effluent appropriate for a diversity of non-potable uses, such as irrigation, industrial processes, and even indirect filtered reuse with extra treatment steps, nanofiltration plays a critical role in wastewater reuse applications. The nanofiltration process removes particles with pore size widths of approximately 0.5–1.5 nm [69]. Arsenic from groundwater and surface water, cations, natural organic matter, biological pollutants, organic pollutants, and nitrates can be removed by using nanofiltration as reviewed by Vander Bruggen and Vandecasteele [72]. Favre-Reguillon et al. (2003) displayed the use of nanofiltration to extract minuscule amounts of U(VI) from seawater [73].

10.4.4 Reverse Osmosis

RO membrane filtration, also known as reverse osmosis, is very effective and can eliminate around 99% of impurities such as pathogens and dissolved solids within the wastewater system. Reverse osmosis is similar to normal osmosis, but the reverse inward solvent flow from higher solute concentration to lower solute concentration is conducted through a membrane. Also known as semi-permeable or selectively permeable, reverse osmosis membranes reject salt. RO membranes are used to remove water hardness, colors, and other organic impurities in chemical industries and agriculture. Applying RO to anaerobic groundwater can eliminate undesirable elements, including iron, manganese, and ammonium, as well as minute amounts of pesticides and medications, commonly referred to as "organic micropollutants" [74].

10.5 NANOMATERIALS IN DISINFECTION

Traditionally, disinfection, removal, or killing of pathogenic microorganisms is commonly accomplished by chlorination. However, the same resulted in the formation of toxic byproducts. The effective innovative disinfection technology involving nanomaterials, due to their exceptional reactivity, high surface area-to-volume ratio, and antibacterial activity, has become an attractive option for disinfection and water purification. The following are some important uses of nanomaterials in water purification through disinfection (Table 10.1).

10.5.1 Photocatalytic Disinfection

The first use of photocatalysis using titanium dioxide for bacterial ablation was documented by Matsunaga et al. in 1985 [89]. Numerous studies on the deactivation of microbes such as bacteria, fungi, algae, viruses, and protozoans have been published since then. Photocatalysis is widely used in the inactivation of microorganisms as reported by Blake et al. in an extensive review (Herrmann 1999) [90]. The use of photocatalysis in decontaminating water tainted with pathogenic bacteria was examined by McCullagh et al. in 2007 [91]. A detailed review of the disinfectant and purification of water using solar photocatalysis was published in 2009 by Malato et al. [92]. The hydroxyl radical is often considered to be the main species implicated in the elimination of microorganisms in photocatalytic disinfection investigations; however, some articles do report the engrossment of additional ROS, such as H_2O_2, O_2^- [93–95]. Nanoparticles like titanium dioxide (TiO_2), zinc oxide (ZnO), and cadmium sulfide (CdS) [96] exhibit photocatalytic properties when exposed to sunlight or UV radiation. These nanoparticles generate Reactive Oxygen Species (ROS) that can effectively inactivate and destroy various pathogens, including bacteria, viruses, and protozoa present in water. It is measured as a sustainable method and environmentally approachable for water treatment [97].

Table 10.1 Application of different types of nanomaterials in water disinfection

Types of Nanomaterials	Application	Ref.
TiO_2 and ZnO nanomaterials	Photocatalytic disinfection	[75]
Noble metals- and carbon nanomaterials- TiO_2 composites	Photocatalytic disinfection of bacteria	[76]
Hydrogel loaded with bimetallic peroxide ($CeCuO_x$)	High-efficiency water disinfection	[77]
CuNPs@PVDF nanocomposite	Eliminating pathogens and reducing organic dye pollution.	[78]
Sol-gel synthesized bismuth titanium oxide ($Bi_4Ti_3O_{12}$)	Water pollution disinfection	[79]
Mo-based nanomaterials and their hybrids	Antiseptic disinfection	[80]
TiO_2-based photocatalytic nanomaterials	Water disinfection	[81]
Iron oxide magnetic nanoparticles	Water disinfection with bacteria.	[82]
Silver-based porous nanocomposites (AgNCs)	Water disinfection	[83]
Chalcogenides-based nanomaterials	Bacterial disinfection and organic contaminants degradation	[84]
Silver nanowire-decorated glass fiber (AgNW-GF) filters	*E. coli* removal from water	[85]
Binary composite heterojunction photocatalyst MoS_2/TiO_2 nanotube arrays (MoS_2/TiO_2 NTAs)	Water decontamination	[86]
Silver nanoparticles (AgNPs)	Inactivation of viruses, fungi, and protozoa	[87]
TiO_2/RGO/MoS_2 nanocomposite	Dye removal and disinfection of wastewater	[88]

10.6 ANTIMICROBIAL NANOPARTICLES

Most waterborne pathogens, like bacteria, viruses, parasites, and some others, contain membranes and cell walls, which serve as a protective barrier against some external resistance. These pathogens pollute the water, which is unsuitable for drinking. To overcome this problem, we use some nanomaterials that show a great response against these pathogens. Nanomaterials like silver, gold, titanium oxide, zinc oxide, platinum, and some others are widely used to remove these pathogens from the water or for water purification. These nanomaterials display numerous means of action-- they can interrupt cell membranes, proteins, enzymes, and mitochondria, affect cellular processes, generate reactive oxygen species, and interact with microbial DNA [98]. These nanomaterials are widely used with filtration membranes, or directly added to water to remove these pathogens or to inactivate them. Antimicrobial nanoparticles are widely used against waterborne pathogens, as these particles can prevent biofilm formation in water treatment systems.

10.7 NANOFILTERATION AND NANOCOMPOSITES MEMBRANES

Nanomembranes are usually categorized by their pore dimensions. They are situated in the region between reverse osmosis and ultrafiltration membranes. Usually, nanofiltration membranes have pore sizes of 2 nm. They eliminate particles with common sizes fluctuating from 0.9 to 10 nm, according to the specific membranes [99]. For water treatment and purification, membranes are combined with nanomaterials such as graphene, fullerenes, nanofibers, nanotubes, and some others [100]. Nanocomposite membranes display excessive mechanical strength and chemical stability, contrary to waterborne pathogens (viruses and bacteria), heavy metals, dye removals, and contaminant elimination.

10.8 CONCLUSION AND FUTURE PROSPECTS

In the treatment and purification of wastewater, innovative nano-based technology combined with biological methods can provide a convenient method. Over the last two decades, different nanomaterials such as nanofibers, carbon nanotubes, titanium dioxide, graphene, fullerenes, and carbon dots have been widely used for wastewater treatment. Additionally, advanced outputs have also been tested through the combination of nanomaterials with biological procedures. Such combined technologies improved the efficiency in terms of contamination removal, pollutant removal, heavy metals removal, dye removal, etc. Nanotechnology can suggest an additional real world and environment-friendly approach to wastewater treatment. However, these methods have restrictions such as the necessity of high end instruments, precise biotic elements for the treatment of diverse pollutants (organic compounds, dyes, nutrients), and the appropriate process environment of each microbe. The application of nanotechnology-based methods for purifying water could be a potential solution to tackle the issues shortly, given the effects on both the environment and human wellness. Furthermore, the green production of nanomaterials for use in wastewater treatment and purification would need significant work and investigation. Enduring research in nanotechnology will prospectively yield advanced solutions for cleaner water. Researchers continue to explore the properties, applications, and limitations of nanomaterials. Nanotechnology is sustainable due to recyclability and cost effectiveness, so it offers valuable advantages in wastewater treatment.

REFERENCES

[1] Singh, K.K., Singh, A. and Rai, S. 2022. A study on nanomaterials for water purification. Materials Today: Proceedings. 51: 1157–1163.

[2] Patanjali, P., Singh, R., Kumar, A. and Chaudhary, P. 2019. Nanotechnology for water treatment: A green approach. In Green Synthesis, Characterization and Applications of Nanoparticles. 485–512. Elsevier.

[3] Kumar, S., Ahlawat, W., Bhanjana, G., Heydarifard, S., Nazhad, M.M. and Dilbaghi, N. 2014. Nanotechnology-based water treatment strategies. Journal of Nanoscience and Nanotechnology. 14(2): 1838–1858.

[4] Report WHO 2014: The Future of Global Water Stress: An Integrated Assessment.

[5] https://www.who.int/teams/environment-climate-change-and-health/water-sanitation-and-health/monitoring-and-evidence/wash-monitoring

[6] Sun, M. and Gupta, A.S. 2020. Vascular nanomedicine: current status, opportunities, and challenges. *In*: Seminars in Thrombosis and Hemostasis. 46(05): 524–544. Thieme Medical Publishers.

[7] Rathod, S., Preetam, S., Pandey, C. and Bera, S.P. 2024. Exploring synthesis and applications of green nanoparticles and the role of nanotechnology in wastewater treatment. Biotechnology Reports. e00830.

[8] Thangavelu, L. and Veeraragavan, G.R. 2022. A survey on nanotechnology-based bioremediation of wastewater. Bioinorganic Chemistry and Applications.

[9] Kim, M., Jee, S.C., Shinde, S.K., Mistry, B.M., Saratale, R.G., Saratale, G.D., et al. 2019. Green-synthesis of anisotropic peptone-silver nanoparticles and their potential application as an anti-bacterial agent. Polymers. 11(2): 271.

[10] Kingsley, J.D., Ranjan, S., Dasgupta, N. and Saha, P. 2013. Nanotechnology for tissue engineering: need, techniques, and applications. Journal of Pharmacy Research. 7(2): 200–204.

[11] Saraswat, P., Singh, S., Prasad, M., Misra, R., Rajput, V.D. and Ranjan, R. 2023. Applications of bio-based nanomaterials in environment and agriculture: a review on recent progress. Hybrid Advances. 100097.

[12] Santhosh, C., Velmurugan, V., Jacob, G., Jeong, S.K., Grace, A.N. and Bhatnagar, A. 2016. Role of nanomaterials in water treatment applications: a review. Chemical Engineering Journal. 306: 1116–1137.

[13] Nyabadza, A., Makhesana, M., Plouze, A., Kumar, A., Ramirez, I., Krishnamurthy, S., et al. 2024. Advanced nanomaterials and dendrimers in water treatment and the recycling of nanomaterials: A review. Journal of Environmental Chemical Engineering. 112643.

[14] Gupta, S. and Babu, B.V. 2009. Removal of toxic metal Cr (VI) from aqueous solutions using sawdust as adsorbent: Equilibrium, kinetics and regeneration studies. Chemical Engineering Journal. 150(2–3): 352–365.

[15] Gangadhar, G., Maheshwari, U. and Gupta, S. 2012. Application of nanomaterials for the removal of pollutants from effluent streams. Nanoscience and Nanotechnology-Asia. 2(2): 140–150.

[16] Yin, J. and Deng, B. 2015. Polymer-matrix nanocomposite membranes for water treatment. Journal of Membrane Science. 479: 256–275.

[17] Kumar, S. 2023. Smart and innovative nanotechnology applications for water purification. Hybrid Advances. 3: 100044.

[18] Prabhakar, V., Bibi, T., Vishnu, P. and Bibi, T. 2013. Nanotechnology, future tools for water remediation. International Journal of Emerging Technology and Advanced Engineering. 3(7): 54–59.

[19] Brame, J., Li, Q. and Alvarez, P.J. 2011. Nanotechnology-enabled water treatment and reuse: emerging opportunities and challenges for developing countries. Trends in Food Science and Technology. 22(11): 618–624.

[20] Bhawana, P. and Fulekar, M. 2012. Nanotechnology: remediation technologies to clean up the environmental pollutants. Research Journal of Chemical Sciences. ISSN. 2231: 606X.

[21] Madhura, L., Singh, S., Kanchi, S., Sabela, M., Bisetty, K. and Inamuddin. 2019. Nanotechnology-based water quality management for wastewater treatment. Environmental Chemistry Letters. 17: 65–121.

[22] Thamarai, P., Kamalesh, R., Saravanan, A., Swaminaathan, P. and Deivayanai, V.C. 2024. Emerging trends and promising prospects in nanotechnology for improved remediation of wastewater contaminants: Present and future outlooks. Environmental Nanotechnology, Monitoring and Management. 100913.

[23] Asghar, N., Hussain, A., Nguyen, D.A., Ali, S., Hussain, I., Junejo, A., et al. 2024. Advancement in nanomaterials for environmental pollutants remediation: a systematic review on bibliometrics analysis, material types, synthesis pathways, and related mechanisms. Journal of Nanobiotechnology. 22(1): 26.

[24] Singh, N., Malik, M.A. and Hashmi, A.A. 2021. Nanotechnology and Its Application in Wastewater Treatment. Sustainable Practices in the Textile Industry. 307–332.

[25] Singh, V., Yadav, P. and Mishra, V. 2020. Recent advances on classification, properties, synthesis, and characterization of nanomaterials. Green Synthesis of Nanomaterials for Bioenergy Applications. 83–97.

[26] Savage, N. and Diallo, M.S. 2005. Nanomaterials and water purification: opportunities and challenges. Journal of Nanoparticle Research. 7: 331–342.

[27] Schwarzenbach, R.P., Egli, T., Hofstetter, T.D., Von Gunten, U. and Wehrli, B. 2010. Global water pollution and human health. Annual Review of Environment and Resources. 35: 109–136.

[28] Qin, B. 2009. Lake eutrophication: Control countermeasures and recycling exploitation. Ecological Engineering. 35(11): 1569–1573.

[29] Tang, M., Lu, S., He, L., Zhu, X., Feng, W. and Zhang, W. 2022. Preparation, Characterization of $ZnTiO_3$/ZnO Composite Materials and Their Photocatalytic Performance. Nanomaterials. 12(8): 1345.

[30] Chaukura, N., Masilompane, T.M., Gwenzi, W. and Mishra, A.K. 2020. Biochar-Based Adsorbents for the Removal of Organic Pollutants from Aqueous Systems. Emerging Carbon-Based Nanocomposites for Environmental Applications. 147–174.

[31] Theron, J., Walker, J.A. and Cloete, T.E. 2008. Nanotechnology and water treatment: applications and emerging opportunities. Critical Reviews in Microbiology. 34(1): 43–69.

[32] Mukhopadhyay, S., Maiti, D., Saha, A. and Devi, P.S. 2016. Shape transition of TiO_2 nanocube to nanospindle embedded on reduced graphene oxide with enhanced photocatalytic activity. Crystal Growth and Design. 16(12): 6922–6932.

[33] Zhang, D.Y., Zheng, Y., Tan, C.P., Sun, J.H., Zhang, W., Ji, L.N., et al. 2017. Graphene oxide decorated with Ru (II)–polyethylene glycol complex for lysosome-targeted imaging and photodynamic/photothermal therapy. ACS Applied Materials and Interfaces. 9(8): 6761–6771.

[34] Mostofizadeh, A., Li, Y., Song, B. and Huang, Y. 2011. Synthesis, properties, and applications of low-dimensional carbon-related nanomaterials. Journal of Nanomaterials. 1–21.

[35] Yu, L., Pengcheng, G., Wen, Y., Shujun, Y., Jian, W. and Xiangke, W. 2017. Adsorption of radionuclide uranium onto carbon-based nanomaterials from aqueous systems. Progress in Chemistry. 29(9): 1062.

[36] Gopinath, K.P., Vo, D.V.N., Gnana Prakash, D., Adithya Joseph, A., Viswanathan, S. and Arun, J. 2021. Environmental applications of carbon-based materials: A review. Environmental Chemistry Letters. 19: 557–582.

[37] Liu, Z., Ling, Q., Cai, Y., Xu, L., Su, J., Yu, K., et al. 2022. Synthesis of carbon-based nanomaterials and their application in pollution management. Nanoscale Advances. 4(5): 1246–1262.

[38] Wang, S., Sun, H., Ang, H.M. and Tadé, M.O. 2013. Adsorptive remediation of environmental pollutants using novel graphene-based nanomaterials. Chemical Engineering Journal. 226: 336–347.

[39] Ali, I., Alharbi, O.M., Tkachev, A., Galunin, E., Burakov, A. and Grachev, V.A. 2018. Water treatment by new-generation graphene materials: hope for a bright future. Environmental Science and Pollution Research. 25: 7315–7329.

[40] Abdelkader, A.M., Cooper, A.J., Dryfe, R.A. and Kinloch, I.A. 2015. How to get between the sheets: a review of recent works on the electrochemical exfoliation of graphene materials from bulk graphite. Nanoscale. 7(16): 6944–6956.

[41] Balandin, A.A. 2011. Thermal properties of graphene and nanostructured carbon materials. Nature Materials. 10(8): 569–581.

[42] Chung, D.D.L. 2016. A review of exfoliated graphite. Journal of Materials Science. 51: 554–568.

[43] Gandhi, M.R., Vasudevan, S., Shibayama, A. and Yamada, M. 2016. Graphene and graphene-based composites: A rising star in water purification, comprehensive overview. Chemistry Select. 1(15): 4358–4385.

[44] Thamaraiselvan, C., Wang, J., James, D.K., Narkhede, P., Singh, S.P., Jassby, D., et al. 2020. Laser-induced graphene and carbon nanotubes as conductive carbon-based materials in environmental technology. Materials Today. 34: 115–131.

[45] Kim, Y.A., Hayashi, T., Endo, M. and Dresselhaus, M.S. 2013. Carbon nanofibers. Springer Handbook of Nanomaterials. 233–262.

[46] Cao, G. 2004. Nanostructures and Nanomaterials: Synthesis, Properties and Applications. Imperial College Press.

[47] Jamil, T., Munir, S., Wali, Q., Shah, G.J., Khan, M.E. and Jose, R. 2021. Water purification through a novel electrospun carbon nanofiber membrane. ACS Omega, 6(50): 34744–34751.

[48] Barhate, R.S. and Ramakrishna, S. 2007. Nanofibrous filtering media: Filtration problems and solutions from tiny materials. Journal of Membrane Science. 296(1–2): 1–8.

[49] Linh, N.T.B., Lee, K.H. and Lee, B.T. 2011. A novel photoactive nano-filtration module composed of a TiO_2-loaded PVA nano-fibrous membrane on sponge Al_2O_3 scaffolds and Al_2O_3-(m-ZrO_2)/t-ZrO_2 composites. Materials Transactions. 52(7): 1452–1456.

[50] Agboola, O., Maree, J. and Mbaya, R. 2014. Characterization and performance of nanofiltration membranes. Environmental Chemistry Letters. 12: 241–255.

[51] Das, R., Bandyopadhyay, R. and Pramanik, P. 2018. Carbon quantum dots from natural resources: A review. Materials Today Chemistry. 8: 96–109.

[52] Shandilya, P., Raizada, P., Sudhaik, A., Saini, A., Saini, R. and Singh, P. 2021. Metal and carbon quantum dot photocatalysts for water purification. Water Pollution and Remediation: Photocatalysis. 81–118.

[53] Lim, S.Y., Shen, W. and Gao, Z. 2015. Carbon quantum dots and their applications. Chemical Society Reviews. 44(1): 362–381.

[54] Zhang, Z., Zheng, T., Li, X., Xu, J. and Zeng, H. 2016. Progress of carbon quantum dots in photocatalysis applications. Particle and Particle Systems Characterization. 33(8): 457–472.

[55] Alshammari, B.H., Lashin, M.M., Mahmood, M.A., Al-Mubaddel, F.S., Ilyas, N., Rahman, N., et al. 2023. Organic and inorganic nanomaterials: fabrication, properties and applications. RSC Advances. 13(20): 13735–13785.

[56] Jeevanandam, J., Barhoum, A., Chan, Y.S., Dufresne, A. and Danquah, M.K. 2018. Review on nanoparticles and nanostructured materials: History, sources, toxicity, and regulations. Beilstein Journal of Nanotechnology. 9(1): 1050–1074.

[57] Dreaden, E.C., Alkilany, A.M., Huang, X., Murphy, C.J. and El-Sayed, M.A. 2012. The golden age: Gold nanoparticles for biomedicine. Chemical Society Reviews. 41(7): 2740–2779.

[58] Ijaz, I., Gilani, E., Nazir, A. and Bukhari, A. 2020. Detailed review on chemical, physical, and green synthesis, classification, characterizations, and applications of nanoparticles. Green Chemistry Letters and Reviews. 13(3): 223–245.

[59] Chavali, M.S. and Nikolova, M.P. 2019. Metal oxide nanoparticles and their applications in nanotechnology. SN Applied Sciences. 1(6): 607.

[60] Balasubramanian, S.B.G.B., Gurumurthy, B. and Balasubramanian, A. 2017. Biomedical applications of ceramic nanomaterials: A review. International Journal of Pharmaceutical Sciences and Research. 8(12): 4950–4659.

[61] Feng, L., Li, W., Bao, J., Zheng, Y., Li, Y., Ma, Y., et al. 2019. Synthesis and photoluminescence properties of silica modified SiO_2@ ANA Si Tb@ SiO_2, SiO_2@ ANA-Si-Tb-L@ SiO_2 core–shell–shell nanostructured composites. Royal Society Open Science. 6(8): 190182.

[62] Thomas, S.C., Kumar Mishra, P. and Talegaonkar, S. 2015. Ceramic Nanoparticles: Fabrication Methods and Applications in Drug Delivery. Current Pharmaceutical Design. 21(42): 6165–6188.

[63] Rao, J.P. and Geckeler, K.E. 2011. Polymer nanoparticles: Preparation techniques and size-control parameters. Progress in Polymer Science. 36(7): 887–913.

[64] Pantic, I. 2010. Magnetic nanoparticles in cancer diagnosis and treatment: novel approaches. Reviews on Advanced Materials Science. 26: 67–73.

[65] Mansha, M., Khan, I., Ullah, N. and Qurashi, A. 2017. Synthesis, characterization, and visible-light-driven photoelectrochemical hydrogen evolution reaction of carbazole-containing conjugated polymers. International Journal of Hydrogen Energy. 42(16): 10952–10961.

[66] Sirohi, R., Kumar, Y., Madhavan, A., Sagar, N.A., Sindhu, R., Bharathiraja, B., et al. 2023. Engineered nanomaterials for water desalination: Trends and challenges. Environmental Technology and Innovation. 30: 103108.

[67] Wang, Z., Horseman, T., Straub, A.P., Yip, N.Y., Li, D., Elimelech, M., et al. 2019. Pathways and challenges for efficient solar-thermal desalination. Science Advances. 5(7): eaax0763.

[68] Qiblawey, H.M. and Banat, F. 2008. Solar thermal desalination technologies. Desalination. 220(1–3): 633–644.

[69] Chadha, U., Selvaraj, S.K., Thanu, S.V., Cholapadath, V., Abraham, A.M., Manoharan, et al. 2022. A review of the function of using carbon nanomaterials in membrane filtration for contaminant removal from wastewater. Materials Research Express. 9(1): 012003.

[70] Maddah, H.A. and Chogle, A.M. 2015. Applicability of low-pressure membranes for wastewater treatment with cost study analyses. Membrane Water Treatment. 6(6): 477–488.

[71] Lafi, R., Gzara, L., Lajimi, R.H. and Hafiane, A. 2018. Treatment of textile wastewater by a hybrid ultrafiltration/electrodialysis process. Chemical Engineering and Processing-Process Intensification. 132: 105–113.

[72] Van der Bruggen, B., Vandecasteele, C., Van Gestel, T., Doyen, W. and Leysen, R. 2003. A review of pressure-driven membrane processes in wastewater treatment and drinking water production. Environmental Progress. 22(1): 46–56.

[73] Favre-Reguillon, A., Lebuzit, G., Foos, J., Guy, A., Draye, M. and Lemaire, M. 2003. Selective concentration of uranium from seawater by nanofiltration. Industrial and Engineering Chemistry Research. 42(23): 5900–5904.

[74] Vingerhoeds, M.H., Nijenhuis-de Vries, M.A., Ruepert, N., van der Laan, H., Bredie, W.L. and Kremer, S. 2016. Sensory quality of drinking water produced by reverse osmosis membrane filtration followed by remineralization. Water Research. 94: 42–51.

[75] Puri, N. and Gupta, A. 2023. Water remediation using titanium and zinc oxide nanomaterials through disinfection and photo catalysis process: A review. Environmental Research. 115786.

[76] He, J., Kumar, A., Khan, M. and Lo, I.M. 2021. Critical review of photocatalytic disinfection of bacteria: from noble metals and carbon nanomaterials-TiO_2 composites to challenges of water characteristics and strategic solutions. Science of the Total Environment. 758: 143953.

[77] Guo, Y., Niu, Z., Huang, J., Ding, Y., Li, X., Song, Y., et al. 2024. Bimetallic peroxide nanocatalytic gel for water disinfection. Journal of Environmental Chemical Engineering. 12(3): 113015.

[78] Abd El-Lateef, H.M., Khalaf, M.M., Kandeel, M. and Gouda, M. 2024. Facile green fabrication of copper/polyvinylidene fluoride nanocomposite for pathogens disinfection and dye degradation in wastewater. Inorganic Chemistry Communications. 161: 112119.

[79] Thiruramanathan, P., Manjula, S., Karthikeyan, N., Srinivasan, R. and Sivakumar, V. 2023. Photodegradation behaviour of sol–gel synthesized $Bi_4Ti_3O_{12}$ for water pollution disinfection. Materials Today: Proceedings.

[80] Liao, J., Wang, L., Ding, S., Tian, G., Hu, H., Wang, Q., et al. 2023. Molybdenum-based antimicrobial nanomaterials: A comprehensive review. Nano Today. 50: 101875.

[81] De Pasquale, I., Porto, C.L., Dell'Edera, M., Curri, M.L. and Comparelli, R. 2021. TiO_2-based nanomaterials assisted photocatalytic treatment for virus inactivation: perspectives and applications. Current Opinion in Chemical Engineering. 34: 100716.

[82] Santos, A.S.G., Ramalho, P.S., Viana, A.T., Lopes, A.R., Gonçalves, A.G., Nunes, O.C., et al. 2021. Feasibility of using magnetic nanoparticles in water disinfection. Journal of Environmental Management. 288: 112410.

[83] Bhardwaj, A.K., Sundaram, S., Yadav, K.K. and Srivastav, A.L. 2021. An overview of silver nanoparticles as promising materials for water disinfection. Environmental Technology and Innovation. 23: 101721.

[84] Enamala, M.K., Chavali, M., Tangellapally, A., Pasumarthy, D., Murthy, M.K., Kuppam, C., et al. 2021. Use of chalcogenide-based nanomaterials for wastewater treatment, including bacterial disinfection and organic contaminants degradation. In Chalcogenide-Based Nanomaterials as Photocatalysts. 243–259. Elsevier.

[85] Bahcelioglu, E., Doganay, D., Coskun, S., Unalan, H.E. and Erguder, T.H. 2020. A point-of-use (POU) water disinfection: Silver nanowire decorated glass fiber filters. Journal of Water Process Engineering. 38: 101616.

[86] Yan, H., Liu, L., Wang, R., Zhu, W., Ren, X., Luo, L., et al. 2020. Binary composite MoS_2/TiO_2 nanotube arrays as a recyclable and efficient photocatalyst for solar water disinfection. Chemical Engineering Journal. 401: 126052.

[87] Hu, Z.T., Chen, Y., Fei, Y.F., Loo, S.L., Chen, G., Hu, M., et al. 2022. An overview of nanomaterial-based novel disinfection technologies for harmful microorganisms: Mechanism, synthesis, devices and application. Science of the Total Environment. 837: 155720.

[88] Mutlag, A.S., Rafiee, E., Khodayari, M. and Eavani, S. 2022. Glass-coated-nanostructure semiconductor $TiO_2/RGO/MoS_2$ for dye removal and disinfection of wastewater: Design and construction of a novel fixed-bed photocatalytic reactor. Materials Science in Semiconductor Processing. 148: 106821.

[89] Matsunaga, T., Tomoda, R., Nakajima, T. and Wake, H. 1985. Photoelectrochemical sterilization of microbial cells by semiconductor powders. FEMS Microbiology Letters. 29(1–2): 211–214.

[90] Herrmann, J.M. 1999. Heterogeneous photocatalysis: fundamentals and applications to the removal of various types of aqueous pollutants. Catalysis Today. 53(1): 115–129.

[91] McCullagh, C., Robertson, J.M., Bahnemann, D.W. and Robertson, P.K. 2007. The application of TiO_2 photocatalysis for disinfection of water contaminated with pathogenic micro-organisms: A review. Research on Chemical Intermediates. 33: 359–375.

[92] Malato, S., Fernández-Ibáñez, P., Maldonado, M.I., Blanco, J. and Gernjak, W. 2009. Decontamination and disinfection of water by solar photocatalysis: recent overview and trends. Catalysis Today. 147(1): 1–59.

[93] Huang, Z., Maness, P.C., Blake, D.M., Wolfrum, E.J., Smolinski, S.L. and Jacoby, W.A. 2000. Bactericidal mode of titanium dioxide photocatalysis. Journal of Photochemistry and Photobiology A: Chemistry. 130(2–3): 163–170.

[94] Sunada, K., Watanabe, T. and Hashimoto, K. 2003. Studies on photokilling of bacteria on TiO_2 thin film. Journal of Photochemistry and Photobiology A: Chemistry. 156(1–3): 227–233.

[95] Cho, M., Chung, H., Choi, W. and Yoon, J. 2004. Linear correlation between the inactivation of *E. coli* and OH radical concentration in TiO_2 photocatalytic disinfection. Water Research. 38(4): 1069–1077.

[96] Chong, M.N., Jin, B., Chow, C.W. and Saint, C. 2010. Recent developments in photocatalytic water treatment technology: a review. Water Research. 44(10): 2997–3027.

[97] Yang, H., He, D., Liu, C., Zhou, X. and Qu, J. 2023. Magnetic photocatalytic antimicrobial materials for water disinfection. Separation and Purification Technology. 124697.

[98] Tang, S. and Zheng, J. 2018. Antibacterial activity of silver nanoparticles: structural effects. Advanced Healthcare Materials. 7(13): 1701503.

[99] Johnson, D.J. and Hilal, N. 2022. Nanocomposite nanofiltration membranes: State of play and recent advances. Desalination. 524: 115480.

[100] Ursino, C., Castro-Muñoz, R., Drioli, E., Gzara, L., Albeirutty, M.H. and Figoli, A. 2018. Progress of nanocomposite membranes for water treatment. Membranes. 8(2): 18.

Chapter **11**

Nanomaterials for Sustainable Agriculture and Food Production

Zaheer Ud Din Sheikh[#1], Neha Kotwal[#1], Deepak Pathania[1], Richa Kothari[1] and Anita Singh[*1,2]

[1]Department of Environmental Sciences, Central University of Jammu, Jammu-181143, Jammu and Kashmir, India

[2]Department of Environmental Studies, Central University of Haryana, Jant-Pali, Mahendergarh, Haryana, India

11.1 INTRODUCTION

Agriculture is fundamental pillar of human civilisation, supporting life, economies, and cultures throughout the world. According to World Bank Open Data, over 26% of the world's workforce is employed in agriculture, making a significant contribution to the Gross Domestic Product (GDP) of many nations [1]. However, this important primary sector faces numerous challenges. Human population is rapidly growing, expected to reach 9.7 billion by 2050, and 10.4 billion by 2100 [2], placing immense pressure on agricultural systems to produce more food. However, this must be achieved against rising environmental challenges, such as soil degradation, water scarcity, and biodiversity loss, in addition to the looming danger of climate change. Furthermore, inequities in access to resources and technology persist, limiting smallholder farmers' ability to adapt and grow. As we manage these challenges, new solutions, including sustainable farming methods,

*For Correspondence: anitasaharan@cuh.ac.in

#Zaheer Ud Din Sheikh and Neha Kotwal have equal contribution as first author

DOI: 10.1201/9781003492184-11

precision agriculture technologies, and equitable policies, are critical to assure food security, preserve natural resources, and create resilient agricultural systems for future generations.

Nanomaterials have received a lot of attention in the agricultural and food production sectors because of their ability to increase crop productivity while reducing environmental impact [3]. Nanomaterials have shown great potential in agriculture and food production, with applications ranging from improving seed germination and growth to enhancing food safety and quality [4, 5]. Nanoagrochemicals, such as nanoformulated biopesticides and nanoclay-reinforced packaging nanocomposites, have been used to protect crops, supply fertilisers, and preserve food [6, 7]. The use of nanoparticles in agroecology, namely silver-based nanomaterials, has shown promise in controlling bacterial and fungal diseases in plants, water, and food [8]. Furthermore, nanosafety issues have become vital in assuring the safe use and disposal of nanomaterials. This chapter deals with the agricultural applications of nanomaterials ranging from soil improvement to pest management and food preservation. Moreover, current trends, future directions and innovations and success stories are also discussed.

11.2 NANOMATERIALS FOR SOIL IMPROVEMENT

Nanofertilizers: Conventional chemical fertilisers though beneficial for crops have poor usage efficiency (losses by volatisation and leaching) making them costly and inefficient. Nanotechnology has brought about modification of these chemical fertilisers by introducing nanofertilizers. They have improved usage efficiency, crop productivity and soil properties. Growth rate of soyabean has increased by about 32% and seed yield by 20% [9]. Plant metabolism grow with increased uptake of nutrients are the other benefits of nanofertilizers. Nanofertilizers work by acting either as carriers/additives for nutrients or by encapsulation of nutrients inside them. They function by controlled or slow release of the fertiliser thus improving usage efficiency. Their transport is a two pathway mechanism i.e. foliar (stomata or trichome roots) and root uptake (root cell membrane). Their absorption and transport in crops are determined by the type or dose of the nanofertilizer, zeta potential of nanoparticles and plant species involved [10]. Various nanofertilizers that are reported to work in agriculture are Zeolites, Nanocomposite, Superabsorbent Fertilisers and Carbon nanotubes [11]. Types of nanofertilizers based on nutrient elements:

Macro-nutrient element nanofertilizer (Nitrogen, P, K)

Micro-nutrient element nanofertilizer (Boron, Molybdenum)

11.2.1 Nanopesticides

Nanopesticides are another application of nanomaterials exceptional usage in agricultural science and technology. They can be defined as products possessing biocidal properties and capabilities containing engineered NPs. Nanopesticides are known to benefit crops by improving their yield, imposing lesser harm to the environment, have higher efficiency and stability with improved pest targeting [12]. Nanoparticles are extensively being used as the formulation of insecticides and pesticides as shown in Figure 11.1. Such nanoparticles work by delivering DNA and other chemicals required by plant tissues to act against insect pests [13]. They manipulate the particle size and structure of pesticides imparting stability and persistence making them last longer on the plant surface. Lipids, Metal Organic Frameworks, Polymers are some forms of nanopesticides and each have different mechanism of pesticide delivery and release [10].

Nano-based inorganic alumina is an inexpensive and reliable option among other insecticides for insect pest control [13]. Recently chiral nanomaterials attracted interest that work by selective adsorption and chiral recognition. They are target specific and thus do not affect useful microorganisms and insects, are environmentally friendly and have better biocompatibility [10].

11.3 SOIL REMEDIATION WITH NANOMATERIALS, BEHAVIOUR AND PERSISTENCE

Understanding the possible behaviour of engineered nanomaterials (ENMs) in the soil matrices, their fate, migration behaviours and transportation aspects are crucial when using ENMs for soil remediation [14]. Their behaviour in a terrestrial environment can be assessed by indices like their adsorption on soil particles, aggregation, agglomeration, indigenous microbial degradation, and dissolution into aquifers [15]. This helps in knowing their persistence and mobility in soil for their consequent bioaccumulation and hazard profile studies. Characters like soil texture (sand, clay and quartz proportions), related SOM, surfactants etc. affect ENMs behaviour in soil [16].

In the soil ENMs may get adsorbed and retarded by soil. SOM, a ubiquitous soil component of the environment gets adsorbed onto soil particles or the ENM surface and affects ENM sorption, transportation, and precipitation behaviour between soil particles. Further SOM affects solubility rates without affecting the dissolution rates of ENMs [17]. Dissolution a critical aspect of fate transformation of ENMs like Cu/CoU, ZnO, nZVI etc. where they are converted to ions causing toxicity to the soil aquifer system (fresh and groundwater), and marine and human health [18]. Factors like ENMs particle size, presence of salts in the environment, coating affect dissolution rates. It has been determined that nZVi coating with stabilisers like chitosan, CMC, polyacrylic acid helps reduce adhesion to surfaces for improved mobility, stability, and reactivity. ENM behaviour in soil still needs more in depth studies for understanding more systematic underground behaviour of ENMs [19].

Nanoparticles usually persist longer in a terrestrial environment and sediments with its metal ions reacting differently in the soil depending on the soil properties and the ageing process [20]. The transformation of NPs in the soil has been studied exhaustively and, it has been reported in a study that the extraction amounts of Cu from the soil changed over time, during different sets of experiments from the soil exposed to CuO NPs or Cu $(NO_3)_2$ and the concentration of Cu and the source were found to affect the behaviour [20]. From another study with silver nanoparticles, it was found that their large clusters in 100–200 nm size were found in tomato roots compared to 1-10 nm size in water suspension indicating the size increase of nanoparticles in plant organs [21]. An enhancement or reduction of the phytotoxicity of NPs is found from different studies like in case of maize, reduced Cu_2O from nano CuO with reduced growth symptoms were found by Servin et al. [22]. Adsorption of Fe_2O_3 NPs on root aquifers like tips, hairs, meristematic zones, and its uptake biomineralised them under the phytochemical influence of roots.

11.4 NANOMATERIALS IN CROP ENHANCEMENT

11.4.1 Nanoencapsulation

Smart delivery methods for nutrients, biofunctional moieties fertilisers and pesticides to plant cells are made possible by nanotechnology. Because of their special qualities, nanoparticles are an excellent choice for a "smart delivery system. One such delivery method that is used is nanoencapsulation, which has the significant eco protective benefit of exposing a relatively small amount of pesticides to the environment, making it safer to use [23]. A nanocapsule structurally contains a shell with elements like lipids, viral capsids, polymers etc. serving as an active compound for the plant to help against diseases and pests. The nanocapsule improves the permeability, solubility and stability of the compound and protects active compound till its released [24]. Further, release of the active compound depends on the shell characteristic features. Nanoencapsulation protects the active compound from premature degradation of enhancing its pest control efficiency for long periods with reduced exposure and dosage of pesticides. The commercialisation still requires proper investigation of their behaviour and fate [25].

11.4.2 Nanobiosensor

A biosensor is such a device with bioreceptor sensitive towards detecting biomolecules and producing quantitative data. Nanotechnology based nanobiosensor possesses outstanding sensitivity and versatility detecting any biochemical and biophysical signal linked to a specific stress for their conversion into a detectable signal. They detect chemicals like pests, abiotic stress, and microbes. They have detected targets using metal nanoparticles (Ag, Au etc), Magnetic NPs, Metal oxide nanoparticles (ZnO, TiO_2) [26].

Nanobiosensors applications in agriculture:

- In assessment of soil quality for detecting soil moisture content, nutrients like nitrates etc. [27]
- In toxic chemical detection like pesticides, insecticides etc. that persist for long time and pose detrimental effects requiring efficient monitoring [28].
- In abiotic stress monitoring like salinity, light, heat, drought, cold etc. that tend to alter the plant biochemical responses and physiology.

Phytohormones like IAA and SA were used for abiotic stress response detection against salinity from soyabean seedlings [29].

- In heavy metal detection among them cadmium, mercury, palladium, arsenic etc. detection with carbon nanotubes and Au nanoparticles are notable [30].

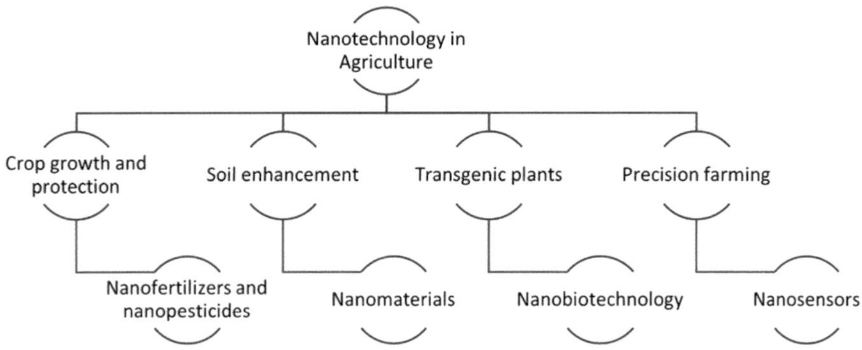

Figure 11.1 Nanotechnological application in agriculture.

Wearable sensors and other smart nanobiosensors are emerging as a novel frontier in hazardous chemical and plant diagnostics, and have the potential to advance both agriculture and plant science research. In a wearable detecting system, such sensors are used directly on the plant elements surfaces like leaves and stems. Wearable sensors with affordability, adaptability, lightweight design, and biocompatibility features have piqued attention in the field of diagnostics. A flexible strain sensor was created by Nassar et al. [31] to measure plant development in addition to the levels of humidity and temperature on the plant leaf surface. Additional intelligent plant leaf sensors, such as graphene sensors (tape-based), have been developed for tracking the behaviour of water transport of the plant in real time [32].

11.4.3 Nanomaterials and Plant Diseases

Identification of the pathogen is the first step in the realm of plant disease, a major limiting factor for crop productivity and strategies like nanosensing, nanodiagnostics have made a significant contribution due to its amazing properties. Nanoparticles with their amazing optical and electrical properties can be fabricated for sensing applications [33]. Various nanostructures for biosensing have been developed with improved detection and performance. Immobilisation of DNA, enzymes etc. has also been achieved through various techniques like encapsulation, covalent bonding or a combination of these [34]. Nanoparticles and antibodies have been integrated to form immunosensors for bacterial detection. Quantum dots based biosensing in disease detection and diagnosis is widely studied e.g. *Candidatus Phytoplasma aurantifolia* based disease of lime named witches broom [35]. Gold nanoparticles with high electroconductivity and activity are used for plant disease detection like AuNPs based electrode in electrochemical detection of a key plant VOC, methyl salicylate, released by plants during infections [36]. DNA-AuNP probes present a promising technological solution for biological sciences precision. They represent a new generation of detection tools that are biosensor-based [37].

Metal oxide nanoparticles, like TiO_2 and SnO_2 can be used for the detection of Volatile Organic Compounds (VOCs), like p-ethylguaiacol, formed from infected strawberries in nanomolar concentrations [38]. Also, in plant disease management nanoparticles like AgNPs, ZnNPs, CuNPs, sulphur and others like nanoforms of magnesium, nickel, iron etc. are reported for antimicrobial activity as shown in Table 11.1. Silver nanoparticles are reported for both antibacterial and antifungal activity, in suppression of plant pathogenic fungi like *Magnaporthe grisea* and bacteria *like Staphylococcus aureus, E. coli.* AgNPs/PVP suppressed different moulds and yeasts like *Candida tropicalis, C. glabrata, Aspergillus brasiliensis* [39].

Table 11.1 Nanomaterials used in plant disease control

S. No.	Nanomaterial	Applications	Ref.
		Silver Nanoparticles	
1.	AgNPs	Inhibitory pathogens such as *Gloeophyllum abietinum* (wood-degrading fungal)	[40]
2.	AgNPs	Disease control in plants	[41]
3	AgNPs	Kills (*Meloidogyne spp.),* root-knot nematodes	[42]
		Copper Nanoparticles	
1.	CuNPs	Inhibits fungal growth and acts against pathogenic fungi Fusarium spp.	[43]
2.	CS-Cu, FQ-Cu, MV-Cu	Acts against *Xanthomonas perforans* (copper-tolerant)	[44]
3.	CuNPs	Effective against *P. hypolateritia* fungi	[45]
4.	CuNPs	They are evaluated for antifungal activity against crop pathogenic *Fusarium* spp.	[46]
		Metal Oxide Nanoparticles	
1.	MgO NPs	Effective for inducing systemic resistance and acts against bacterial wilt disease	[47]
2.	ZnO	Antimicrobial Zincide helping in foliage and fruit protection	[48]
3.	CeO_2	Potential to suppress Fusarium wilt disease in tomato and increased yield	[23]
		Nanoformulations	
1.	Graphene oxide/silver nanoparticles	Acts antibacterial for *Xanthomonas oryzae pv. oryzae*	[49]
2.	Nanosilica	Antifungal effect against phytopathogens	[50]
3.	CNMs	Yield and growth induction improvement of crop. Other functions such as smart delivery system, micronutrient carrier and biosensor	[51]
4.	Engineered NP	Crop yield and nutrient utilisation improvement with disease resistance	[52]

11.5 NANOMATERIALS FOR WATER MANAGEMENT IN AGRICULTURE

Alternative technology for wastewater treatment that is more sustainable and greener is a considerable need as the conventional remediation procedures require pumping wastewater and moving contaminated soil for treatment, which unfortunately results in polluted waste that needs to be disposed of eventually. This makes the process time consuming. It is possible to remediate soils and wastewater on the spot itself without the need for transportation or pumping by using *in situ* nano-remediation techniques [53].

The higher surface-to-volume aspect of nanomaterials have made them the choice of consideration in the area of environmental cleaning and remediation [54]. In order to detect and eliminate the pollutants, like pharmaceutical waste, heavy metals, cyanides and pesticides (and intermediates), microbes, antibiotics in wastewater, viruses, and parasites, use of new nanomembranes, metal or metal oxide nanoparticles and other nano based materials is a promising approach [55]. Nanomaterials like nanosilver and nanoclays are used to enhance water filtration. A unique method for identifying bacterial, viral, and fungal pathogens in soils, water, and plants is the use of nanosensors [26]. There are many modes of nanomaterials that can work in water and wastewater remediation as shown in Figure 11.2.

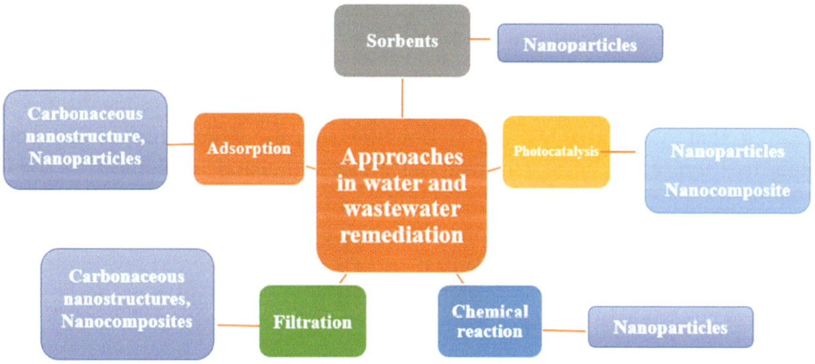

Figure 11.2 Approaches in remediation of wastewater.

The various types of nanomaterials that play a role in water remediation are described below:

11.5.1 Graphene and Derivatives

Graphene-based metals, graphene oxide, reduced graphene oxide, and graphene based metal oxide show a high absorption capacity as they prevent the trapped electrons from reuniting. They also have a large surface area, increased purity, and excellent environmental compatibility and selectivity [56]. An additional significant discovery in recent years was the identification of oxy acids. They are less expensive than metal oxides and have greater photoactivity and surface. The creation of an oxy acid salt occurs when an acid and a base react. In the process of organic pollutants degradation (such as dyes, phenol, etc.), $BiPO_4$ (Eg = 3.85 eV) possess a unique structure and is shown for twice the photocatalytic activity, one tenth of the BET (Brunauer, Emmett, and Teller) surface than TiO_2 (P25) and one-sixth of the cost [57].

11.5.2 Composite Adsorbents

They were created by impregnating metal nanoparticles (NPs) into porous substrates, such as polymers and C_3N_4, in order to improve the use of these NPs in actual wastewater treatment [58]. It has become more important to integrate many nanoparticles into a nanocomposite to investigate any beneficial modifications to their functionality brought about by the synergistic effects of association. Components like methanol and 4-nitrophenol (4NP) have been effectively reduced using sodium borohydride (SBH) through the use of noble metal NMs (mainly Ag, Pd, and Pt) having modifiable internal framework and compositions of shell [57].

11.5.3 Carbon-Based Materials

A class of nanomaterials that has received extensive research is carbon based, which includes carbon dots, fullerenes, nanodiamonds, and nanofoams. Simply put, carbon nanotubes are graphene layers shaped like tubes; depending on how many layers there are, they can have one or more walls [59]. Microplastics (MPs) extraction was explored using magnetic carbon nanotubes (M-CNTs) [60]. The effective extraction of MPs from solutions and wastewater samples has been found, indicating that M-CNTs are a potential material for MP pollution prevention. Owing to their enticing mechanical, electrical, thermal, and physicochemical characteristics, carbon nanotube membranes have demonstrated enormous potential in separation applications. Using batch sorption tests, for the rapid removal of methylene blue, MGOLS i.e., magnetic graphene oxide-functionalised lauryl sulphate when employed demonstrated a remarkable Langmuir loading capacity, reaching a maximum of 624.42 mg/g [61].

11.5.4 Metal Oxide Nanoparticles

Trending nanoparticles prepared recently comprises of metal oxides like titanium dioxide, iron oxide and nanosilver and nanogold like nanoparticles. Recently iron oxide has been usually employed for magnetic adsorbents preparation in water detoxification [62].

11.6 APPLICATIONS OF NANOTECHNOLOGIES IN FOOD INDUSTRY

In the last decade, a staggering number of items have been developed and utilised in the food industry with several food businesses focused on increasing manufacturing efficiency, food qualities, flavour, and safety, as well as developing and promoting innovative products based on nanomaterials. Numerous nanomaterials have demonstrated great potential in all facets of the food industry, from farming to plates.

11.6.1 Food Processing and Packaging

Food packaging, food preservation, and additives are merely a few of the areas where food nanotechnology has permeated consumer goods. Recognition of this innovative technique has progressed food processing and storage to ensure food safety [63]. It has also been discovered that several traditional compounds utilised in food additives and packaging materials partially exist at the nanoscale. For instance, it has been discovered that food-grade TiO_2 NPs exist in the nanoscale range up to about 40% [64]. Long term exposure to nanoparticles like TiO_2 NPs can result in negative effects, even though they are usually acknowledged to be minimally hazardous at ambient conditions. So far, only two items containing unique nanomaterials have been added directly to food: titanium dioxide and iron oxide, which are already employed as food pigments and colourants, respectively.

The prime reason is the lack or very little regulation and legislation pertaining to nanofood, specifically due to the complexity of nanomaterials and the case by case legislative processes. The lack of understanding regarding the potential toxic effects and risks linked with new nanomaterials is to further investigating the reason for the restricted regulation [65]. There is a lack of information regarding the long term effects of nanomaterials, particularly metal nanoparticles (NPs), and most studies concentrate on the toxicity of nanomaterials *in vitro* [66]. The following need to be addressed: transport of nanomaterials into food; fate of nanomaterials; bioaccumulating potentials and their effect on ecosystems; toxicity to tissues and organs as well as chronic effects on the human body. Public approval is another crucial factor that is frequently overlooked by authorities, producers, and academics [67].

Nanomaterials are and have been extensively used in many facets of the food industry, including production, processing, storage, and delivery of food. In the food sector, tested nanomaterials include metal and metal oxide nanoparticles i.e., inorganic and organic and combination (such as clay) [68]. Due to its antibacterial action, silver nanoparticles (NPs) are the most widely manufactured and utilised commercially, whereas gold NPs are extensively researched for use as sensors and detectors. The uses of titanium dioxide nanoparticles (NPs) as a food additive (white colour pigment), flavour enhancer, and disinfectant have all been thoroughly researched [69]. Natural product NPs are typically used in the food business as components or supplements in addition to being designed as a delivery system. Engineered nanostructures have special qualities that make them excellent additives or components in food production. The FDA also approves certain inorganic oxide compounds like MgO (E530), SiO_2 (E551), and TiO_2 (E171) for use as food colour additives, flavour carriers, and anti-caking agents [63]. For example, TiO_2 is frequently added to gum, cake frosting, puddings and candies. As a nutritional dietary supplement in animal feeds, copper oxide, iron oxide, and ZnO are currently classified by the U.S. FDA as "generally recognised as safe" (GRAS) [70]. Dicopper oxide (Cu_2O or copper (I) oxide) usage as a nutritional feed for animal species is determined to be safe for consumer use by the "European Food Safety Authority (EFSA) Panel on Additives and Products or Substances used in Animal Feed (FEEDAP Panel)" [71].

Extension of product shelf life and prevention of rotting and nutrient loss lead to increased food safety. Active packaging serves like an inert shield against external factors in addition to preserving food. The packaging techniques that are susceptible to climate change indicate this. These systems either scavenge toxic gases or create beneficial antioxidant and antibacterial chemicals. Antimicrobial agents, enzyme immobilisation systems, as well as oxygen scavenging techniques are utilised to promote food stability. By using nanocomposites, active packaging extends to deliver functional components including microorganisms, vitamins, and minerals.

In food packaging materials, silver nanoparticles, capable of eradicating bacteria in less than six minutes, are employed to maintain the freshness of food products. Alcohol bottles are another common use for multi-layer polyethylene terephthalate (PET) bottles. In food packaging, nanoparticles of metals and metal-oxides are used in nanocomposites as antibacterial agents. Copper oxide, silver-based nanofillers, titanium dioxide, and zinc oxide are utilised. Since silver nanoparticles are powerful against a broad range of microorganisms, they are specifically popular kind of nanoparticle. It has been found that antibacterial drugs containing silver ions bind to the surface of cells, rupture the membrane, damage DNA, and release silver ions [72]. Many factors, including the nanostructures' shape, huge surface area, particle envelopment, and functionalisation, have a significant role in their antibacterial properties [72]. These nanostructures have been shown to be able to cross both the inner and outer membranes of bacteria. Certain subclasses of nanoclays, such as montmorillonite, kaolinite, hectorite, halloysite etc., are formed based on the physiochemical characteristics of the nanomaterials.

The U.S. FDA has notified effective "Food Contact Substances (FCS)" of montmorillonite and bentonite, listing them as GRAS. The potential of edible coatings containing nanoparticles for food preservation and storage has also been demonstrated. Fruits and vegetables that have been coated, stay edible throughout the storage and transit phases. The highly sensitive characteristics of nanomaterials make them suitable for other uses in food contact packaging, such as the detection of pesticides [73], diseases, and toxins [74]. ZnO Quantum Dots (QD) have been reported by Sahoo et al. [73] to be helpful in the detection of a number of pesticides, including atrazine, glyphosate, tetradifon, and aldrin. Moreover, ZnO QD may potentially degrade pesticides by photochemical reaction. In recent times, "smart packing" and "active packaging" is becoming a recent choice in nanomaterials based sensor utilisation for monitoring the food quality [75].

11.6.2 Nanotechnology Based Solutions for Food Production, Quality, and Safety

Food often loses its quality due to several chemical interactions that occurs between food ingredients and the environment. It is possible to use a variety of nanomaterials to prevent these unwanted processes. An imbalanced redox state results from the presence of nanotoxins in metal and metal oxide nanoparticles, which function by reactive oxygen species generation and oxidative stress induction [76]. Because of this, reduced reactivity nanoparticles are used as carriers of antioxidants [77]. Vitamins and flavonoids that have been encapsulated can be released into an environment that is slightly acidic, like the stomach [78].

Dietary components can have their detrimental effects eliminated with the use of nanotechnology. To assist the body in minimising cholesterol accumulation, certain nutraceuticals are incorporated into the carrier [79]. Most people are unaware that nanostructures in food can affect their health. For example, green tea has nanoselenium, found to improve the absorption of selenium in the body [80]. Since nanoencapsulation involves wrapping the components, guaranteeing product function, and the controlled release of the core, it is an essential part of nanotechnology. Encapsulated chemicals, which possess greater stability, longer shelf life, and a pH-triggered controlled release, are frequently made up of several active molecules [81]. Research has demonstrated that nanoparticles can greatly enhance food materials and packaging. Silicate stacks were used to create nanocomposites with many advantageous characteristics, including as flame and UV light resistance. A plethora of novel natural ingredients have been formulated to improve the appearance and flavour of food, including colour and visual appearance [82].

11.7 ENVIRONMENTAL IMPLICATIONS AND SAFETY CONCERNS

Although there is rising concern about the widespread presence of nanoparticles (NPs) in the environment and associated potential influence on ecosystems, little is known about how many NPs infiltrate agricultural soil systems via direct or indirect paths. Research on the fate, movement, accessibility, and potential environmental risks of metal-based NPs is still in its early stages [83].

11.7.1 Ecotoxicological Implications of the Nanoparticles

Yang and Watts (2005) reported that, alumina nanoparticles tend to inhibit the growth in cabbage, carrot, maize, cucumber, and soyabean roots. Increased amounts of nanosized ferrofluids slowed popcorn development by altering oxidative stress within leaf cells, lowering photosynthesis and metabolic rate [84]. Lin and Xing [85] studied how *Lolium perenne* (ryegrass) internalises and transfers ZnO nanoparticles. Various microscopy techniques revealed that ZnO nanoparticles lowered ryegrass biomass, shrank root tips, and caused considerable vacuolation or collapse of root epidermal and cortical cells. Despite this, Zn migration from roots to shoots was modest when exposed to ZnO nanoparticle treatments. The toxic effects of nanoparticulate ZnO, bulk ZnO, and $ZnCl_2$ to freshwater microalgae was investigated using *Pseudokirchneriella subcapitata*. The study found that nanoparticulate ZnO, bulk ZnO, and $ZnCl_2$ all had similar levels of toxicity, with an IC50 value of approximately 60 μg Zn/L after 72 hours [86]. Karlsson et al. [87] studied several nanoparticles and nanotubes to determine their cytotoxic effect and potential for damaging DNA and oxidative stress. They investigated nanoparticles of metal oxides including CuO, $CuZnFe_2O_4$, Fe_2O_3, Fe_3O_4, TiO_2, and ZnO, as well as carbon nanoparticles and multiwalled carbon nanotubes (MWCNT). The findings revealed considerable differences in the harmful effects of various nanoparticles. CuO nanoparticles caused the highest cytotoxicity and DNA impairment, whereas ZnO influenced both cell survival and DNA integrity. TiO_2 particles largely damaged DNA, but iron oxide particles (Fe_2O_3, Fe_3O_4) exhibited low to zero toxicity, with the exception of $CuZnFe_2O_4$ particles, which were very damaging. Carbon nanotubes caused cytotoxicity and DNA damage at low concentrations. Xia et al. [88] examined the oxidative stress and cytotoxic effect of CeO_2, TiO_2, and ZnO, in RAW 264.7 and BEAS-2B cell lines. The findings revealed that ZnO caused toxicity in both cell types, resulting in reactive oxygen species (ROS) production, oxidant damage, inflammatory response, and the ultimate death of a cell. On the other hand, CeO_2 nanoparticles were internalised into BEAS-2B and RAW 264.7 macrophages, without inducing an inflammatory response or cytotoxicity. The CeO_2 nanoparticles also reduced ROS generation and increased cellular resilience to external oxidative stresses.

Various nanoparticle-mediated transgenic delivery approaches are emerging in the crowded field of plant biotechnology procedures. Along with established methods, modern technologies such as CRISPR provide fascinating opportunities for crop manipulation. However, substantial problems remain, which could be solved by a variety of ways, such as efficient delivery systems for various genomes, new hybrid nanoparticle designs, and advances in pollen magneto function and CRISPR techniques. Regardless of the potential of nanotechnology applications, ongoing support and understanding of these issues are required to enable positive growth in the sector without setbacks [89].

Copper-based nanopesticides, such as Cu $(OH)_2$, can harm spinach plants by hindering their metabolic pathways and lowering antioxidant levels such 4-hydroxybutyric acid, α-tocopherol, ascorbic acid, and ferulic acid. Additionally, crops treated with these nanopesticides showed lower levels of phenolic chemicals [90]. Researchers and toxicologists have expressed concern about nanoparticles' various applications, as well as their build up and incorporation into everyday goods. They have also observed multiple effects, which include the generation of reactive oxygen species (ROS), DNA impairment, genotoxic effects, and organ damage in people [91].

11.7.2 Fate and Behaviour of Nanomaterials in Agricultural Systems

The behaviour of metal-based nanoparticles (NPs) after release into the environment is critical in defining their mobilisation, reactivity, toxic effects, and possible dangers in agricultural systems. Metal-based nanoparticles (NPs) can form and precipitate on the soil matrix, diffuse in the soil solutions, or experience physico chemical reactions with their surrounding components, resulting into different properties. Metal-based NPs' environmental behaviour in agricultural settings is governed by both their inherent qualities and the soil's pore water properties. The size-dependent quantum phenomenon cause NPs to have different physico chemical characteristics. Furthermore, metal-based NPs are frequently produced with specific properties, such as chemical constitution, shape, surface layering, and functionality, to improve their effectiveness, which can have a substantial impact on their environmental behaviour. Furthermore, the soil matrix, which includes solids, solution, and gas phases, is a complicated system that complicates our understanding of metal-based NP behaviour in the environment. The intricacy of metal-based NPs makes it difficult to analyse and predict their environmental behaviour [83, 92–96].

11.7.2.1 Fate in Water

Nanoparticles (NPs) undergo several mechanisms once they enter the water systems, including aggregation, dispersion, solubilisation, settling, photochemical reactions, conversion events, decomposition by living organisms, and reactions with natural colloids and other water constituents. These mechanisms influence the fate, bioavailability, and ecotoxicity of designed nanoparticles in water [97]. Metal-based nanoparticles, for example, can aggregate or disperse depending on pH, ionic strength, and the presence of organic matter. Organic macromolecules can coat NP surfaces, preventing aggregation and increasing their persistence. However, aggregation can still occur in the presence of organic matter and calcium ions, resulting in flocculation. Surface charge also influences NP behaviour, as electrostatically functionalised NPs become unstable in watery settings. Furthermore, NPs dissolve and undergo chemical alteration in response to environmental conditions, which are primarily regulated by redox and pH. Zinc oxide (ZnO) NP dissolution increases in low conditions, producing zinc ions that may bind to ligands and accelerate dissolving in saltwater [98]. Temperature also influences NP solubility, with lower temperatures leading to more solubility due to negative enthalpy [99]. Additionally, photooxidation processes can occur for carbon-based NPs, such as carbon nanotubes (CNTs), when exposed to sunlight, resulting in surface alteration [100]. Overall, knowing these complicated processes is critical for determining the environmental fate and dangers of manufactured nanoparticles in aquatic systems.

effectiveness across various agricultural domains, from soil enrichment to seed treatment, fertilisation to pest management, and water conservation to food preservation. However, there are a few concerns related to application of nanomaterials. Their potential toxicity and environmental impact should be thoroughly understood and addressed before mass production and utilisation. Despite these concerns, nanoparticles offer promising prospects for sustainable agriculture, presenting innovative solutions to improve crop yield, optimise nutrient absorption, and mitigate environmental pressures.

REFERENCES

[1] World Bank. World Bank Open Data. [cited 2024 Apr 30]. Employment in agriculture (% of total employment) (modeled ILO estimate). Available from: https://data.worldbank.org/indicator/SL.AGR.EMPL.ZS?locations=IN#:~:text=by CountryIndicator-,Employment%20in%20agriculture%20(%25%20of%20total%20employment)%20(modeled%20ILO%20estimate,.ilo.org/data.

[2] United Nations. United Nations: Population. [cited 2024 Apr 30] Available from: https://www.un.org/en/global-issues/population

[3] Liu, Y., Yue, L., Wang, Z. and Xing, B. 2019. Processes and mechanisms of photosynthesis augmented by engineered nanomaterials. Environmental Chemistry. 16(6): 430.

[4] Khot, L.R., Sankaran, S., Maja, J.M., Ehsani, R. and Schuster, E.W. 2012. Applications of nanomaterials in agricultural production and crop protection: A review. Crop Protection. 35: 64–70.

[5] Tuantranont, A., Wisitsoraat, A., Kiatpathomchai, W., Khumwan, P., Karuwan, C., Primpray, V., et al. 2021. Nanomaterials in agricultural and food applications. *In*: Lau, W.J., Faungnawakij, K., Piyachomkwan, K., Ruktanonchai, U.R. (Eds). Handbook of Nanotechnology Applications [Internet]. Elsevier. [cited 2024 Apr 30]. 383–404. Available from: https://linkinghub.elsevier.com/retrieve/pii/B9780128215067000168

[6] Eghonghon Ukhurebor, K. 2021. The Role of Biosensor in Climate Smart Organic Agriculture toward Agricultural and Environmental Sustainability. In: Swaroop Meena R. (Ed.). Agrometeorology [Internet]. IntechOpen. [cited 2024 Apr 30]. Available from: https://www.intechopen.com/books/agrometeorology/the-role-of-biosensor-in-climate-smart-organic-agriculture-toward-agricultural-and-environmental-sus

[7] Jampílek, J. and Kráľová, K. 2019. Nanobiopesticides in agriculture: State of the art and future opportunities. *In*: Koul, O. (Ed.). Nano-Biopesticides Today and Future Perspectives [Internet]. Elsevier. [cited 2024 Apr 30]. 397–447. Available from: https://linkinghub.elsevier.com/retrieve/pii/B9780128158296000188

[8] Abd-Elsalam, K.A. 2021. Silver-based nanomaterials for sustainable applications in agroecology: A note from the editor. *In*: Abd-Elsalam, K.A. (Ed.). Silver Nanomaterials for Agri-Food Applications [Internet]. Elsevier. [cited 2024 Apr 30]. 1–14. Available from: https://linkinghub.elsevier.com/retrieve/pii/B9780128235287000019

[9] Liu, R. and Lal, R. 2014. Synthetic apatite nanoparticles as a phosphorus fertilizer for soybean (Glycine max). Scientific Reports. 4(1): 5686.

[10] Tang, Y., Zhao, W., Zhu, G., Tan, Z., Huang, L., Zhang, P., et al. 2023. Nano-pesticides and fertilizers: solutions for global food security. Nanomaterials. 14(1): 90.

[11] Usman, M., Farooq, M., Wakeel, A., Nawaz, A., Cheema, S.A., Rehman, H.U., et al. 2020. Nanotechnology in agriculture: Current status, challenges and future opportunities. Science of the Total Environment. 721: 137778.

[12] Fraceto, L., Castro, V.L., Grillo, R., Ávila, D.S., Oliveira, H. and Lima, R. (Eds). 2020. Nanopesticides: From Research and Development to Mechanisms of Action and Sustainable Use in Agriculture [Internet]. Cham: Springer International Publishing. [cited 2025 May 8]. Available from: https://link.springer.com/10.1007/978-3-030-44873-8

[13] Rautela, I., Dheer, P., Thapliyal, P., Shah, D., Joshi, M., Upadhyay, S., et al. 2021. Current Scenario and Future Perspectives of Nanotechnology in Sustainable Agriculture and Food Production. Plant Cell Biotechnology and Molecular Biology. 22: 99–121.

[14] Marcon, L., Oliveras, J. and Puntes, V.F. 2021. *In situ* nanoremediation of soils and groundwaters from the nanoparticle's standpoint: A review. Science of the Total Environment. 791: 148324.

[15] Perea Vélez, Y.S., Carrillo-González, R. and González-Chávez, MaDCA. 2021. Interaction of metal nanoparticles–plants–microorganisms in agriculture and soil remediation. Journal of Nanoparticle Research. 23(9): 206.

[16] Qian, Y., Qin, C., Chen, M. and Lin, S. 2020. Nanotechnology in soil remediation—Applications vs. implications. Ecotoxicology and Environmental Safety. 201: 110815.

[17] Gao, X., Rodrigues, S.M., Spielman-Sun, E., Lopes, S., Rodrigues, S., Zhang, Y., et al. 2019. Effect of soil organic matter, soil pH, and moisture content on solubility and dissolution rate of CuO NPs in soil. Environmental Science and Technology. 53(9): 4959–4967.

[18] Garner, K.L. and Keller, A.A. 2014. Emerging patterns for engineered nanomaterials in the environment: a review of fate and toxicity studies. Journal of Nanoparticle Research. 16(8): 2503.

[19] Mishra, D., Pandey, V. and Khare, P. 2021. Engineered Nanoparticles in Agro-ecosystems: Implications on the Soil Health. *In*: Singh, P., Singh, R., Verma, P., Bhadouria, R., Kumar, A., Kaushik, M. (Eds). Plant-Microbes-Engineered Nano-particles (PM-ENPs) Nexus in Agro-Ecosystems [Internet]. Cham: Springer International Publishing. [cited 2025 May 10]. 103–118. (Advances in Science, Technology & Innovation). Available from: http://link.springer.com/10.1007/978-3-030-66956-0_7

[20] Ahmed, B., Rizvi, A., Ali, K., Lee, J., Zaidi, A. and Khan, M.S., et al. 2021. Nanoparticles in the soil–plant system: a review. Environmental Chemistry Letters. 19(2): 1545–1609.

[21] Siddiqi, K.S. and Husen, A. 2022. Plant response to silver nanoparticles: a critical review. Critical Reviews in Biotechnology. 42(7): 973–90.

[22] Servin, A.D., Pagano, L., Castillo-Michel, H., De La Torre-Roche, R., Hawthorne, J., Hernandez-Viezcas, J.A., et al. 2017. Weathering in soil increases nanoparticle CuO bioaccumulation within a terrestrial food chain. Nanotoxicology. 11(1): 98–111.

[23] Gour, N., Upadhyaya, P. and Patel, J. 2019. Nanomaterials as therapeutic and diagnostic tool for controlling plant diseases. *In*: Comprehensive Analytical Chemistry [Internet]. Elsevier. [cited 2025 May 10]. 225–261. Available from: https://linkinghub.elsevier.com/retrieve/pii/S0166526X19300236

[24] Desai, S., Singh, M., Chavan, A., Wagh, N.S. and Lakkakula, J. 2022. Micro- and nanoencapsulation techniques in agriculture. *In*: Agricultural Nanobiotechnology [Internet]. Elsevier. [cited 2025 May 10]. 297–323. Available from: https://linkinghub.elsevier.com/retrieve/pii/B9780323919081000092

[25] Mondal, R., Dam, P., Chakraborty, J., Paret, M.L., Katı, A., Altuntas, S., et al. 2022. Potential of nanobiosensor in sustainable agriculture: the state-of-art. Heliyon. 8(12): e12207.

[26] Singh, R.P. 2021. Recent Trends, Prospects, and Challenges of Nanobiosensors in Agriculture. *In*: Pudake, R.N., Jain, U., Kole, C. (Eds). Biosensors in Agriculture: Recent Trends and Future Perspectives [Internet]. Cham: Springer International Publishing. [cited 2025 May 10]. 3–13. (Concepts and Strategies in Plant Sciences). Available from: http://link.springer.com/10.1007/978-3-030-66165-6_1

[27] Bharti, A., Jain, U. and Chauhan, N. 2024. From lab to field: Nano-biosensors for real-time plant nutrient tracking. Plant Nano Biology. 9:100079.

[28] Panoth, D., Manikkoth, S.T., Jahan, F., Thulasi, K.M., Paravannoor, A. and Vijayan, B.K. 2022. Nanosensors for pesticide detection in soil. *In*: Nanosensors for Smart Agriculture [Internet]. Elsevier. [cited 2025 May 10]. 237–258. Available from: https://linkinghub.elsevier.com/retrieve/pii/B9780128245545000021

[29] Hu, Y., Wang, X., Wang, C., Hou, P., Dong, H., Luo, B., et al. 2020. A multifunctional ratiometric electrochemical sensor for combined determination of indole-3-acetic acid and salicylic acid. RSC Advances. 10(6): 3115–3121.

[30] Guo, Z., Kang, Y., Liang, S. and Zhang, J. 2020. Detection of Hg(II) in adsorption experiment by a lateral flow biosensor based on streptavidin-biotinylated DNA probes modified gold nanoparticles and smartphone reader. Environmental Pollution. 266: 115389.

[31] Nassar, J.M., Khan, S.M., Villalva, D.R., Nour, M.M., Almuslem, A.S. and Hussain, M.M. 2018. Compliant plant wearables for localized microclimate and plant growth monitoring. NPJ Flexible Electronics. 2(1): 24.

[32] Oren, S., Ceylan, H., Schnable, P.S. and Dong, L. 2017. High-Resolution Patterning and Transferring of Graphene-Based Nanomaterials onto Tape toward Roll-to-Roll Production of Tape-Based Wearable Sensors. Advanced Materials Technologies. 2(12): 1700223.

[33] Kalia, A., Abd-Elsalam, K.A. and Kuca, K. 2020. Zinc-Based Nanomaterials for Diagnosis and Management of Plant Diseases: Ecological Safety and Future Prospects. JoF. 6(4): 222.

[34] Kralova, K. and Jampilek, J. 2013. Applications of nanomaterials in plant disease management and protection. *In*: Nanotechnology in Agriculture and Agroecosystems [Internet]. Elsevier. [cited 2025 May 10]. 239–296. Available from: https://linkinghub.elsevier.com/retrieve/pii/B9780323994460000131

[35] Rad, F., Mohsenifar, A., Tabatabaei, M., Safarnejad, M.R., Shahryari, F., Safarpour, H., et al. 2012. Detection of Candidatus Phytoplasma aurantifolia with a quantum dots fret-based biosensor. Journal of Plant Pathology. 94(3): 525–534.

[36] Umasankar, Y. and Ramasamy, R.P. 2013. Highly sensitive electrochemical detection of methyl salicylate using electroactive gold nanoparticles. Analyst. 138(21): 6623.

[37] Hu, L., Fu, X., Kong, G., Yin, Y., Meng, H.M., Ke, G., et al. 2020. DNAzyme–gold nanoparticle-based probes for biosensing and bioimaging. Journal of Materials Chemistry B. 8(41): 9449–9465.

[38] Fang, Y., Umasankar, Y. and Ramasamy, R.P. 2014. Electrochemical detection of p-ethylguaiacol, a fungi infected fruit volatile using metal oxide nanoparticles. Analyst. 139(15): 3804–3810.

[39] Khan, M.R., Rizvi, T.F. and Ahamad, F. 2019. Application of Nanomaterials in Plant Disease Diagnosis and Management. *In*: Abd-Elsalam, K.A., Prasad, R. (Eds). Nanobiotechnology Applications in Plant Protection [Internet]. Cham: Springer International Publishing. [cited 2025 May 10]. 19–33. (Nanotechnology in the Life Sciences). Available from: http://link.springer.com/10.1007/978-3-030-13296-5_2

[40] Alghuthaymi, M.A., Almoammar, H., Rai, M., Said-Galiev, E. and Abd-Elsalam, K.A. 2015. Myconanoparticles: synthesis and their role in phytopathogens management. Biotechnology and Biotechnological Equipment. 29(2): 221–236.

[41] Ardakani, A.S. 2013. Toxicity of silver, titanium and silicon nanoparticles on the root-knot nematode, Meloidogyne incognita, and growth parameters of tomato. Nematol. 15(6): 671–677.

[42] Narayanan, K.B. and Park, H.H. 2014. Antifungal activity of silver nanoparticles synthesized using turnip leaf extract (Brassica rapa L.) against wood rotting pathogens. European Journal of Plant Pathology. 140(2): 185–192.

[43] Viet, P.V., Nguyen, H.T., Cao, T.M. and Hieu, L.V. 2016. Fusarium Antifungal Activities of Copper Nanoparticles Synthesized by a Chemical Reduction Method. Journal of Nanomaterials. 2016: 1–7.

[44] Strayer-Scherer, A., Liao, Y.Y., Young, M., Ritchie, L., Vallad, G.E., Santra, S., et al. 2018. Advanced Copper Composites Against Copper-Tolerant Xanthomonas perforans and Tomato Bacterial Spot. Phytopathology®. 108(2): 196–205.

[45] Ponmurugan, P., Manjukarunambika, K., Elango, V. and Gnanamangai, B.M. 2016. Antifungal activity of biosynthesised copper nanoparticles evaluated against red root-rot disease in tea plants. Journal of Experimental Nanoscience. 11(13): 1019–1031.

[46] Bramhanwade, K., Shende, S., Bonde, S., Gade, A. and Rai, M. 2016. Fungicidal activity of Cu nanoparticles against Fusarium causing crop diseases. Environmental Chemistry Letters. 14(2): 229–235.

[47] Imada, K., Sakai, S., Kajihara, H., Tanaka, S. and Ito, S. 2016. Magnesium oxide nanoparticles induce systemic resistance in tomato against bacterial wilt disease. Plant Pathology. 65(4): 551–560.

[48] Graham, J.H., Johnson, E.G., Myers, M.E., Young, M., Rajasekaran, P., Das, S., et al. 2016. Potential of Nano-Formulated Zinc Oxide for Control of Citrus Canker on Grapefruit Trees. Plant Disease. 100(12): 2442–2447.

[49] Liang, Y., Yang, D. and Cui, J. 2017. A graphene oxide/silver nanoparticle composite as a novel agricultural antibacterial agent against Xanthomonas oryzae pv. oryzae for crop disease management. New Journal of Chemistry. 41(22): 13692–13699.

[50] Goswami, P., Mathur, J. and Srivastava, N. 2022. Silica nanoparticles as novel sustainable approach for plant growth and crop protection. Heliyon. 8(7): e09908.

[51] Verma, S.K., Das, A.K., Gantait, S., Kumar, V. and Gurel, E. 2019. Applications of carbon nanomaterials in the plant system: A perspective view on the pros and cons. Science of the Total Environment. 667: 485–499.

[52] Verma, S.K., Das, A.K., Patel, M.K., Shah, A., Kumar, V. and Gantait, S. 2018. Engineered nanomaterials for plant growth and development: A perspective analysis. Science of the Total Environment. 630: 1413–1435.

[53] Raja, M.A. and Husen, A. 2020. Role of nanomaterials in soil and water quality management. *In*: Nanomaterials for Agriculture and Forestry Applications [Internet]. Elsevier. [cited 2025 May 10]. 491–503. Available from: https://linkinghub.elsevier.com/retrieve/pii/B9780128178522000202

[54] Chenab, K.K., Sohrabi, B., Jafari, A. and Ramakrishna, S.. 2020. Water treatment: functional nanomaterials and applications from adsorption to photodegradation. Materials Today Chemistry. 16: 100262.

[55] Tripathy, J., Mishra, A., Pandey, M., Thakur, R.R., Chand, S., Rout, P.R., et al. 2024. Advances in Nanoparticles and Nanocomposites for Water and Wastewater Treatment: A Review. Water. 16(11): 1481.

[56] Nisa, M.U., Nadeem, N., Yaseen, M., Iqbal, J., Zahid, M., Abbas, Q., et al. 2023. Applications of graphene-based tungsten oxide nanocomposites: A review. Journal of Nanostructure in Chemistry. 13(2): 167–196.

[57] Taghipour, S., Hosseini, S.M. and Ataie-Ashtiani, B. 2019. Engineering nanomaterials for water and wastewater treatment: review of classifications, properties and applications. New Journal of Chemistry. 43(21): 7902–7927.

[58] Khodakarami, M. and Bagheri, M. 2021. Recent advances in synthesis and application of polymer nanocomposites for water and wastewater treatment. Journal of Cleaner Production. 296: 126404.

[59] Elgarahy, A.M., Elwakeel, K.Z., Akhdhar, A. and Hamza, M.F. 2021. Recent advances in greenly synthesized nanoengineered materials for water/wastewater remediation: an overview. Nanotechnology for Environmental Engineering. 6(1): 9.

[60] Tang, Y., Zhang, S., Su, Y., Wu, D., Zhao, Y. and Xie, B. 2021. Removal of microplastics from aqueous solutions by magnetic carbon nanotubes. Chemical Engineering Journal. 406: 126804.

[61] Yakout, A.A., Shaker, M.A., Elwakeel, K.Z. and Alshitari, W. 2019. Lauryl sulfate@ magnetic graphene oxide nanosorbent for fast methylene blue recovery from aqueous solutions. Journal of Dispersion Science and Technology. 40(5): 707–715.

[62] Abdel Maksoud M.I.A., Elgarahy A.M., Farrell, C., Al-Muhtasebm A.H., Rooney, D.W. and Osman, A.I. 2020. Insight on water remediation application using magnetic nanomaterials and biosorbents. Coordination Chemistry Reviews. 403: 213096.

[63] Siddiqui, S. and Alrumman, S.A. 2021. Influence of nanoparticles on food: An analytical assessment. Journal of King Saud University—Science. 33(6): 101530.

[64] Dudefoi, W., Terrisse, H., Richard-Plouet, M., Gautron, E., Popa, F., Humbert, B., et al. 2017. Criteria to define a more relevant reference sample of titanium dioxide in the context of food: a multiscale approach. Food Additives and Contaminants: Part A. 1–13.

[65] He, X., Deng, H. and Hwang, H min. 2019. The current application of nanotechnology in food and agriculture. Journal of Food and Drug Analysis. 27(1): 1–21.

[66] Wyser, Y., Adams, M., Avella, M., Carlander, D., Garcia, L., Pieper, G., et al. 2016. Outlook and Challenges of Nanotechnologies for Food Packaging. Packaging Technology and Science. 29(12): 615–648.

[67] Jagtiani, E. 2022. Advancements in nanotechnology for food science and industry. Food Frontiers. 3(1): 56–82.

[68] Hoseinnejad, M., Jafari, S.M. and Katouzian, I. 2018. Inorganic and metal nanoparticles and their antimicrobial activity in food packaging applications. Critical Reviews in Microbiology. 44(2): 161–181.

[69] Dikshit, P., Kumar, J., Das, A., Sadhu, S., Sharma, S., Singh, S., et al. 2021. Green Synthesis of Metallic Nanoparticles: Applications and Limitations. Catalysts. 11(8): 902.

[70] Lehto, S., Buchweitz, M., Klimm, A., Straßburger, R., Bechtold, C. and Ulberth, F. 2017. Comparison of food colour regulations in the EU and the US: a review of current provisions. Food Additives and Contaminants: Part A. 34(3): 335–355.

[71] EFSA Panel on Additives and Products or Substances used in Animal Feed (FEEDAP). Safety and efficacy of dicopper oxide as feed additive for all animal species. EFS2 [Internet]. 2016 Jun [cited 2025 May 10];14(6). Available from: https://data.europa.eu/doi/10.2903/j.efsa.2016.4509

[72] Yinn I.X., Zhang, J., Zhao, I.S., Mei, M.L., Li, Q. and Chu, C.H. 2020. The Antibacterial Mechanism of Silver Nanoparticles and Its Application in Dentistry. IJN. 15: 2555–2562.

[73] Sahoo, D., Mandal, A., Mitra, T., Chakraborty, K., Bardhan, M. and Dasgupta, A.K. 2018. Nanosensing of Pesticides by Zinc Oxide Quantum Dot: An Optical and Electrochemical Approach for the Detection of Pesticides in Water. Journal of Agricultural and Food Chemistry. 66(2): 414–423.

[74] Stephen Inbaraj, B. and Chen, B.H. 2016. Nanomaterial-based sensors for detection of foodborne bacterial pathogens and toxins as well as pork adulteration in meat products. Journal of Food and Drug Analysis. 24(1): 15–28.

[75] Mustafa, F. and Andreescu, S. 2020. Nanotechnology-based approaches for food sensing and packaging applications. RSC Advances. 10(33): 19309–19336.

[76] Biswas, R., Alam, M., Sarkar, A., Haque, M.I., Hasan, Md M. and Hoque, M. 2022. Application of nanotechnology in food: processing, preservation, packaging and safety assessment. Heliyon. 8(11): e11795.

[77] Kumar, H., Bhardwaj, K., Nepovimova, E., Kuča, K., Singh Dhanjal, D., Bhardwaj, S., et al. 2020. Antioxidant Functionalized Nanoparticles: A Combat against Oxidative Stress. Nanomaterials. 10(7): 1334.

[78] Wang, Y, Nowack, B. 2018. Environmental risk assessment of engineered nano-SiO_2, nano iron oxides, nano-CeO_2, nano-Al_2O_3, and quantum dots. Environmental Toxicology and Chemistry. 37(5): 1387–1395.

[79] Ndlovu, N., Mayaya, T., Muitire, C. and Munyengwa, N. 2020. Nanotechnology Applications in Crop Production and Food Systems. International Journal of Plant Breeding and Crop Science. 7(1): 603–613.

[80] Xiang, J., Rao, S., Chen, Q., Zhang, W., Cheng, S., Cong, X., et al. 2022. Research Progress on the Effects of Selenium on the Growth and Quality of Tea Plants. Plants. 11(19): 2491.

[81] Zhang, L., Yu, D., Regenstein, J.M., Xia, W. and Dong, J. 2021. A comprehensive review on natural bioactive films with controlled release characteristics and their applications in foods and pharmaceuticals. Trends in Food Science and Technology. 112: 690–707.

[82] Sun, R., Lu, J. and Nolden, A. 2021. Nanostructured foods for improved sensory attributes. Trends in Food Science and Technology. 108: 281–286.

[83] Chen, H. 2018. Metal based nanoparticles in agricultural system: behavior, transport, and interaction with plants. Chemical Speciation and Bioavailability. 30(1): 123–134.

[84] Răcuciu, M. and Creanga, D. 2007. TMA-OH coated magnetic nanoparticles internalized in vegetal tissue. Romanian Journal of Physics. 52.

[85] Lin, D. and Xing, B. 2008. Root Uptake and Phytotoxicity of ZnO Nanoparticles. Environmental Science and Technology. 42(15): 5580–5585.

[86] Franklin, N.M., Rogers, N.J., Apte, S.C., Batley, G.E., Gadd, G.E. and Casey, P.S. 2007. Comparative toxicity of nanoparticulate ZnO, bulk ZnO, and $ZnCl_2$ to a freshwater microalga (Pseudokirchneriella subcapitata): The Importance of particle solubility. Environmental Science and Technology. 41(24): 8484–8490.

[87] Karlsson, H.L., Cronholm, P., Gustafsson, J. and Möller, L. 2008. Copper Oxide Nanoparticles Are Highly Toxic: A Comparison between Metal Oxide Nanoparticles and Carbon Nanotubes. Chemical Research in Toxicology. 21(9): 1726–1732.

[88] Xia, T., Kovochich, M., Liong, M., Mädler, L., Gilbert, B., Shi, H., et al. 2008. Comparison of the Mechanism of Toxicity of Zinc Oxide and Cerium Oxide Nanoparticles Based on Dissolution and Oxidative Stress Properties. ACS Nano. 2(10): 2121–2134.

[89] Kumari, R., Suman, K., Karmakar, S., Mishra, V., Lakra, S.G., Saurav, G.K., et al. 2023. Regulation and safety measures for nanotechnology-based agri-products. Frontiers in Genome Editing. 5: 1200987.

[90] Upadhyay, S.K., Kumar, S., Rout, C., Vashistha, G. and Aggarwal, D. 2022. Risks, and concerns of use of nanoparticles in agriculture. *In*: Rajputm V.D., Verma, K.K., Sharma, N., Minkina, T. (Eds). The Role of Nanoparticles in Plant Nutrition under Soil Pollution [Internet]. Cham: Springer International Publishing. [cited 2024 May 5]. 371–394. (Sustainable Plant Nutrition in a Changing World). Available from: https://link.springer.com/10.1007/978-3-030-97389-6_16

[91] Fröhlich, E. and Fröhlich, E. 2016. Cytotoxicity of Nanoparticles Contained in Food on Intestinal Cells and the Gut Microbiota. IJMS. 17(4): 509.

[92] Deb, S.K. and Shukla, M.K. 2011. A Review of Dissolved Organic Matter Transport Processes Affecting Soil and Environmental Quality. Journal of Environmental and Analytical Toxicology. [Internet]. [cited 2024 May 5]. 01(02). Available from: https://www.omicsonline.org/a-review-of-dissolved-organic-matter-transport-processes-affecting-soil-and-environmental-quality-2161-0525.1000106.php?aid=2306

[93] Hotze, E.M., Phenrat, T. and Lowry, G.V. 2010. Nanoparticle Aggregation: Challenges to Understanding Transport and Reactivity in the Environment. Journal of Environmental Quality. 39(6): 1909–1924.

[94] Lin, D., Tian, X., Wu, F. and Xing, B. 2010. Fate and transport of engineered nanomaterials in the environment. Journal of Environmental Quality. 39(6): 1896–1908.

[95] Praetorius, A., Labille, J., Scheringer, M., Thill, A., Hungerbühler, K. and Bottero, J.Y. 2014. Heteroaggregation of Titanium Dioxide Nanoparticles with Model Natural Colloids under Environmentally Relevant Conditions. Environmental Science and Technology. 48(18): 10690–10698.

[96] Schmid, G. 2011. Nanoparticles: From Theory to Application. John Wiley & Sons.

[97] Peng, C., Zhang, W., Gao, H., Li, Y., Tong, X., Li, K., et al. 2017. Behavior and Potential Impacts of Metal-Based Engineered Nanoparticles in Aquatic Environments. Nanomaterials. 7(1): 21.

[98] Miao, A.J., Zhang, X.Y., Luo, Z., Chen, C.S., Chin, W.C., Santschi, P.H., et al. 2010. Zinc oxide–engineered nanoparticles: Dissolution and toxicity to marine phytoplankton. Environmental Toxicology and Chemistry. 29(12): 2814–2822.

[99] Reed, R.B., Ladner, D.A., Higgins, C.P., Westerhoff, P. and Ranville, J.F. 2012. Solubility of nano-zinc oxide in environmentally and biologically important matrices. Environmental Toxicology and Chemistry. 31(1): 93–99.

[100] Deline, A.R., Frank, B.P., Smith, C.L., Sigmon, L.R., Wallace, A.N., Gallagher, M.J., et al. 2020. Influence of Oxygen-Containing Functional Groups on the Environmental Properties, Transformations, and Toxicity of Carbon Nanotubes. Chemical Reviews. 120(20): 11651–11697.

[101] Bondarenko, L., Kahru, A., Terekhova, V., Dzhardimalieva, G., Uchanov, P. and Kydralieva, K. 2020. Effects of Humic Acids on the Ecotoxicity of Fe_3O_4 Nanoparticles and Fe-Ions: Impact of Oxidation and Aging. Nanomaterials. 10(10): 2011.

[102] Zykova, M.V., Volikov, A.B., Buyko, E.E., Bratishko, K.A., Ivanov, V.V., Konstantinov, A.I., et al. 2023. Enhanced Antioxidant Activity and Reduced Cytotoxicity of Silver Nanoparticles Stabilized by Different Humic Materials. Polymers (Basel). 15(16): 3386.

[103] Yusuf, E.O., Amber, I., Officer, S. and Oluyemi, G.F. 2024. Transport of nanoparticles in porous media and associated environmental impact: A review. Journal of Engineering Research. S2307187724000063.

[104] Jackson, P., Jacobsen, N.R., Baun, A., Birkedal, R., Kühnel, D., Jensen, K.A., et al. 2013. Bioaccumulation and ecotoxicity of carbon nanotubes. Chemistry Central Journal. 7: 154.

[105] Pereira, A.D.E.S., Oliveira H.C. and Fraceto L.F. 2019. Polymeric nanoparticles as an alternative for application of gibberellic acid in sustainable agriculture: A field study. Scientific Reports. 9(1): 7135.

[106] Acharya, P., Jayaprakasha, G.K., Crosby, K.M., Jifon, J.L. and Patil, B.S. 2020. Nanoparticle-mediated seed priming improves germination, growth, yield, and quality of watermelons (Citrullus lanatus) at multi-locations in Texas. Scientific Reports. 10(1): 5037.

[107] Gao, Y., Li, D., Li, D., Xu, P., Mao, K., Zhang, Y., et al. 2020. Efficacy of an adhesive nanopesticide on insect pests of rice in field trials. Journal of Asia-Pacific Entomology. 23(4): 1222–1227.

[108] Gao, X., Shi, F., Peng, F., Shi, X., Cheng, C., Hou, W., et al. 2021. Formulation of nanopesticide with graphene oxide as the nanocarrier of pyrethroid pesticide and its application in spider mite control. RSC Advances. 11(57): 36089–36097.

[109] El-Ghany, M.F.A., El-Kherbawy, M.I., Abdel-Aal, Y.A., El-Dek, S.I. and Abd El-Baky, T. 2021. Comparative Study between Traditional and Nano Calcium Phosphate Fertilizers on Growth and Production of Snap Bean (Phaseolus vulgaris L.) Plants. Nanomaterials. 11(11): 2913.

[110] Abd El-Azeim, M.M., Sherif, M.A., Hussien, M.S., Tantawy, I.A.A. and Bashandy, S.O. 2020. Impacts of nano- and non-nanofertilizers on potato quality and productivity. Acta Ecologica Sinica. 40(5): 388–397.

Chapter **12**

Life Cycle Assessment of Nanomaterials

Mansi Sharma*[1] and Garima Slathia[2]

[1]Assistant Professor, Department of Physics & Environmental Sciences, Sharda School of Engineering & Science, Sharda University, Greater Noida-201310, Uttar Pradesh, India

[2]Analyst, Agro Tech Aromatics Pvt. Ltd., SIDCO Complex, Bari Brahmna, Samba-181133, Jammu and Kashmir, India

12.1 INTRODUCTION

12.1.1 The Rise of Nanomaterials and their Applications

The concept of "nanotechnology" was first introduced in 1950s by an American physicist, Richard Feynman in his talk on "*There's Plenty of Room at the Bottom*". The key elements of his speech revolved around the idea that human work abilities are not guided by the laws of nature, but rather by research limitations at the atomic or molecular levels in terms of incompetent equipment or methods. This eventually paved the way for sowing seeds of modern technology in scientific research minds' [1]. Nanotechnology, slowly surpassing its infancy stage, has witnessed impressive strides in the last century. This burgeoning field continues to offer innovative solutions for the development of novel materials with the hope that these nanomaterials can find applications in various sectors (e.g., agriculture, food, medicine, energy storage, environmental monitoring, etc.). Nanomaterials, precisely categorized as tiny particles measuring 1–100 nanometers in size, have been incorporated into various applications, albeit unknowingly by humans, for

*For Correspondence: mansi.sharma@sharda.ac.in

DOI: 10.1201/9781003492184-12

ages [2]. This traces back to 4,500 years ago, when the use of asbestos nanofibers by mankind was evident in reinforced ceramic mixtures [3]. Likewise, lead sulfide nanoparticles were used by Egyptians about 4,000 years ago in the formulation of hair dyes [4]. These materials, crafted at the nanoscale, harbor a repertoire of unique properties, including electrical, optical, magnetic, etc., mainly owing to their extensive surface area, exceptionally high surface energy, and behavior governed by quantum mechanics. Nanomaterials (NMs) or nanoparticles (NPs) act as a connecting link between bulky materials and atomic/molecular structures, making them a topic of intense scientific interest. Based on the composition, NMs can be grouped into: (a) inorganic (e.g., metal/metal oxide-based NPs), (b) organic, (c) carbon-based (e.g., graphene, hollow tubes, carbon nanofibers), and composite-based (e.g., nano copper-carbon chitosan alginate). A multitude of applications of NMs in various fields like chemical industries, medicine, electrical devices, photocatalysis, cosmetics, environmental clean up have been reported (Figure 12.1), thanks to their remarkable physical and chemical properties [5]. For example, food packaging materials are increasingly incorporating NMs like polymer-clay nanocomposites to create barriers that shield products from environmental damage [6]. Cutting edge tools like nanosensors are being utilized in rapid and accurate detection of microbial pathogens in food storage and manufacturing [7]. Interestingly, NMs are making waves in environmental sensing applications, particularly on site cleaning of polluted areas, treatment of wastewater, enhancing energy efficiency of solar cells, and water and air purification [8].

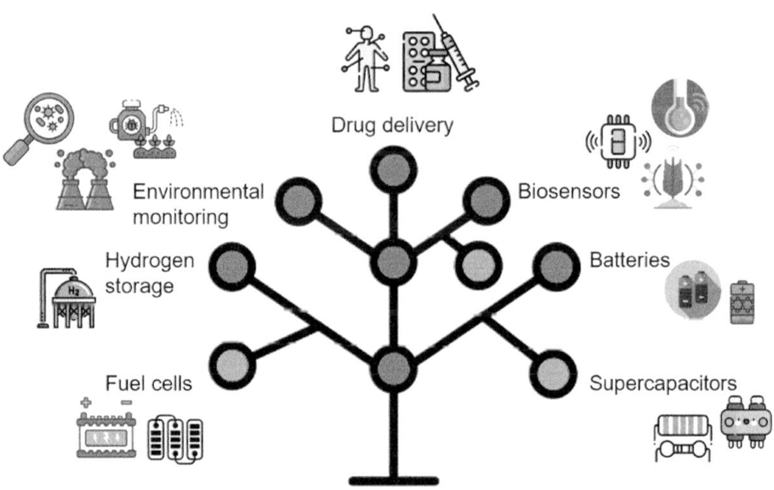

Figure 12.1 A pictorial representation of the various applications of nanomaterials.

Although NMs have exciting applications in diverse fields, their inadvertent use poses a significant risk for health and the environment. The most commonly used methods for NM synthesis rely on bottom-up approaches (e.g., carbonization, vapor deposition and activation, self-assembly, etc.), which are energy- and resource-intensive. This often leads to the release of pollutants in soil, water, and air in the form of emissions or effluents [9]. In the past, various studies have highlighted the exceptional functionalities of NMs across various applications; however, a wide lacuna still exists in identifying the potential environmental impacts of NMs throughout their life span. Moreover, present NM pathways raise questions about the sustainability of their production and use, due to their potential to cause unforeseen environmental damage [10]. In this regard, the Life Cycle Assessment (LCA) tool is a key to unlocking sustainability, closing the gap in our understanding, and creating eco-friendly NMs.

12.1.2 An Introduction of Life Cycle Assessment (LCA)

Life cycle assessment (LCA) evaluates the environmental footprint of NMs from cradle-to-grave stages, including the type of raw materials used, energy consumed, and environmental emissions. As a standardized approach that follows ISO (International Organization for Standardization) 14040 guidelines, LCA aims to investigate a product's environmental impact by a four stage framework: (i) description of the goals and scope of assessment, (ii) creation of a product's life cycle inventory, (iii) assessment of the environmental and health risks of product, and (iv) interpretation of the findings and analysis of the product's overall environmental impact [11] (Figure 12.2). Despite previous attempts at LCA for NMs, there are several roadblocks. These may include, for instance, difficulty capturing all the unique characteristics of NMs in a single analysis, withholding information regarding material or energy inputs during NM production by the companies owing to secrecy concerns, and failure of current methods of NM characterization to accurately assess their environmental impacts once released [12].

The rise of nanotechnology has made it imperative to investigate the potential environmental risks of NMs using LCA. In general, LCA helps in quantifying the environmental impacts, including energy use and emissions released. By acknowledging the disadvantages of a product (e.g., NMs), LCA can empower scientists, product designers, and consumers to make informed decisions for a sustainable future. However, due to inconsistencies in research in terms of methods, data collection, characterization, and generation of inconclusive or contradictory results, a critical appraisal of current LCA practices in nanotechnology is deemed necessary [13].

12.2 LIFE CYCLE STAGES OF NANOMATERIALS

Life cycle assessment of a product takes into account the resource requirements and emissions released in the entire span of a product's life cycle in order to evaluate and better understand the environmental and human health effects connected to that system or process. Finding any relevant life cycle stage is the first step in conducting LCA, since it allows for the evaluation of every scenario that may pose a risk. Because every product has unique uses, manufacturing processes, and waste management procedures thus, unique hotspots for nanomaterial release and

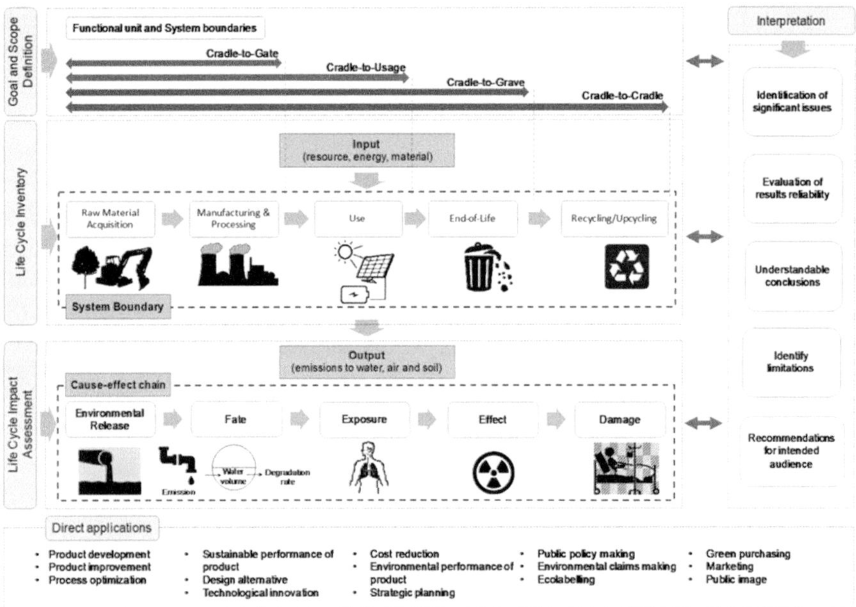

Figure 12.2 Environmental impact of a product: The Life Cycle Assessment (LCA) Framework (Reproduced from Nizam et al. [12] https://doi.org/10.3390/nano11123324).

related risks, and the life cycle are entirely product-dependent. Regardless of the perceived environmental benefits at the "use" stage of NMs, to get a complete and accurate picture of environmental impacts associated with NMs, all the life cycle stages including material processing and manufacturing must be involved [14]. The life cycle of a product is divided into various stages from resource extraction and production to end of life of the product. Broadly, the life cycle stages of NMs are classified into three types, i.e., upstream, middle stream, and downstream.

12.2.1 Upstream

The upstream stage consists of resource extraction, production, or synthesis of NMs.

12.2.1.1 Resource Extraction and Processing

The first life cycle stage, i.e., extraction of raw materials and manufacturing of NMs involves the integration of multidisciplinary branches, including chemistry, engineering, material sciences, and biology so that the unique properties of nanoscale materials for different applications can be harnessed. Some of the most common sources of NMs include metals or their oxides (e.g., NPs), carbon-based NMs (e.g., graphene, nanotubes), hybrid materials or their composites, and biological sources (e.g., plant extracts, yeast, bacteria, fungi). The most widely used approach for NM development is the synthetic method, which enables the creation of NMs via chemical, physical, or biological processes. Other sustainable methods incorporated in the resource extraction for NMs include recycling of waste materials to extract NMs, thereby minimizing environmental impacts and resource depletion [15].

12.2.1.2 Manufacturing of Nanomaterials

All strategies of NM synthesis fall into two categories: (a) bottom-up and (b) top-down (Figure 12.3). In the bottom-up approach, tiny building blocks are self assembled into NPs to create new particles of nanoscale dimensions, using biological or chemical methods. On the contrary, top-down strategy involves the breakdown of bulky materials into tiny NPs using physical methods. Table 12.1 provides a detailed description of the various top-down and bottom-up methods used for the specific type of NM synthesis.

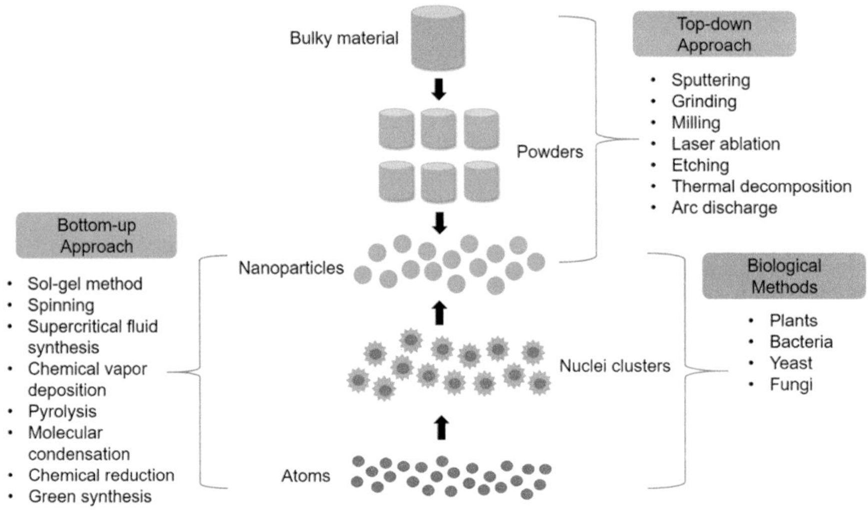

Figure 12.3 A schematic diagram of the synthesis of nanomaterials via different approaches.

Table 12.1 Various top-down and bottom-up approaches adopted for the synthesis of special nanomaterials

Method	Description	Type of Nanomaterial	Ref.
Top-down Methods			
Mechanical Milling	Produces nanomaterials from bulk material and mixtures of various phases	Nanocomposites	[2]
Sputtering	Bombards hard solid surfaces with the particles carrying high energy (e.g., plasma, gas) using magnetron, radio-frequency, or DC diode	Thin nano-films	[16]

Method	Description	Type of Nanomaterial	Ref.
Lithography	Uses electrons or a focused beam of light	Nanoarchitectures	[17]
Arc discharge method	Uses two graphite rods positioned inside a chamber to create an electric current and produce heat that vaporizes carbon from rods	Fullerenes, carbon nanohorns, carbon nanotubes, graphene, amorphous spherical carbon nanoparticles	[2]
Thermal decomposition	Uses heat for the chemical breakdown of materials at a particular temperature	Metal nanoparticles, Carbon quantum dots	[18]
Laser ablation	Uses plasma condensation and a laser beam to vaporize the source material as a result of the high energy of laser irradiation	Fullerenes	[19]
Bottom-up Methods			
Pyrolysis	Burning of precursor (vapor or liquid state) at high pressure in a furnace	Carbon-based nanomaterials	[20]
Sol-gel	Dissolving the alkoxide molecular precursors in water or alcohol and heating and stirring to cause hydrolysis and convert precursor into gel	Metal oxide nanoparticles	[21]
Chemical vapor deposition	Involves the heating of a substrate at high temperature and introducing a gas into the chamber as a precursor to decompose the gas and release atoms which recombine to form nanomaterials	Carbon-based nanomaterials (e.g., carbon nanotubes, graphene)	[22]
Soft and hard templating	Involves the interaction between soft templates (e.g., flexible organic molecules, block copolymers, and surfactants [cationic anionic, non-ionic]) and precursors via hydrogen bonding, Van der Waals or electrostatic interactions. On the other hand, hard templating or nanocasting involves injecting the precursor molecules into solid template pores	Nanoporous materials	[23]
Reverse micelle	Involves water-in-oil emulsion to produce reverse micelles where hydrophilic heads point towards a water-containing core. The core serves as a nanoreactor for nanoparticle synthesis	Nanoparticles	[24]

12.2.2 Midstream

The midstream life cycle stage includes the integration of NMs in products along with their applications and interactions after getting released into the environment.

12.2.2.1 Nanomaterial Use and Integration in Products

Nanomaterials are endowed with a plethora of unique properties, including chemical, magnetic, or optical properties, attracting their wide utility in diverse fields ranging from agriculture, electronics, and displays to coatings, films, textiles, energy storage, biomedical devices, automotive industries, and environmental monitoring. For instance, the precise delivery of bioactive compounds or nutrients via nanoencapsulation, pathogen detection by nanosensors, fruit or vegetable preservation using edible nanofilms, and nanomedicines for fighting viral infections are some of the applications of NMs in agriculture or food sector [25, 26]. Besides this, other useful NM applications are seen in water pollution remediation (e.g., using nanoadsorbents), manufacture of energy storage devices such as supercapacitors (e.g., using carbon-based electrode NMs), and enhancing the efficiency of solar cells (e.g., using nanocoatings) [27–29].

Two key factors are described to govern the unique characteristics of NMs: (i) NM incorporation into diverse matrices (e.g., food, composites, cement, etc.) and (ii) NM interaction with the surroundings. It is necessary to optimize the functionalities of NMs while designing NM-based products. This is especially useful for complex NMs, like polymers and metal-organic frameworks, that require integration of more than one unique property at once. To maximize the functionality of NMs, "Multifunctional Efficiency" that builds on the idea of atom economy in green chemistry, is a powerful tool that helps design NMs by minimizing the presence of inactive components and creating a simple, sustainable process for NM synthesis [30]. Besides possessing increased functionalities, NMs or nano-enabled products are also expected to be safe for humans as well as the environment throughout their life cycle. There are two key concepts that need to be addressed while considering NM safety: (i) hazard, i.e., the potential of NMs to cause damage and (ii) exposure, i.e., the dose encountered by humans or the environment over time. Thus, safe applications of NMs require a comprehensive analysis of potential hazards and risks associated with them, right from manufacturing to disposal. Taken together, it can be said that NMs with a low hazard potential, but frequent exposure may be equally risky as NMs with a high hazard potential and limited exposure [31].

12.2.2.2 Potential Release and Environmental Interactions

Although NMs are now an essential part of many applications, there are significant environmental and health concerns which limit their widespread use. This is mainly ascribed to the fact that NMs are susceptible to atmospheric release, landfill waste, soil, and water resources from various chemical, engineering, electronics, and medical industries. Therefore, it is essential to comprehend the behavior of NMs in diverse environments, their routes of exposure, and effects on human health due to their uncontrolled release into the environment. Figure 12.4 below depicts the interactions of NMs with different biota when released into the environment.

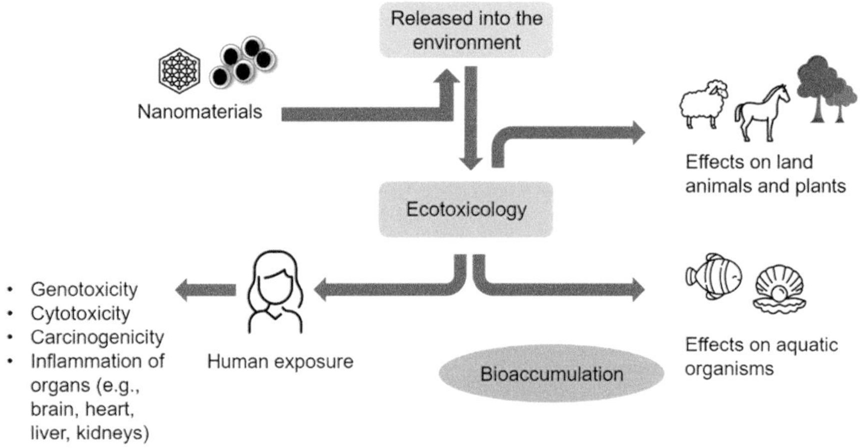

Figure 12.4 A schematic illustration of the interactions of nanomaterials with the environment biota.

The release of NMs into the environment can be through various routes: (a) industrial effluents, (b) accidental exposure or (c) decomposition of materials containing NMs in them. A tangible container or a product that contains embedded or used NMs can be the way by which nanowaste is released from its source. At any point in the lifetime, these NMs can become nanowaste and end up in the environment. For example, NMs can be released into the environment in a number of ways during the manufacturing process. During use, they may be released as a result of normal wear and tear such as abrasion of painted walls. Lastly, they might be released during disposal or the end of life in a landfill, or during incineration and recycling processes [32]. Exposure of NMs to humans can occur through various routes, like injection, ingestion, inhalation, and skin penetration. Some of the possible release mechanisms of NMs contained in different products are listed in Table 12.2. It is obvious and alarming that NMs are present in large quantities in water supplies, air, and soil used to grow food crops. One of the important factors impeding our understanding of the fate of NMs in the environment is their nanodimension. To forecast long term effects, a better comprehension of the possibilities of toxicity of newly introduced NMs, particularly emerging nanomaterials (ENMs), is necessary [33]. By taking proactive steps, the effects of NMs on the ecosystem, environment, and human health, can be mitigated.

Table 12.2 Possible release mechanisms of nanomaterials from different sources

Sources	Release Mechanism	Environmental Matrix	Route of Exposure
Nanomaterial manufacture	Released directly	Air	Inhalation
Vehicular traffic	Released directly	Air	Inhalation
Stationary combustion sources	Released directly	Air	Inhalation
Sprays/Aerosols	Released directly	Air	Inhalation
Environmental remediation processes	Released directly	Soil, sediments, surface water, and wastewater	Ingestion, Dermal contact
Airstreams	Direct deposition (dry/wet)	Soil and water	Inhalation, Ingestion
Food/ Food packaging	Product application, Leaching	Wastewater and food	Ingestion, Dermal contact
Paint dissolution/Land application of sludge sites	Subsurface runoff	Soil and water	Ingestion, Dermal contact
Wastewater sludge	Application as a biosolid, Leaching	Soil, subsurface water, and groundwater	Ingestion, Dermal contact
Wastewater discharge	Released directly	Surface water	Ingestion
Coatings	Product application, Photodegradation, Weathering	Air, soil, and water	Inhalation, Ingestion, Dermal contact
Textiles	Washing	Wastewater	Ingestion
Cosmetic products	Product application	Water and wastewater	Ingestion
Medical products	Product application	Wastewater	Ingestion
Composites/Plastics	Weathering, Photodegradation, Abrasion	Soil and air	Inhalation, Ingestion, Dermal contact
Tires	Abrasion	Soil and air	Inhalation
Household appliances	Leaching	Water	Ingestion
Electronics	Recycling, Disposal	Air and wastewater	Inhalation, Ingestion, Dermal contact
Automotives	Recycling, Disposal	Air and wastewater	Inhalation, Ingestion, Dermal contact

12.2.3 Downstream

The downstream stage of the life cycle consists of the end-of-life management and recycling of NMs. Nanomaterials need to have their environmental impact evaluated, their recycling options investigated, and safe disposal techniques taken into consideration for responsible end of life management.

12.2.3.1 End of Life Management

There is a continuous stream of new products containing NMs being introduced to the market. Due to this, an increasing number of NM-enabled products reach the end of their useful lives and are disposed of. Common methods for managing solid nanowaste include landfilling, incineration, and recycling (Figure 12.5). However, if the waste is not properly handled, NMs have the potential to pollute soil, groundwater, and other environmental compartments. After combustion of NMs in waste incineration plants, the waste ash emitted is likely to be composed of NMs. Although incineration of NMs can increase the chances of leftover NPs, but there is no need of discharging nanowaste into the atmosphere. For instance, cerium oxide NPs were recovered during incineration process by facilitating their binding to the sludge in the exhaust phase and extracted from flue gas using special filters [34]. These NPs are transferred to dumping sites where incinerated waste is further collected and processed. However, in some cases, a significant problem often encountered is the generation of various toxic pollutants, including polyaromatic hydrocarbons, dioxins, and furans during incineration. Therefore, proper handling of ash generated in incineration plants is critical while conducting the end of life cycle management of NMs [35].

12.2.3.2 Recycling and Disposal of Nanomaterials

Another strategy of NM treatment involves the disposal of NMs in landfills, especially the waste products, but there are some limitations attached to this method. For example, addition of NMs to landfills influences waste degradation, processing of leachates, and potential release of NM into the environment. This can lead to continued emergence of NMs in soil and groundwater if they are mostly contained in waste dumping and disposal leachates [36]. As a result, researchers are now interested in synthesizing and manufacturing sustainable nanocomposites utilizing green nanocomposites, biodegradable nanocomposites made with renewable raw materials, and magnetic NPs. These approaches aim to prevent the disposal of non-recyclable NMs, particularly man made engineered NMs, and mitigate their negative environmental effects. Thus far, recycling is thought to be a sustainable NM treatment strategy, highly preferred over landfilling [37]. The process mainly utilizes physical, mechanical, or chemical processes, like cleaning, grinding, sorting, pulping, extrusion, melting, desiccation, reforming, and refining. A number of factors during recycling may affect the release of NMs into the environment: (a) type of mechanism, (b) stiffness of waste matrix, (c) phase temperature, and (d) NM affinity to various phases (e.g., Sair, solid, liquid). For example, the release of nano-/macro-scale particles through grinding of carbon nanotube-polypropylene composite, however, the same nanocomposite reported drastically lower emissions using the shredding process [38]. Likewise, recycling and remelting of glass waste may release the NMs into water or exhaust gas. In particular, the techniques for solvent evaporation and extraction offer versatile approaches for the preparation and recycling of NMs, due to their simplicity, scalability, and lower residual solvent potential [39]. For example, in the era of precision medicine, continued developments in recycled NM design are broadening the diagnostic and therapeutic roles of these materials for many biomedical applications [40].

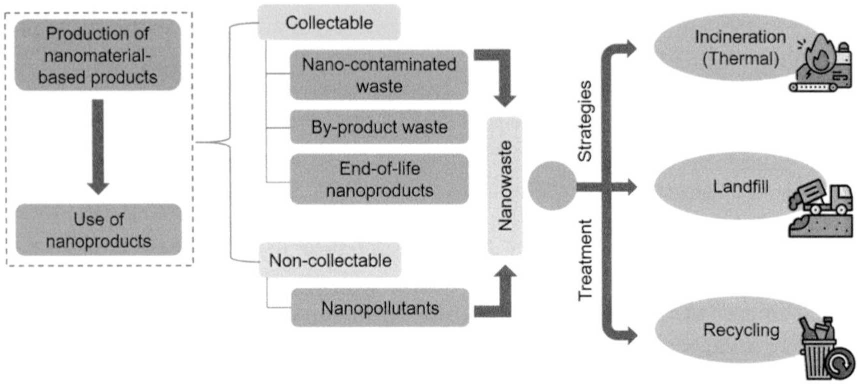

Figure 12.5 A pictorial representation of nanowaste generation and its possible treatment strategies.

12.3 LCA METHODOLOGY FOR NANOMATERIALS

There are several challenges that need to be taken into consideration while conducting the LCA of engineered NMs. These include: (i) defining the purpose, function, and benefits of NM in the "Goal and Scope Definition" phase, (ii) gathering complete data concerning the entire life cycle of NM in the "Inventory Analysis" phase, and (iii) categorizing the factors affecting LCA and developing ways to measure the specific environmental impacts of NMs in the "Impact Assessment" phase.

12.3.1 Definition of Goals and Scope

As stated by the ISO, the "scope" of LCA needs to be properly defined in terms of detailed, indepth, and sufficient information pertaining to the goal that needs to be addressed. The key goal of most of the LCA studies is to review the environmental impacts of NMs, from the acquisition of raw materials to manufacturing and processing phases. In LCA, this initial stage defines the aim of the study, the product under consideration, its system boundaries, and functional unit [12]. For example, engineered NMs often outperform conventional materials owing to their exceptional characteristics and functionalities, eventually terminating in high profits. In order to correctly assess their environmental benefits, consideration of their useful properties is an essential part of the scope. This specific phase considers the ways to handle the various pathways of NMs in the environment. A fixed value for the amount of NM, i.e., functional unit is also set as a reference standard in the entire LCA study. The results on environmental impacts from various impact categories are based on this functional unit, and any limitations in the study (e.g., data uncertainties, margins of error) need to be clearly identified [41]. A functional unit only based on weight of the material is relatively insufficient. Thus, a comprehensive LCA should be inclusive of all the linearly scaled parameters, like functioning and performance of systems, input and output materials, and energy consumed [42].

The system boundary covers all the processes and activities related to the life cycle stages of NMs [9]. Most studies on metal-/carbon-based NMs and nanocomposites only cover cradle-to-gate approach, i.e., a LCA including extraction and transportation of raw materials to manufacturing of the desired product and its use. This often neglects the disposal of product and potential toxicity of NMs released into the environment. This is quite necessary because combustion of the product might release NMs into the environment and thus, affect the various ecosystems, natural resources, and human health. Future LCA studies should therefore focus on creation until disposal phase of NMs, i.e., cradle-to-grave approach, which also involves NM recycling or reuse [43].

12.3.2 Inventory Analysis: Identification of Inputs and Outputs

Life cycle inventory (LCI) phase requires the collection of data and is the most vital, but demanding step in LCA. This is mainly attributed to its prolonged time and high labor needed for the collection of detailed data on the various processes included in NM scoping. Gathering high-quality data for LCA studies of NMs is by itself a challenging task. Unfortunately, the data outlining clear explanations of assumptions, pros, and cons of data collection methods, and meeting the standards of transparency and reliability, are often limited. To fill the gaps, researchers rely on estimations and secondary or generic data, that are highly uncertain and limit the scope of LCA [44]. While a number of up-to-date databases exist that work with LCA software, the main limitation is their genericity and applicability only to processes that are not specific to a particular product. "Ecoinvent" is a popular choice among the various inventory databases, storing information on the inputs (e.g., raw materials, energy, and water consumption) and outputs (e.g., products, co-products, environmental emissions) of LCA [41]. Other databases, including BEES and ILCD also exist, which use specific models according to the scope and objectives of LCA, assessing their environmental impact, fate, and risk of exposure [14]. To date, most of the studies are lacking in sufficient output data coverage in relation to emissions released into the environment throughout their life cycle. Furthermore, NM synthesis or manufacturing processes are primarily retrieved from secondary sources, i.e., literature reviewing or laboratory data.

Ideal LCA studies should thus, focus on gathering LCI data at each stage of life cycle, for example, the origin and composition of NMs, transformation or alteration of properties of NMs, etc., after their release into the environment [12].

12.3.3 Impact Assessment: Categorization and Evaluation

Life cycle impact assessment (LCIA) phase is aimed at analyzing the LCI data to identify the environmental impacts of NMs in different impact categories. Several steps involved in this phase include classifying the impacts, categorizing their severity, normalizing the results, evaluating, and weighing the impacts on specific environmental compartments [45]. The impact score of each impact category (e.g., climate change, eutrophication, ecotoxicity, etc.) is calculated by multiplication of the released mass (e.g., into air) with a characterization factor [46]. Assessment of the environmental performance of NMs and their impacts at the mid-point level (e.g., eutrophication, formation of smog, depletion of ozone, climate change, etc.) has been a focus of majority of previous LCA studies. However, only a quarter of the studies have considered the full impact, reaching the end-point level of LCIA. Commonly focused areas in these studies were global warming, acidification, and ecotoxicity due to their significant environmental concerns, such as bioaccumulation and negative effects on human health or other living beings [12]. One of the studies [47] that carried out LCIA throughout the life cycle of NMs (i.e., end-point level), until disposal explored the environmental impacts of nano-titanium dioxide coatings. Although improved environmental performance of these coatings was reported, but their long-term effects after dumping in landfills have not been well-researched. Therefore, careful compilation of LCIA data on the end-of-life stage is imperative for a LCA approach to reduce uncertainties into the overall results of various studies.

There are three major factors that govern the flow of toxicity of NMs in humans, namely, the fate factor, exposure factor, and effect factor (Figure 12.6). Accurately predicting the NM environmental fate and behavior requires a consideration of potential NM sources, distribution after their release, transformation, and environmental persistence [48]. Unfortunately, many limitations exist in current models as they cannot track NM transport owing to their inconclusive and unquantifiable nature [41]. It is highly useful to study the different processes that affect the movement and behavior of NMs using accurate models. However, there exists a major lacuna in most of the LCA studies of NMs, lacking information about their release and environmental effects on various ecosystems over time [14, 49]. Although the analysis of the fate factor of NMs remains untapped, but this area certainly holds significant promise. For example, Barton et al. [50] investigated the fate factor and transformation (e.g., hetero-aggregation) of cerium dioxide NPs during wastewater treatment that might affect their redox abilities.

The value of exposure factor greatly varies with input parameters of fate and movement of NMs. Moreover, the value is based on assumptions and might only consider the possible highest exposure in worst scenarios. Eckelman et al. [51] were the first to study the exposure factor to evaluate carbon nanotubes in a water body, taking into account estimated concentrations of dissolved organic carbon, suspended solids, and aquatic organisms. Quantification of potential exposure of NMs can be difficult due to our limited understanding of the various mechanisms of NM exposure in the environment. To accurately assess NM environmental exposure, gaining data on the magnitude, frequency, and duration of NM exposure is quintessential, but till date, reliable methods to quantify NM concentrations in the environment have not been developed. Besides this, accurately differentiating between natural and manufactured NMs is another challenge. Some existing models, like MFA might predict how much NM is present in a certain geographical area, but a high uncertainty level in data and lack of reliable methods to confirm their outcomes, is most likely [49]. Several NMs might also adopt alternate routes of disposal, consequently ending in various exposure routes [52].

The effect factor mainly covers the impact of NMs on various ecosystems, i.e., the ecotoxicity aspect of NMs. The factor is essentially based on toxicological data. For instance, researchers estimate the effect factor in freshwater ecotoxicity by calculating the chronic effective concentrations in aquatic organisms, whereas the effect factor for toxicity in humans might be based on lethal or effective doses in animals [53]. Although a wide variety of methods exist to assess NM toxicity for both the environment and human health, there is still room for improvement. Significant challenges that make the process difficult include the tiny size of NMs for conducting toxicity tests and varied toxicity of different NMs (e.g., size, thickness, surface properties, surface functionalities). Therefore, more direct studies on the chronic effects of NMs are needed to strengthen the data used to determine effect factors [9, 54].

12.3.4 Life Cycle Interpretation: Results and Conclusions

Selection of the right assessment tool for interpretation of results in LCA studies is necessary because different frameworks offer different data types and insights. Choosing a wrong LCA framework or misinterpreting the results can lead to misleading outcomes. To avoid this, approaches such as multicriteria decision analysis and comparative, sensitivity, and perturbation analyses can be employed [42]. Although current LCA studies in this direction are largely lacking, but integrating these decision-making techniques can strengthen LCA interpretations, especially for cradle-to-gate and cradle-to-grave stages [55]. A few studies have proposed possible suggestions for the reduction of environmental impacts of NMs. This can be achieved by optimization of extraction step as NM synthesis relies heavily on electricity during extraction. Nevertheless, using less energy, i.e., lower temperatures might also decrease NM production. According to Gilbertson et al. [56], the development of carbon nanotubes results in higher impacts mainly during lithography stages (e.g., etching trenches, contact leads, deposition of metals). Interestingly, the impacts on both the environment and human health were similar throughout the LCA, from the mid-point to the end-point. In general, energy consumption stands out as the most significant factor that impacts the environment and human health in the mid-point or end-point assessment. For example, there are higher chances of ozone depletion, formation of particulate matter, toxicity risks for humans, and production of ionizing radiations in the end-point stage [12].

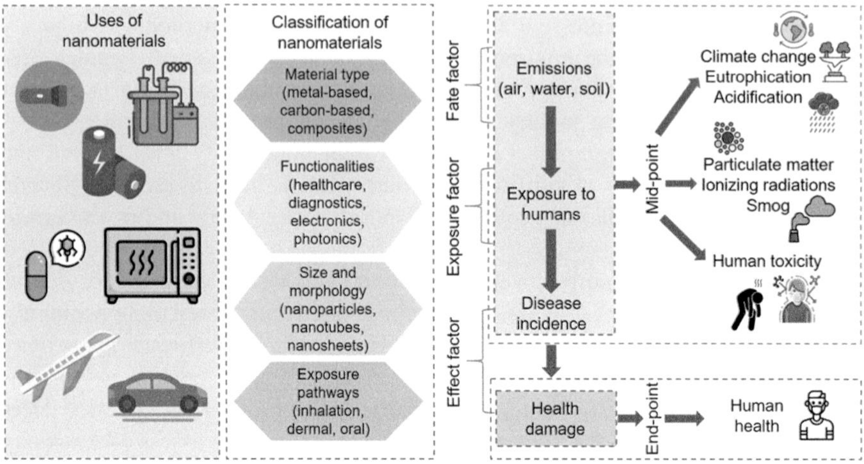

Figure 12.6 An illustration depicting the uses of nanomaterials and their toxicological effects on environment and human health for conducting a life cycle assessment.

12.4 CHALLENGES AND CONSIDERATIONS

12.4.1 Data Availability and Uncertainty

The LCA studies for NMs are inherently uncertain due to a multitude of factors, which include model structure, deciding parameters, variation in impacts on the environment across space and time, and the nature of uncertainty itself [49]. The uncertainty in selection of model and framework for conducting LCA studies on NMs can produce inconsistent and conflicting results. Nevertheless, researchers can help in reducing these uncertainties by improving data availability through incorporating probabilistic distributions, quantifying the effects of different distributions, and ranking uncertainties arising from the model outputs. Increased simplicity and prioritization of environmental uncertainties and hotspots is often observed in latest LCA models for NMs. The parameter choice is a crucial step in LCA, constrained by inconclusive or unrepresentative data. This can be improved by the development of more robust models designed for scaling NM production, manufacture, and performance, alongside their fate prediction and toxicity assessment. Such models inherit great precision in emission values and offer reliable environmental impact assessments [57, 58].

Uncertainty can also be generated due to temporal and spatial non-representativeness in few LCA studies. For example, challenges of understanding NM environmental release and flow. These can be overcome by adopting

Probabilistic Material Flow Analysis (PMFA) and Bayesian networks. While the PMFA method studies the range of values calculated for NM flows throughout the life cycle, Bayesian networks determine the relationship between parameters through construction of probability tables. Though these approaches are useful, but encountered by limitations such as uncertainties owing to lack of knowledge and increased data randomness. This demands increased research efforts targeted towards improvement in data collection and development of more robust models [59]. Limited availability of accurate inventory data for NMs is most likely ascribed to the manufacturers' practice of not disclosing detailed information on materials or energy inputs. Similarly, there is paucity of data concerning NM emissions from manufacture to disposal. In most cases, these emissions are neither measured by manufacturers nor governmental organizations [46]. Inconsistencies in laboratory data often adds another layer of uncertainty to LCA studies. Approaches like Monte Carlo simulations can calculate the uncertainties in methods, but the inconsistencies in lab-scale methods are usually challenging in terms of production of detailed quality results in few cases [12].

12.4.2 Characterization and Transformation of Nanomaterials

A majority of LCA studies struggle to face extreme difficulty in accurately predicting NM release and flow throughout their life cycle, particularly during reuse and recycling. This challenge is further amplified by our limited knowledge on the types of emissions released during various waste disposal methods. Due to this knowledge gap, most studies make simplifying assumptions, treating NM products like regular products at their end-of-life, overlooking the possibility of NM transformation or aging over time [42, 46]. Conducting LCIA for NMs necessitates overcoming a key hurdle, i.e., the absence of characterization factors tailored to these materials. Comprehensive risk assessments of NMs that evaluate their environmental fate, exposure pathways for living organisms, and toxicity, are important for creating relevant and representative characterization factors. Future research that incorporates these essential LCIA elements will lead to a more comprehensive understanding of the environmental risks posed by various NMs [49].

12.4.3 Modeling Environmental Interactions

Understanding the fate and behavior of NMs in different environmental settings is critical for assessing their exposure risks and impacts on environment and human health. While quantification and direct measurement of NMs in complex environmental matrices remains a challenge, environmental fate models can help us estimate NM distribution in the environment [60]. The environmental fate and behavior of NMs can be investigated using simulation programs, leading to accurate LCA assessments of these materials [61].

12.5 CONCLUSION

With nanotechnology rapidly booming into a multi-trillion-dollar industry, a comprehensive evaluation of its impacts on the environment and human health across a product's life cycle is imperative. Although LCA is a valuable tool, evaluating the environmental impacts of engineered NMs using it is still under development due to several hurdles, and in many cases, there is a dearth of fully optimized methodologies. A major challenge for conducting LCA studies of NMs is the limited availability of actual inventory data on the raw materials used and produced at each stage of the NM life cycle. Accurate quantification of the NPs released throughout the life cycle stages of NMs presents another major gap. Lastly, there is limited understanding of the various mechanisms controlling the fate and behavior of NMs once they are released into the environment. Therefore, to strengthen the LCA studies of NMs, it is imperative to identify and assess the various factors governing the fate, movement, transformation, and effect of NMs in the environment. Furthermore, it is of utmost importance to develop LCA environmental impact characterization factors and better LCI databases for NMs. While LCAs have been conducted for nanoproducts in the past, however, they were primarily focused on evaluating the accountable emissions generated during production phase. In future, mapping out a more comprehensive LCA approach and validating its effectiveness for NMs by collecting accurate inventory data, calculating characterization factors, and analyzing the environmental and toxicological impacts of NMs throughout their life cycle, warrants further investigation. For example, a more thorough evaluation of impacts of NMs on humans can be done using "Nano LCRA" and "Comprehensive Environmental Assessment" frameworks that effectively integrate LCA with risk assessment methodologies. Moreover, to ensure nanotechnology fosters a positive impact, several solutions are recommended: (a) waste minimization using green chemistry principles during NM manufacturing process, (b) using green chemistry metrics to assess the environment-friendliness of NMs or their synthesis methods, and (c) adopting a proactive approach while designing novel nanoproducts.

12.6 FUTURE DIRECTIONS AND OPPORTUNITIES

12.6.1 Standardization and Harmonization of LCA Methods

While a wealth of LCA studies exist in academia, there is a significant gap in adoption of this approach by industries for assessing, minimizing, and communicating the impacts of products or processes on the environment. To overcome this, it is necessary to harmonize the LCA methodology into a user-friendly assessment tool. LCA harmonization should include the following key concepts: (a) inventory generation, (b) hypothesis development, (c) impact assessment method, and (d) communication indicators for communicating their environmental footprint in a clear and concise manner. The harmonization procedure facilitates the comparison of results across various LCA studies, regardless of the environmental impacts assessed at the mid-point or end-point level. Moreover, this harmonization is a crucial initial step towards achieving full standardization, which involves the establishment of mandatory rules by recognized organizations, dictating a specific framework for conducting LCA studies [62].

12.6.2 Integration of LCA with Design and Innovation

Developing a novel conceptual framework that addresses the impacts of NMs on the environment and human health is currently needed for the safe application of NMs. Recently the Safe(r) by Design (SbD) framework is garnering increased attention in many industries. This approach mainly focuses on manufacturing NMs without compromising their functionalities and characteristics essential for intended applications [63]. Likewise, Safe(r) Innovation Approach (SIA) builds on the idea of SbD and Socio-Economic Analysis (SEA) to minimize the environmental and human health risks or uncertainties associated with NMs. The origin of SbD concept underlies the NANoREG project and integrates health, environmental safety, and functionality when designing NMs or products thereof [64]. The core of the concept is the Cooper Stage-Gate model, which involves a series of clearly defined stages, guiding development from initial ideas to market launch, including NM recycling and reuse aspects. Specific tasks are integrated at each stage of the model and movement between different stages is guarded by "gates" where innovation projects are critically assessed on the basis of predetermined criteria. These decisions weigh potential risks, costs, and anticipated benefits, ensuring only promising projects advance to the next stage. In future, extensive research needs to be carried out on investigating the unique NM properties, allowing their effective implementation by manufacturers and regulators using appropriate tools [65]. The SbD concept is also evolving to embrace a plethora of tools or methods directed towards hazard assessment, risk of exposure, and overall NM safety. This can favor safe incorporation of nanotechnological advancements after complying with established NM safety regulations. A list of limitations and recommendations for LCA practitioners working with NMs has been depicted in Figure 12.7, with an aim to close the gaps in data inconsistency, ambiguity, and unavailability of complete data.

12.6.3 Life Cycle Optimization for Sustainable Nanotechnologies

To ensure truly sustainable technologies, a Triple Bottom Line (TBL) approach is needed to consider all the three pillars of sustainability, i.e., environmental, social, and economic using specialized tools and techniques tailored to address their impacts on each other [66]. Sustainability of innovation needs to take into account the levels of product, production methods, industry, project, sector, supply chain, and industrial policies. Two main tools relevant for designing a sustainable product are ecological and human health Risk Analysis (RA) and LCA. Multi-Criteria

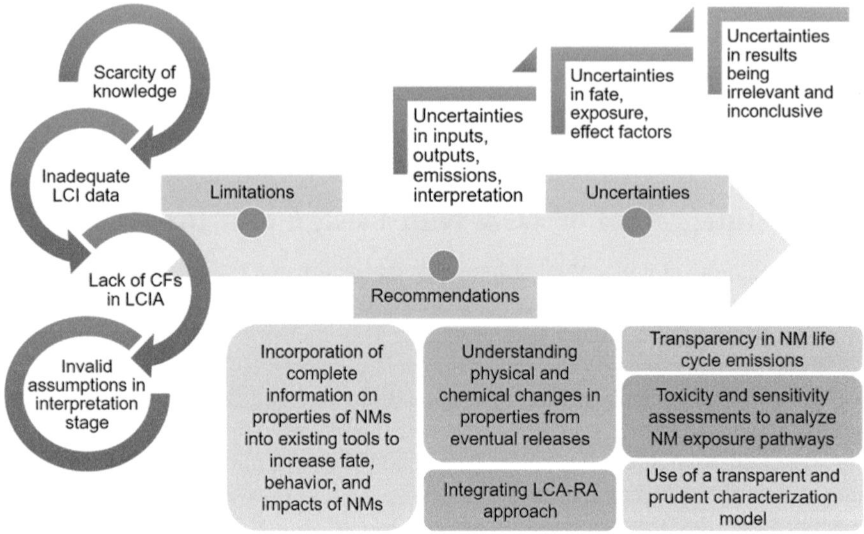

Figure 12.7 An overview of limitations, uncertainties, and recommendations to the life cycle assessment (LCA) practitioners working with nanomaterials. *Abbreviations:* LCI: life cycle inventory; LCIA: life cycle impact assessment; NMs: nanomaterials; RA: risk analysis behavior.

Decision Analysis (MCDA) is a powerful quantification tool that involves various methods (e.g., outranking, analytic hierarchy process, elimination and choice expressing reality, multi-attribute utility/value theory, etc.) to evaluate and rank multiple alternatives based on a pre-determined goal [67]. A novel approach, namely, Safe-and-Sustainable-by-Design (SSbD) considers NM or nanoproduct safety and sustainability at the early stages of material or product design. This approach addresses the functionality, safety, and circularity of NMs throughout their life cycle. Most of the assessment tools in previously reviewed studies have highlighted the importance of Life Cycle Thinking (LCT) for assessing a product's sustainability. To achieve a more nuanced understanding of sustainability of products, there has been a large spur in the emergence of economic assessments (through Life Cycle Costing; LCC) and social assessments (throughSocial-LCA; S-LCA). LCSA integrates three methods, i.e., S-LCA, environmental LCA, and LCC for accounting all the three pillars of sustainability [68]. Integration of these approaches, albeit resource-intensive and expert-oriented, can provide systematic and meaningful results on the overall sustainability of NMs.

REFERENCES

[1] Santamaria, A. 2012. Historical Overview of Nanotechnology and Nanotoxicology. *In*: Reineke, J. (Eds). Nanotoxicity. Totowa, NJ: Humana Press. 1–12. (Methods in Molecular Biology. Vol. 926). Available from: https://link.springer.com/10.1007/978-1-62703-002-1_1

[2] Baig, N., Kammakakam, I. and Falath, W. 2021. Nanomaterials: A review of synthesis methods, properties, recent progress, and challenges. Materials Advances. 2(6): 1821–1871.

[3] Heiligtag, F.J. and Niederberger, M. 2012. The fascinating world of nanoparticle research. Materials Today. 16(7–8): 262–271.

[4] Walter, P., Welcomme, E., Hallégot, P., Zaluzec, N.J., Deeb, C, Castaing, J., et al. 2006. Early Use of PbS Nanotechnology for an Ancient Hair Dyeing Formula. Nano Letters. 6(10): 2215–2219.

[5] El-Khawaga, A.M., Zidan, A. and El-Mageed, A.I.A.A. 2023. Preparation methods of different nanomaterials for various potential applications: A review. Journal of Molecular Structure. 1281: 135148.

[6] Huang, S., Tyagi, V., Bhattacharya, B. and Wang, Y. 2023. Nanotechnology: A new approach to advanced food packaging. *In*: Sharma, A., Vijayakumar, P.S., Prabhakar, P.K., Kumar, R. (Eds). Nanotechnology Applications for Food Safety and Quality Monitoring. Elsevier. [cited 2025 May 20]. 241–64. Available from: https://linkinghub.elsevier.com/retrieve/pii/B9780323857918000033

[7] Biswas, R., Alam, M., Sarkar, A., Haque, MI., Hasan, Md. M. and Hoque, M. 2022. Application of nanotechnology in food: Processing, preservation, packaging and safety assessment. Heliyon. 8(11): e11795.

[8] Arvidsson, R., Peters, G., Hansen, S.F. and Baun A. 2022. Prospective environmental risk screening of seven advanced materials based on production volumes and aquatic ecotoxicity. NanoImpact. 25: 100393.

[9] Pallas, G., Vijver, M.G., Peijnenburg, W.J.G.M. and Guinée, J. 2020. Life cycle assessment of emerging technologies at the lab scale: The case of nanowire-based solar cells. Journal of Industrial Ecology. 24(1): 193–204.

[10] Ali, F., Neha, K. and Parveen, S. 2023. Current regulatory landscape of nanomaterials and nanomedicines: A global perspective. Journal of Drug Delivery Science and Technology. 80: 104118.

[11] Gallagher, M.J., Allen, C., Buchman, J.T., Qiu, T.A., Clement, P.L., Krause, M.O.P., et al. 2017. Research highlights: Applications of life-cycle assessment as a tool for characterizing environmental impacts of engineered nanomaterials. Environmental Science: Nano. 4(2): 276–281.

[12] Nizam, N.U.M., Hanafiah, M.M. and Woon, K.S. 2021. A content review of life cycle assessment of nanomaterials: Current practices, challenges, and future prospects. Nanomaterials. 11(12): 3324.

[13] Lee, J., Ryu, K.H., Ha, H.Y., Jung, K.D. and Lee, J.H. 2020. Techno-economic and environmental evaluation of nano calcium carbonate production utilizing the steel slag. Journal of CO_2 Utilization. 37: 113–121.

[14] Miseljic, M. and Olsen, S.I. 2014. Life-cycle assessment of engineered nanomaterials: a literature review of assessment status. Journal of Nanoparticle Research. 16(6): 2427.

[15] Ahmed, S.F., Mofijur, M., Rafa, N., Chowdhury, A.T., Chowdhury, S., Nahrin, M., et al. 2022. Green approaches in synthesising nanomaterials for environmental nano-bioremediation: Technological advancements, applications, benefits and challenges. Environmental Research. 204: 111967.

[16] Rane, A.V., Kanny, K., Abitha, V.K. and Thomas, S. 2018. Methods for synthesis of nanoparticles and fabrication of nanocomposites. *In*: Bhagyaraj, S.M., Oluwafemi, O.S., Kalarikkal, N., Thomas, S. (eds). Synthesis of Inorganic Nanomaterials.

Elsevier. 121–139. Available from: https://linkinghub.elsevier.com/retrieve/pii/B9780081019757000051

[17] Alhamad, A.A., Zeghoud, S., Amor, I.B., Zaater, A., Amor, A.B., Aouadif, A., et al. 2023. A short review of Nanomaterials: Synthesis methods, properties, and applications: Review. Algerian Journal of Chemical Engineering AJCE. 1(01): 01–7.

[18] Ijaz, I., Gilani, E., Nazir, A. and Bukhari, A. 2020. Detail review on chemical, physical and green synthesis, classification, characterizations and applications of nanoparticles. Green Chemistry Letters and Reviews. 13(3): 223–245.

[19] Zhang, D., Li, Z. and Sugioka, K. 2021. Laser ablation in liquids for nanomaterial synthesis: diversities of targets and liquids. Journal of Physics: Photonics. 3(4): 042002.

[20] Zahid, M.U., Pervaiz, E., Hussain, A., Shahzad, M.I. and Niazi, M.B.K. 2018. Synthesis of carbon nanomaterials from different pyrolysis techniques: A review. Materials Research Express. 5(5): 052002.

[21] Bokov, D., Turki Jalil, A., Chupradit, S., Suksatan, W., Javed Ansari, M., Shewael, I.H., et al. 2021. Nanomaterial by sol-gel method: synthesis and application. Wang, Z. (Eds). Advances in Materials Science and Engineering. 2021(1): 5102014.

[22] Hachem, K., Ansari, M.J., Saleh, R.O., Kzar, H.H., Al-Gazally, M.E., Altimari, U.S., et al. 2022. Methods of chemical synthesis in the synthesis of nanomaterial and nanoparticles by the chemical deposition method: A review. BioNanoScience. 12(3): 1032–1057.

[23] Malgras, V., Tang, J., Wang, J., Kim, J., Torad, N.L., Dutta, S., et al. 2019. Fabrication of nanoporous carbon materials with hard- and soft-templating approaches: A review. Journal of Nanoscience and Nanotechnology. 19(7): 3673–3785.

[24] Lone, I.H., Radwan, N.R.E., Aslam, J. and Akhter, A. 2019. Concept of reverse micelle method for the synthesis of nano-structured materials. CNANO. 15(2): 129–136.

[25] Nikaeen, G., Abbaszadeh, S. and Yousefinejad, S. 2020. Application of Nanomaterials in Treatment, Anti-Infection and Detection of Coronaviruses. Nanomedicine (Lond). 15(15): 1501–1512.

[26] Shafiq, M., Anjum, S., Hano, C., Anjum, I. and Abbasi, B.H. 2020. An overview of the applications of nanomaterials and nanodevices in the food industry. Foods. 9(2): 148.

[27] Saleem, H. and Zaidi, S.J. Developments in the application of nanomaterials for water treatment and their impact on the environment. Nanomaterials. 10(9): 1764.

[28] Jun, L.Y., Khan, F.S.A., Mubarak, N.M., Yon, L.S., Bing, C.H., Khalid, M., et al. 2021. Importance of nanomaterials in engineering application. *In*: Mubarak, N.M., Khalid, M., Walvekar, R. and Numan, A. (Eds). Contemporary Nanomaterials in Material Engineering Applications. Cham: Springer International Publishing. 2021. 1–20. (Engineering Materials). Available from: http://link.springer.com/10.1007/978-3-030-62761-4_1

[29] Khan, Y., Sadia, H., Ali Shah, S.Z., Khan, M.N., Shah, A.A., Ullah, N., et al. 2022. Classification, synthetic, and characterization approaches to nanoparticles, and their applications in various fields of nanotechnology: A review. Catalysts. 12(11): 1386.

[30] Freund, R., Lächelt, U., Gruber, T., Rühle, B. and Wuttke, S. 2018. Multifunctional efficiency: Extending the concept of atom economy to functional nanomaterials. ACS Nano. 12(3): 2094–2105.

[31] Kraegeloh, A., Suarez-Merino, B., Sluijters, T. and Micheletti, C. 2018. Implementation of safe-by-design for nanomaterial development and safe innovation: Why we need a comprehensive approach. Nanomaterials. 8(4): 239.

[32] Zahra, Z., Habib, Z., Hyun, S. and Sajid, M. 2022. Nanowaste: Another future waste, its sources, release mechanism, and removal strategies in the environment. Sustainability. 14(4): 2041.

[33] Malakar, A., Kanel, S.R., Ray, C., Snow, D.D. and Nadagouda, M.N. 2021. Nanomaterials in the environment, human exposure pathway, and health effects: A review. Science of the Total Environment. 759: 143470.

[34] Walser, T., Limbach, L.K., Brogioli, R., Erismann, E., Flamigni, L., Hattendorf, B., et al. 2012. Persistence of engineered nanoparticles in a municipal solid-waste incineration plant. Nature Nanotech. 7(8): 520–524.

[35] Suman, T.Y. and Pei, D.S. 2022. Nanomaterial waste management. *In*: Rai, M., Nguyen, T.A. (Eds). Nanomaterials Recycling. Elsevier. 21–36. Available from: https://linkinghub.elsevier.com/retrieve/pii/B9780323909822000020

[36] Bolyard, S.C., Reinhart, D.R. and Santra, S. 2013. Behavior of engineered nanoparticles in landfill leachate. Environmental Science and Technology. 130710152553007.

[37] Behera, A., Sahini, D. and Pardhi, D. 2022. Procedures for recycling of nanomaterials: A sustainable approach. *In*: Rai, M., Nguyen, T.A. (Eds). Nanomaterials Recycling. Elsevier. 175–207. Available from: https://linkinghub.elsevier.com/retrieve/pii/B9780323909822000093

[38] Zhang, J., Panwar, A., Bello, D., Jozokos, T., Isaacs, J.A., Barry, C., et al. 2016. The effects of recycling on the properties of carbon nanotube-filled polypropylene composites and worker exposures. Environmental Science: Nano. 3(2): 409–417.

[39] Haleema, Munir M.U., Phan, D.N. and Khan, M.Q. 2022. Recycling of nanomaterials by solvent evaporation and extraction techniques. *In*: Rai, M. and Nguyen, T.A. (Eds). Nanomaterials Recycling. Elsevier. 209–222. Available from: https://linkinghub.elsevier.com/retrieve/pii/B978032390982200010X

[40] Abd Elkodous, M., El-Husseiny, H.M., El-Sayyad, G.S, Hashem, A.H., Doghish, A.S., Elfadil, D., et al. 2021. Recent advances in waste-recycled nanomaterials for biomedical applications: Waste-to-wealth. Nanotechnology Reviews. 10(1): 1662–1739.

[41] Mullen, E. and Morris, M.A. 2021. Green nanofabrication opportunities in the semiconductor industry: A life cycle perspective. Nanomaterials. 11(5): 1085.

[42] Cucurachi, S. and Blanco Rocha, C.F. 2019. Life-cycle assessment of engineered nanomaterials. *In*: Pacheco-Torgal, F., Diamanti, M.V., Nazari, A., Granqvist, C.G., Pruna, A., Amirkhanian, S. (Eds). Nanotechnology in Eco-efficient Construction. 2nd ed. Elsevier. 815–846. Available from: https://linkinghub.elsevier.com/retrieve/pii/B9780081026410000311

[43] Bartolozzi, I., Daddi, T., Punta, C., Fiorati, A. and Iraldo, F. 2020. Life cycle assessment of emerging environmental technologies in the early stage of development: A case study on nanostructured materials. Journal of Industrial Ecology. 24(1): 101–115.

[44] Moloi, M.S., Lehutso, R.F., Erasmus, M., Oberholster, P.J. and Thwala, M. 2021. Aquatic environment exposure and toxicity of engineered nanomaterials released from nano-enabled products: Current status and data needs. Nanomaterials. 11(11): 2868.

[45] Huijbregts, M.A.J., Steinmann, Z.J.N., Elshout, P.M.F., Stam, G., Verones, F., Vieira, M., et al. 2016. ReCiPe2016: A harmonised life cycle impact assessment method at midpoint and endpoint level. International Journal of Life Cycle Assessment. 22(2): 138–147.

[46] Salieri, B., Turner, D.A., Nowack, B. and Hischier, R. 2018. Life cycle assessment of manufactured nanomaterials: Where are we? NanoImpact. 10: 108–120.

[47] Hischier, R., Nowack, B., Gottschalk, F., Hincapie, I., Steinfeldt, M. and Som, C. 2015. Life cycle assessment of façade coating systems containing manufactured nanomaterials. Journal of Nanoparticle Research. 17(2): 68.

[48] Ludin, N.A., Mustafa, N.I., Hanafiah, M.M., Ibrahim, M.A., Asri Mat Teridi, M., Sepeai, S., et al. 2018. Prospects of life cycle assessment of renewable energy from solar photovoltaic technologies: A review. Renewable and Sustainable Energy Reviews. 96: 11–28.

[49] Hischier, R., Salieri, B. and Pini, M. 2017. Most important factors of variability and uncertainty in an LCA study of nanomaterials—Findings from a case study with nano titanium dioxide. NanoImpact. 7: 17–26.

[50] Barton, L.E., Auffan, M., Olivi, L., Bottero, J.Y. and Wiesner, M.R. 2015. Heteroaggregation, transformation and fate of CeO_2 nanoparticles in wastewater treatment. Environmental Pollution. 203: 122–129.

[51] Eckelman, M.J., Mauter, M.S., Isaacs, J.A. and Elimelech, M. 2012. New perspectives on nanomaterial aquatic ecotoxicity: Production impacts exceed direct exposure impacts for carbon nanotubes. Environmental Science and Technology. 46(5): 2902–2910.

[52] Boldrin, A., Hansen, S.F., Baun, A., Hartmann, N.I.B. and Astrup T.F. 2014. Environmental exposure assessment framework for nanoparticles in solid waste. Journal of Nanoparticle Research. 16(6): 2394.

[53] Temizel-Sekeryan, S. and Hicks, A.L. 2020. Global environmental impacts of silver nanoparticle production methods supported by life cycle assessment. Resources, Conservation and Recycling. 156: 104676.

[54] Arvidsson, R., Kushnir, D., Sandén, B.A. and Molander, S. 2014. Prospective life cycle assessment of graphene production by ultrasonication and chemical reduction. Environmental Science and Technology. 48(8): 4529–4536.

[55] Hanafiah, M.M., Leuven, R.S.E.W., Sommerwerk, N., Tockner, K. and Huijbregts, M.A.J. 2013. Including the introduction of exotic species in life cycle impact assessment: The case of inland shipping. Environmental Science and Technology. 47(24): 13934–13940.

[56] Gilbertson, L.M., Busnaina, A.A., Isaacs, J.A., Zimmerman, J.B. and Eckelman, M.J. 2014. Life cycle impacts and benefits of a carbon nanotube-enabled chemical gas sensor. Environmental Science and Technology. 48(19): 11360–11368.

[57] Parsons, S., Murphy, R.J., Lee, J. and Sims, G. 2015. Uncertainty communication in the environmental life cycle assessment of carbon nanotubes. IJNT. 12(8/9): 620.

[58] Lacirignola, M., Blanc, P., Girard, R., Pérez-López, P. and Blanc, I. 2017. LCA of emerging technologies: addressing high uncertainty on inputs' variability when performing global sensitivity analysis. Science of the Total Environment. 578: 268–280.

[59] Igos, E., Benetto, E., Meyer, R., Baustert, P. and Othoniel, B. 2019. How to treat uncertainties in life cycle assessment studies? International Journal of Life Cycle Assess. 24(4): 794–807.

[60] Suhendra, E., Chang, C.H., Hou, W.C. and Hsieh, Y.C. 2020. A review on the environmental fate models for predicting the distribution of engineered nanomaterials in surface waters. IJMS. 21(12): 4554.

[61] Ross, B.N. and Knightes, C.D. 2022. Simulation of the environmental fate and transformation of nano copper oxide in a freshwater environment. ACS EST Water. 2(9): 1532–1543.

[62] Jourdaine, M., Loubet, P., Trebucq, S. and Sonnemann, G. 2020. A detailed quantitative comparison of the life cycle assessment of bottled wines using an original harmonization procedure. Journal of Cleaner Production. 250: 119472.

[63] Chen, C., Leong, D.T. and Lynch, I. 2020. Rethinking nanosafety: Harnessing progress and driving innovation. Small. 16(21): 2002503.

[64] Gottardo, S., Alessandrelli, M., Amenta, V., Atluri, R., Barberio, G., Bekker, C., et al. 2017. NANoREG framework for the safety assessment of nanomaterials. [Internet]. Italy: European Commission; 2017 [cited 2025 May 20]. Available from: https://data.europa.eu/doi/10.2760/245972

[65] Gottardo, S., Mech, A., Drbohlavová, J., Małyska, A., Bøwadt, S., Riego Sintes, J., et al. 2021. Towards safe and sustainable innovation in nanotechnology: State-of-play for smart nanomaterials. NanoImpact. 21: 100297.

[66] Raza, S.S., Janajreh, I. and Ghenai, C. 2014. Sustainability index approach as a selection criteria for energy storage system of an intermittent renewable energy source. Applied Energy. 136: 909–920.

[67] Linkov, I. and Seager, T.P. 2011. Coupling multi-criteria decision analysis, life-cycle assessment, and risk assessment for emerging threats. Environmental Science and Technology. 45(12): 5068–5074.

[68] Stoycheva, S., Zabeo, A., Pizzol, L. and Hristozov, D. 2022. Socio-Economic Life Cycle-Based Framework for Safe and Sustainable Design of Engineered Nanomaterials and Nano-Enabled Products. Sustainability. 14(9): 5734.

Chapter **13**

Environmental and Health Safety on Nanomaterials for Zero Carbon Emission

Ashish Kumar[1], Lavi Aluna[2], Atin Kumar Pathak[1], Pramod Kumar Singh[2], Neelu Raina[3], Shamshad Ahmad[4], Abhishek Anand[5] and Har Mohan Singh*[2]

[1]RINA Classification and Certification India Private Limited, 505, KM Trade Tower, Kaushambi, Ghaziabad-201010, Uttar Pradesh, India

[2]Department of Physics & Environmental Sciences, Sharda University, Greater Noida-201310, Uttar Pradesh, India

[3]Department of Chemical Engineering, Faculty of Engineering, King Mongkut's University of Technology North Bangkok (KMUTNB), Bangkok, Thailand

[4]District Water Testing Laboratory, Uttar Pradesh Jal Nigam, Bahraich-271801, Uttar Pradesh, India

[5]Institute of Power Engineering, Universiti Tenaga Nasional, 43000, Kajang, Selangor, Malaysia

13.1 INTRODUCTION

As per the European Commission, nanomaterial can be explained as material containing particles, aggregately or in an unbound state where around 50% of the particles or more than that are in the range of 1–100 nanometres in one or

*For Correspondence: harenergy19@gmail.com

DOI: 10.1201/9781003492184-13

more external dimensions [1]. A nanomaterial is one-millionth of a millimetre-around 100,000 times smaller than the diameter of a human hair. These materials can be engineered, incidental, or naturally occurring. Nanomaterials showcase unique chemical and physical properties due to their small size. Nanomaterials are mechanically strong as compared to large structures because of their compact size which makes them free from internal imperfections. They show electrical conductivity and catalytic activity because of their exceptionally small size [2]. Nanomaterials can rearrange their crystalline structure when dispersed and have an unusually high surface area as compared to their volume; this makes them more reactive and improves interactions with other materials, for example, biological molecules liquids, and gases. Nanomaterials showcase improved catalytic and thermal properties, the large surface area of nanomaterials allows them to be highly reactive, making them efficient in adsorption, sensing of molecules, and catalysis. This is very useful in biological diagnostics [3]. The quantum property of nanomaterial changes with the decrease in their sizes. At the nanoscale, quantum effects become significant, leading to changes in the magnetic electronic and optical properties of nanomaterials [4].

Nanomaterials are the building blocks of nanosciences and nanotechnology, common types of nanomaterials include nanotubes, nanocomposites, nanoparticles, nanowires, and quantum dots [5]. Nanostructured sciences and technology represent a vast and interdisciplinary field of research and development that has experienced explosive growth globally in recent years. This area holds immense potential to transform the methods by which materials and products are fabricated, as well as the range and characteristics of functionalities that can be attained (Figure 13.1). Its current commercial influence is already significant and is assured to expand further in the years ahead.

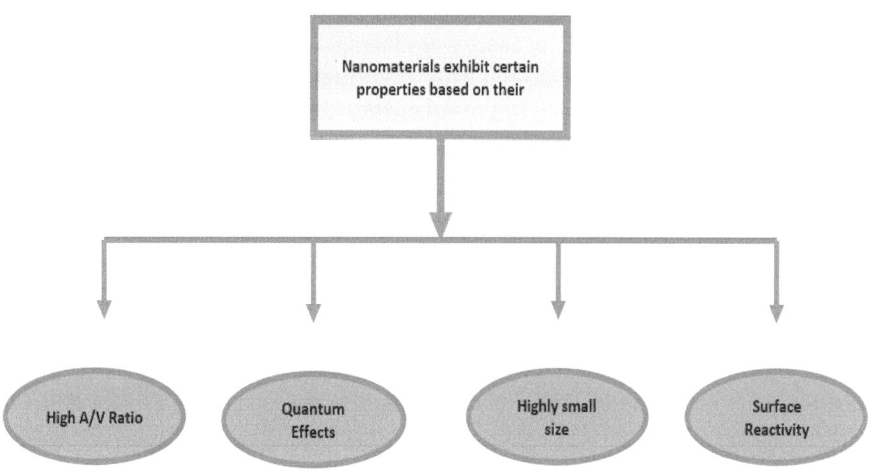

Figure 13.1 Unique characteristic of features of nanomaterial.

13.2 APPLICATIONS OF NANOMATERIALS

Nanomaterials possess very minute dimensions and structures that give rise to several unique properties hence developing a wide range of applications in various fields such as healthcare, manufacturing, electronics, energy, environmental remediation, textile and fabrics, construction, etc., [6]. In healthcare and medicine, nanomaterials are useful in drug delivery, medical imaging, and biosensors; for example, nanomaterials such as dendrimers, liposomes, and other polymeric nanoparticles are being used to encapsulate drugs and target specific cells and tissue hence improving the drug effectiveness and reducing the side effects [7]. Nanomaterials in medical imaging are employed as contrast agents in imaging technologies such as Magnetic Resonance Imaging (MRI), Computed Tomography (CT) scan and fluorescence imaging, improving imaging accuracy and giving high-resolution images [8]. Some nanostructured materials are used as biosensors for the detection of biomolecules like proteins, pathogens, and deoxyribonucleic acid with high specificity and sensitivity. Cosmetics and personal care nanoparticle-based sunscreen developed for better ultraviolet radiation (UV) protection as well as nanoemulsions are used to provide improved quality skincare ingredients improve texture and feel, enhance sun protection, peptides, and hyaluronic acid showcase anti-aging effects, used in skin brightening and depigmentation, acne treatment, and skin barrier repair [9]. Electronic applications of nanoparticles include the use of transistors, semiconductors, and displays. In the fabrication of transistors and semiconductors nanomaterials like carbon graphene and nanotubes are used to provide better, fast, improved, and more effective electronic components. Quantum dots are employed in displays of mobile phones and computers as well as television to produce vibrant colours with high efficiency, giving rise to enhanced display quality in electronic devices [10].

Nanomaterials are used in energy-producing materials such as solar cells, fuel cells, and batteries. Lithium-ion batteries utilize nanostructured electrodes to increase surface area, providing improved energy storage capacity, charging rates, and better life cycle. In fuel cells nanoparticles serve as catalysts, enhancing the speed of chemical reactions required for the energy conversion, leading to higher performance and durability of fuel cell devices [11, 12]. In solar cells materials used in fabrication are nanostructured, they enhance light absorption, electron transport, and charge separation [12], hence improving energy conversion efficiency. In textiles nanomaterials like nanocoating and nanofibres are incorporated into textiles for the addition of properties like stain resistance (fluoropolymers and silicones), water repellence, antibacterial activity (silver nanoparticles), ultraviolet (UV) protection (zinc oxide and titanium dioxide), improving the durability (titanium dioxide and silica), thermal regulation (phase change materials), wrinkles resistance, moisture management, electrical conductivity to fit biosensors and the performance of fabrics [13]. Environmental remediation is applicable in water purification and air filtration where nanomaterials are widely adopted to improve technologies. In water purification nanomaterials for example nanoscale membranes, nanoparticles, and nanocomposites are used to process treatment of water for the removal of pathogens, heavy metals, and other pollutants from contaminated water sources [14]. Nanomaterials are employed in air filters they capture particulate matter, pollutants, and allergens and improve indoor air quality along with a reduction in respiratory-related health risks. Nanoparticles are employed in the remediation of soil to manage soil contamination from numerous sources like heavy metals, pesticides, and hydrocarbons. Nanomaterials are also in use in developing sensors and monitoring devices [14, 15]. In construction and infrastructure nanomaterials are used to create self-cleaning coatings for glass surfaces, infrastructural material, and building facades reducing maintenance costs and enhancing the look. In construction, nanostructured materials are incorporated into coating, concrete, and composites to improve mechanical strength, resistance to environmental damage, and durability of the building [16]. In the automotive industry nanocomposite materials are used for the fabrication of lightweight and strong automotive parts, and nanocoatings are used to make the car parts scratch-resistant and self-healing [17]. Apart from nanoparticles such as nanoclays are used in improving fuel efficiency, some nanoparticles are utilized as additives in lubricants which enhance engine performance, and nanoparticles like graphene and carbon nanotubes are used in the enhancement of thermal management and in reducing emissions [17]. In agriculture, nanoencapsulation is used for targeting specific delivery of nutrients in food. Nanoparticles can be utilized to deliver fertilizers and nutrients to plants with more efficacy because nanoparticles have a high surface area to volume ratio, nanoparticles can also be used as nanopesticides for targeted delivery minimizing environmental contamination, nanosensors can be deployed in fields to get better crop yields and resource efficacy, nanomaterial-based crop protection coatings and films which can serve as a physical barrier against disease, pests, and environmental stress, nanoparticles can also enhance

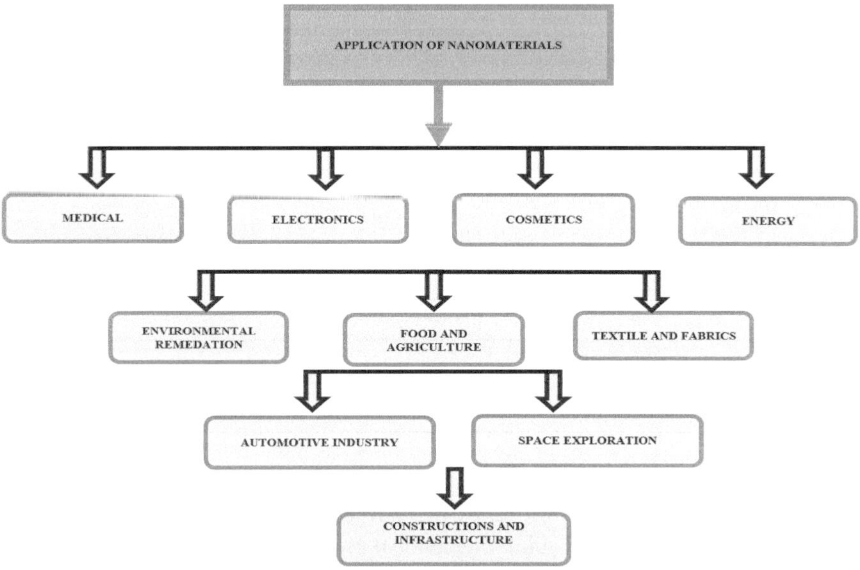

Figure 13.2 Various applications of nanomaterials.

plant genetics [18]. In space exploration lightweight and durable spacecraft components, as well as nanoscale sensors for monitoring the environmental conditions are created by the use of nanomaterials. Nanoparticles like nanofuels and nanocomposites are being used in propulsion systems in spacecrafts, radiation shielding by the use of nanoparticles is also being explored, and nanoparticles are used in composite materials to build spacecraft structures [19]. They facilitate better thermal management, strong, strengthened, durable, and lightweight spacecraft. Apart from that instrumentations and sensors are also made with nanoparticles. Thus the application of nanotechnology revolutionized the contemporary era by covering agriculture to spacecraft sectors (Figure 13.2).

13.3 SOURCES OF NANOMATERIALS

Nanomaterials and living things have been peacefully co-evolving because nanomaterials are the components of the earth's system. Nanomaterials may be categorized into three primary types: "natural, incidental, and engineered nanomaterials" depending on where they originate from [20]. The existing equilibrium has been disrupted by the rise in accidental, engineered, and anthropogenic nanomaterials. Figure 13.3 illustrates the various sources of nanomaterials.

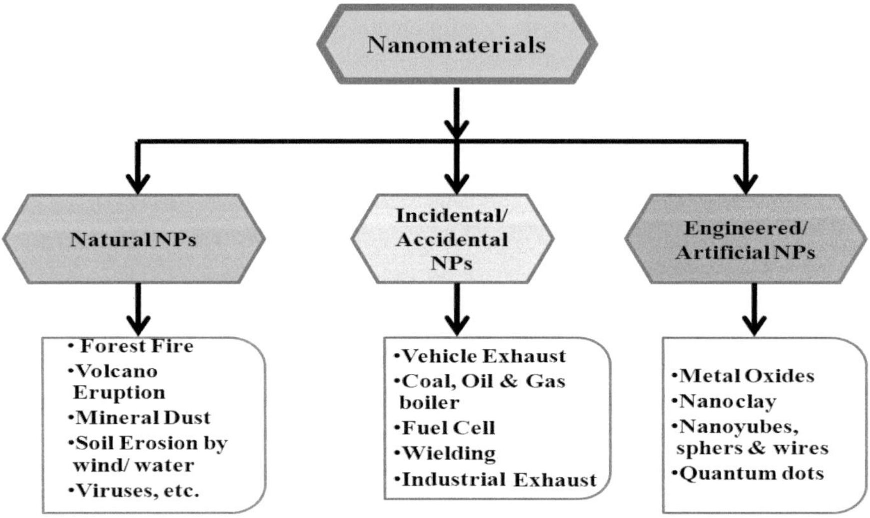

Figure 13.3 Various types of nanoparticles.

13.3.1 Natural Nanomaterials

Natural nanomaterials are not a result of man-made actions or processes; rather, they are the result of natural (bio), mechanical processes, or geochemical process. A few examples are the wax crystals covering nasturtium leaves or lotus leaves, spider and silk spider mites, some butterfly wing scales, horn materials (claws, skin, feathers, hair), natural colloids (milk, blood), corals, nacre, and the human bone matrix. Other examples include foraminifera (mainly chalk) and virus (capsid, protein) structures [21].

In the varied chemical environments of the Earth's crust, crystal formation produces naturally occurring inorganic nanomaterials. Despite being natural, these particles are often not biological. Volcanic ash clouds, for instance, can result from volcanic activity and include a diverse range of polydisperse micro and nanoparticles. The size of these particles ranges from 100 to 200 nm, and their main chemical composition is iron and silicate compounds. Natural organic nanomaterials are produced via crystallization under a variety of environmental and natural circumstances. These are organic particles, most often originating from biological. For instance, the architecture of viruses is generally nanoscale, with sizes varying from 20 to 300 nm. They participate in several biological processes and are made up of proteins and nucleic acids. Other sources of natural nanomaterials include ocean spray, radon gas release, rock weathering that contains metals or anions, acid mine drainage sites, and forest fires (combustion materials) [22].

13.3.2 Incidental Nanomaterials

The incidental nanomaterials usually occur due to man-made processes different processes, nanomaterials that are created accidently or byproducts of different processes, either directly or indirectly by anthropogenic influences (e.g., mechanical, or industrial) processes, as opposed to intentionally engineered or synthesized for specific applications. These materials can originate from human activity or occur naturally, and their nanoscale size may have special qualities and possible consequences [23].

Incidental nanomaterials can be found in a variety of setting, including automobile exhaust fumes, welding fumes, solid fuel heating (heather grown at home) materials can originate from human activity or occur naturally, and their nanoscale size may have special qualities and possible including automobile exhaust fumes, welding fumes, solid fuel heating (heather grown at home), natural mineral weathering and erosion, and metalworking operations [24]. Concerns regarding these nanoparticles' potential effects on the environment may arise if they are found in treated water.

Unintentionally created during a purposeful process, incidental atmospheric nanoparticles may worsen air pollution. Numerous nanomaterials are produced by forest fires, including pigments, cement, and fume silica. Byproducts of water treatment processes, such as the use of certain chemicals or filtering techniques, may include nanomaterials [25]. Concerns regarding these nanoparticles' potential effects on the environment may arise if they are found in treated water.

13.3.3 Engineered Nanomaterials

Based on their dimensions and unique qualities (such as nanostructured medical implants), tailored nanoparticles and nanostructured materials are created for specific uses [26]. For these nanomaterials to have as few unintended negative consequences as possible, they need to be well characterized. The aerosol (fumed silica) that was created in the 1940s was the source of the first commercial nanomaterials. Natural and accidental nanoparticles could possess a variety of forms [27]. Engineered nanomaterials can be used in medicine, electronic transistors, surfaces that repel water and dirt. More sustainable nanomaterials are needed to develop for ecological balance.

13.3.4 Anthropogenic Nanomaterials

Anthropogenic nanomaterials are either produced intentionally or accidentally by human activities. These particles may potentially cause adverse health effects and ecosystem impairment. This category of particles includes both accidental/incidental and artificial/engineered nanoparticles, Tire and Road Wear Particles (TRWP), paint flakes, soot, and plastic particles (micro and nano) as well as other combustion and friction particulate byproducts products [28]. These nanomaterials can often not be designed with special properties in mind, so ideally they are considered as pollutants.

13.4 SAFETY CONSIDERATION OF NANOMATERIALS

While nanosafety is undoubtedly neither a simple setup nor an easy fix, there should be a methodical approach that evaluates the dangers and suggests reasonable actions to ensure both environmental and human security [29, 30]. The advantages of new technology may outweigh the drawbacks of exposing humans to novel substances. There are three basic contexts in which to consider safety:

13.4.1 Environmental Pollution

Working with nanomaterials requires an atmosphere suitable for the kind of materials being used. These settings might be anything from fully equipped clean rooms with masks and safety clothes to negative pressure clean benches (negative air pressure, which means they suck air inward and this design prevents the operator from exposure to aerosolized infectious agents). It may be necessary to use self-contained breathing equipment in very specific situations. Maintaining the material in the proper liquid constantly is one technique to regulate nanomaterials and lower some danger [30]. This reduces the possibility of material or airborne particles sticking to clothing. The semiconductor sector has extensive knowledge of clean room technology, which reduces and controls airborne particulates. High-Efficiency Particulate Air (HEPA) filters with specialized exhaust processing are installed in clean rooms. In scenarios involving chemical fume hoods, vented closures, and other specialized equipment, it is necessary to process Local Exhaust Ventilation (LEV) rather than just vent it to the atmosphere. Using HEPA filters alone is not always enough. The size of the nanoparticles being employed and the anticipated percentage of particles that can be caught should be taken into account when choosing a HEPA filter [31].

13.4.1.1 Exposure in the Working Environment

When people unintentionally interact with nanoparticles often through activities connected to industrial processes, consumer items, or environmental sources unintentional human exposure to nanomaterials can take place [32]. Skin contact, ingestion, or inhalation are the possible routes of exposure. Workers always face challenges while working with nanomaterials. The particulars of the material being used will determine how the works and the environment may be affected. As a result, employees need to recognize that there might be unintended consequences from their work environment. The semiconductor industry is no stranger to treating hazardous materials in a way that reduces worker danger [33]. When working with nanomaterials, the same kind of methodology needs to be applied. As the nanomaterials are advancing, more research will be required to determine and understand the health effects on workers.

13.4.1.2 Deliberate Human Exposure

People's exposure while working with nanomaterials in the workplace, among consumers, and in the environment is most likely going to increase noticeably (Figure 13.4). Even though a lot of information has already been discovered about the impact of accidentally created environmental nanoparticles (like

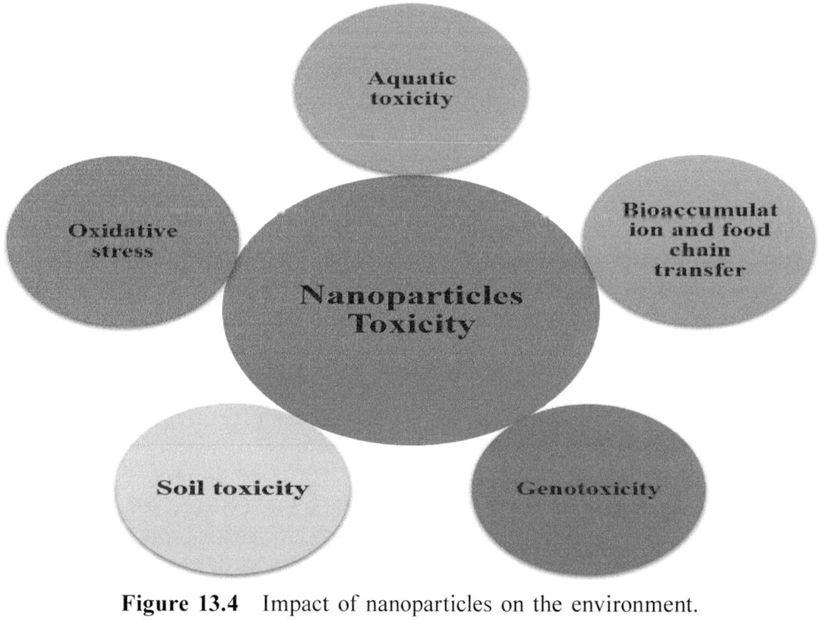

Figure 13.4 Impact of nanoparticles on the environment.

those produced by combustion) on human health, primarily through respiratory pathways, there is still a need for an appropriate risk assessment method to evaluate the safety of engineered nanomaterials [34]. Research and Development (R&D) of nanomaterial products should incorporate safety evaluation, which includes risk assessment and risk management, in the same way as other new technologies. Potential exposure scenarios should guide thoughtful, case-explicit thinking in risk assessment and management.

13.5 IMPACT OF NANOMATERIALS

Nanotechnology has advanced noticeably in the past two decades new and varied uses for nanoparticles are being discovered daily. On the other side, there is less amount of knowledge about the fate of nanoparticles in biological systems and their toxicity [34]. The possible modes of exposure for nanoparticles include ingestion, inhalation, and skin penetration. On the other hand, the combined effects of particle surface area, shape, size, surface chemistry, and surface area determine the overall toxicity of nanoparticles [35]. Growing manufacturing and deliberate (drug administration, sunscreens) or accidental (environmental, occupational) exposure to nanoparticles are expected to raise the likelihood of harmful health impacts. To guarantee the risk-free and sustainable application of nanotechnology, innovative nanomaterials must be physiologically characterized for potential health risks [36]. Table 13.1 listed a broad spectrum of studies on various effects of nanomaterials on human health and environment.

Table 13.1 Studies on various effects of nanomaterials on human health and environment

S. No.	Nanoparticles	Adverse Impact	References
1.	Zinc oxide (ZnO), and Copper (CuO)	Profoundly impacts the dermal cells beneath the epidermal tissue after trans-epidermal penetration.	[37]
2.	Silicon dioxide (SiO_2)	Metabolism is disrupted with the production of oxidative stress. Reactive oxygen species are produced GSH metabolism, pantothenate, and coenzyme A (CoA) production are significantly disrupted. Abnormalities in mitochondrial structure and functioning are caused.	[38]
3.	Silicon dioxide (SiO_2)	Increased formation of reactive oxygen species, destabilizes the potential of the mitochondrial membrane and encourages cytotoxicity	[30]
4.	Copper oxide (CuO), and Zinc oxide (ZnO)	Particles from diesel exhaust present novel physicochemical interactions that have the potential to decrease CuO NPs' cytotoxicity while increasing ZnOs.	[39]

5.	Titanium dioxide–(TiO_2), Copper oxide–(CuO), Zinc oxide–(ZnO), Iron oxide –(Fe_2O_3), Manganese oxide–(Mn_2O_3), Chromium oxide–(Cr_2O_3), Nickel oxide–(NiO)	The vitality of all NPs, with the exception of Cr_2O_3 and Fe_2O_3, decreased over time and in response to dosage. Every NP dramatically reduced the rate of cellular growth.	[40]
6.	Iron oxide nanoparticles with silica-coating (SiO_2)	Cytotoxicity, rare genotoxic effects, with no difference in the repair process of DNA.	[41]
7.	Titanium dioxide and Zinc oxide (TiO_2 and ZnO)	Higher persistence is shown by TiO_2 in uptake and cell surface. Induce generation of reactive oxygen species. ZnO demonstrates mild response and a low number of cell surface damage to cells and lung function.	[42]
8.	Copper oxide (CuO), Silver (Ag), Titanium dioxide (TiO_2), and Zinc oxide (ZnO)	Neutrophils are activated by both Ag and CuO, while TiO_2 does not induce neutrophil responses.	[43]
9.	Graphene oxide (GO)	In 28 days triggers apoptotic pathways, genes from the Inhibitor of the Apoptosis Protein (IAP) family are downregulated as a result.	[44]
10.	Reduced graphene oxide (rGO)	Continuous exposure to rGO does not significantly alter human lung epithelial cells' DNA methylation patterns.	[45]
11.	Graphene oxide (GO)	Prolonged exposure increases the risk of lung illnesses and/or pulmonary infections. It also induces autophagosome accumulation (coming from inhibition of autophagy flux) and elicits a pro- inflammatory response after two weeks of exposure	[46]

Despite the fact that nanoparticles have always existed in nature, recent advancements have been made in the creation and utilization of engineered nanoparticles, which have applications in a variety of fields such as water treatment, bioremediation, paints, coatings, food packaging, medicine, cosmetics, and electronics [47]. In addition, novel nanomaterials are being investigated for possible therapeutic and diagnostic uses in the detection and treatment of cancer, where the nanoscale characteristics allow entrance and intracellular transit to certain target locations [47, 48]. Additionally, nanomaterials have the potential to significantly reduce waste generation, and industrial pollution, and improve the efficiency of energy, production, and consumption [49]. There is a tremendous need to get these applications into the market because of the seemingly endless potential of nanotechnology and the possibility of making money from developing and promoting these innovations. However, there are releases into the air, soil, and aquatic systems as a result of the creation, use, and disposal of produced nanoparticles. Investigating their entry into, passage through, and effects on the environment's health is therefore essential. Unknown toxicological qualities of manufactured nanoparticles when used carelessly might create number of risks to the environment, animals, and public health. There are currently no rules specifically pertaining to the use of nanoparticles, and our understanding of the detrimental impacts of these particles is still quite restricted. As nanotoxicology and nanomedicine are indistinguishable, their respective values are contingent upon their sensible use. In their life cycle, they hold a potential risk for humans and the environment upon exposure. It is not dangerous in and of itself to have nanomaterials (materials containing nanoparticles) present [50]. They are only dangerous in specific situations, especially because of their enhanced responsiveness and movement. A true hazard would only arise if certain characteristics of specific nanoparticles were detrimental to the environment or to living things.

13.5.1 Environment and Health

The possible impacts of nanomaterials on the environment, while exposed for a longer period of time, consist of positive as well as some negative impacts (Figure 13.4). The positive ones include pollution remediation via using nanoparticles such as titanium dioxide, carbon nanotubes and iron are used to clear pollutants from soil, water, and air, energy efficiency, for improvements in agriculture, etc. Environmental risks that the exposure of nanoparticles results after a long period of time. Toxicity to aquatic life is a major negative effect that is studied, the nanoparticles can be easily taken up by the organisms present in water because of their tiny size, for example, silver (Ag) nanoparticles are highly toxic for fishes causing stress on oxygen availability and cellular damage [51]. In contamination of soil, nanoparticles affect those microbes that are crucial for the recycling of nutrients in the soil for example, nanoparticles like titanium dioxide inhibit the growth of bacteria and fungi that are beneficial for soil [52]. The plant can also uptake the nanoparticles and affects their growth. Nanoparticles can spread in the air and be inhaled by living organisms, due to their small size they can penetrate deep into the respiratory systems of living beings in the environment causing respiratory as well as cardiovascular issues in animals and humans. After translocation from one place to another they can also impact the functioning of other organs. They can generate oxidative stress if accumulated in high quantity inside the body in fauna leading to damage of cells or even deaths of the animals and plants [53]. Their accumulation can also become a cause of genotoxicity and create a mutation in the DNA which can have a long-term impact [54]. Some nanoparticles last for a very long period in the environment, for example carbon nanotubes in sediments of soil, can accumulate over time leading to bioaccumulation, and can be transferred through food chains from one trophic level to others. While nanoparticles provide numerous technological advantages, the adverse effect they have on environment cannot be ignored and required an effective mitigation.

The potential impacts of nanomaterials on human health, whether derived from manmade or natural sources, need to be carefully studied. Like any other foreign substance, nanomaterials are recognized by the immune system in the human body, which then initiates a biological process that ends with their removal (Figure 13.5). Toxic effects of nanoparticles occur when these unwanted or excessive biological reactions occur. Nonetheless, several forms of extremely biocompatible nanomaterials can interact with the body without leading to any unpleasant or undesired events (such as toxicity, immunological response, thrombosis, or cancer) [55, 56]. The characteristics of the nanomaterial and its interactions with different bodily components (such as immune cells, proteins, and lipids) can have an impact on the kind and strength of the host reaction. However, risk factors in humans are ruled by exposure level, concentration, and the size, shape, surface area, structure, etc., of the nanomaterial (Figure 13.6).

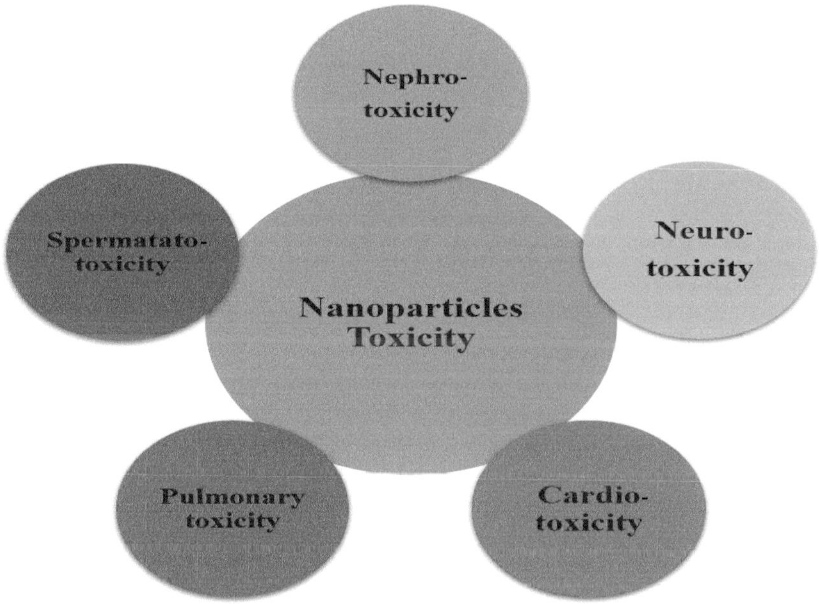

Figure 13.5 Impact of nanoparticles on human health.

- **Effect of dose and exposure duration:** The quantity of nanoparticles that enter the cells is directly proportional to the molar concentration of nanoparticles in the surrounding medium times the duration of exposure [57].
- **Particle shape effect:** Different toxicity levels at various aspect ratios are exhibited by nanoparticles due to their shape-dependent toxicity. For instance, 10 μm asbestos fibres have been linked to lung cancer, 5–10 μm asbestos fibres have been linked to mesothelioma, and 2 μm asbestos fibres have been linked to asbestosis [58].

- **Surface area impact:** Generally speaking, as surface area and particle size decrease, so does the toxicological effect of nanoparticles. Furthermore, it should be emphasized that human cells respond differently to nano and microparticles at the same mass dosage [59].
- **Effect of crystal structure:** Nanoparticles may display distinct sub-cellular localization, oxidative pathways, and cellular uptake depending on their crystal structure [60]. For instance, the toxicity of the two crystalline polymorphs of TiO_2, rutile and anatase, varies. Anatase nanoparticles (200 nm) do not cause DNA damage in the dark, while rutile nanoparticles (200 nm) do cause DNA damage by oxidation [61].
- **Pre-exposure effect:** Shorter exposure times or pre-exposure to lower nanoparticles concentrations can both increase cellular phagocytic activity. As a result of this pre-exposure, the human body becomes somewhat more resilient to nanoparticles [62].

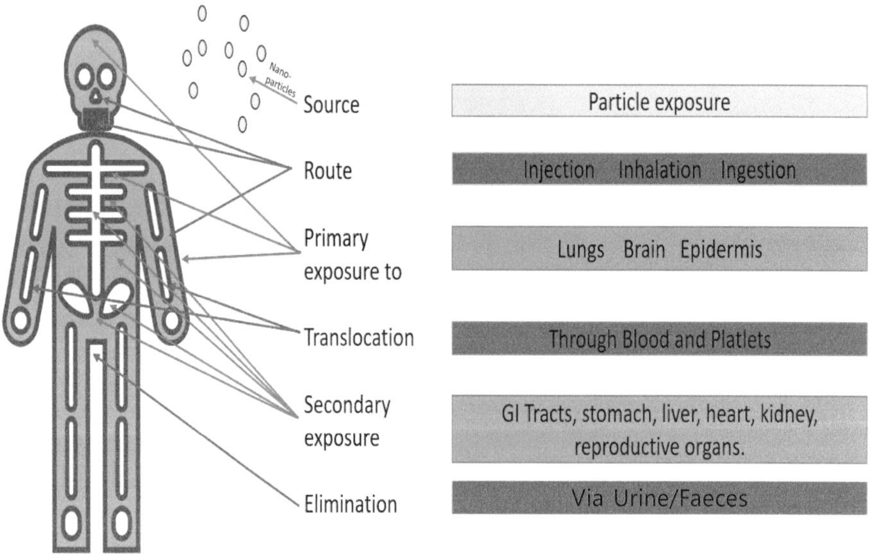

Figure 13.6 Harmful consequences of nanoparticles after injection, inhalation, and ingestion.

Human skin often blocks foreign compounds, although the lungs and digestive system are among the organs that are vulnerable to them. In terms of size, nanoparticles are like viruses. The Human Immunodeficiency Virus (HIV) particle, for example, has a diameter of around 100 nm. When breathed, nanoparticles can easily enter human circulatory pathway and reach to many other parts of the body, such as the kidney, heart, liver, blood cells. It is important to note that the origin of nanoparticles affects their toxicity. While some appear to be beneficial to health, many of them appear to be harmless [63]. Thus, depending on their composition, nanoparticles may induce asthma, cancer, irreversible oxidative stress, and damage to organelles [64].

Common chronic harmful consequences of nanoparticles include uptake by the nucleus, neuronal tissue, reticuloendothelial system, and the production of neoantigens that may induce organ malfunction and hypertrophy [65]. Additionally, research has shown that nanomaterials may interfere with the epigenetic mechanism, which includes methylation, histone tail adaptation, and micro RNA mechanisms in addition to adaptation in gene expression grades without alterations in the actual DNA itself [66]. Cancers, complicated cardiovascular systems, physiological abnormalities, autoimmune diseases, neurological diseases, and mental illnesses have all been linked to epigenetic modification. Because of their tiny size and sharp edges, nanomaterials based on graphene can also penetrate the cytoplasm. They may readily pierce cell membranes, causing damage to the membrane and cytoplasmic leakage. When they settle in live cells, they can emit Reactive Oxygen Species (ROS), which can lead to mitochondrial dysfunction by decreasing the potential of the mitochondrial membrane (MMP) and damage cell membranes by releasing lactate dehydrogenase (LDH). In the latter instance, ROS can cause lipid peroxidation by combining with membrane lipids' unsaturated fatty acids to produce lipid peroxides like malondialdehyde (MDA). These findings demonstrate that GO may cause the production of intracellular and extracellular ROS in a dose-dependent and time-dependent way, even at concentrations which are very low [67].

13.6 NANOMATERIALS IN REDUCING CARBON EMISSION

The current trend of research involves exploring various nanomaterials and their wide-ranging application, particularly in sustainable practices aimed at mitigating climate change and reducing carbon emissions. Nanotechnology offers promising solutions to numerous environmental challenges, suggesting that traditional methods should give way to nano-based technologies. Presently, the world faces a multitude of environmental issues that have persisted for decades, with climate change emerging as a predominant threat. Over the past 40–50 years, there has been a documented increase in surface temperatures by 1.5–2°C, largely attributed to global warming. Primarily carbon emissions are a major source for global warming, while nanotechnology, exemplified by advancements like multi-walled carbon nanotubes holds the potential for effectively combatting these emissions [68]. Multi-walled carbon nanotubes attached with different functional groups along with the addition of an amino group enhance their separation and adsorption properties [71]. Amino-modified materials exhibit CO_2 selectivity during CO_2 and N_2 adsorption and separation trials. Enhanced CO_2 uptake profiles are associated with high nitrogen content and microporosity [71].

Nanomaterials and hybrid nanocomposites can be used for Carbon Capture, Utilization, and Storage (CCUS) technology and CCUS has a high potential to mitigate industrial CO_2 emission which mainly comes from burning of fossil fuels [72]. Apart from this, the major CO_2 emitting sources can be modulated

Table 13.2 Nanomaterials applications for low-carbon emission

S. No.	Sector	Applications of Nanomaterials	Nanotechnology	Insights on Carbon Reduction	References
1.	Healthcare and Medicine	Drug delivery, Medical imaging (MRI, CT scan, biosensors)	Dendrimers, liposomes, polymeric nanoparticles	Durable, efficient equipment, potential reduction in carbon emission	[69]
2.	Cosmetics and Personal care	UV protection cream	Nanoemulsions, nanoliposomes, fullerene, gold, and silver nanoparticles, nanohydroxyapatite	High potential in carbon emission reduction as compared to conventional products.	[70]
3.	Electronics Applications	Transistors, semiconductors, displays	Nanotubes, carbon graphene, quantum dots	Highly efficient equipment, less carbon emission.	[71]
4.	Energy Production	Solar cells, fuel cells, lithium-ion batteries	Nanoelectrodes, nano-titanium dioxide, nano perovskite, nano-platinum, nano-iron oxide	Composites combined with electroactive materials showcase high efficiency of devices.	[72]
5.	Textiles	Water-repellent, antibacterial-protected, and durable clothing	Nanocoating, Nanofibres, nano zinc oxide, nano copper, nano-silica, nanoencapsulation	Increased durability leads to reduce carbon emission	[13]
6.	Environmental Remediation	Water purification, air filtration, monitoring devices	Nanoscale membranes, nanocomposites	High efficiency and low carbon emission	[13, 37]
7.	Construction and infrastructure	Self-cleaning glass, and incorporated into coating, concrete, and composites	Nanosilica, nano clays, nano alumina, nano copper, nanotubes, nano-titanium dioxide	Increase longevity of infrastructure leads to reduce maintenance which leads low carbon emission	[73]
8.	Automotive industry	Shining surfaces, lubricants, lightweight parts, scratch-resistant parts	Nanoclays nanocomposite, carbon nanotubes, nanosilver, graphene	Lightweight parts require less fuel energy to drive hence less carbon emission	[74]
9.	Agriculture	Specific delivery of nutrients, fertilizers, pesticides	Nano chitosan, nanozeolites, nano-emulsion, nanocarbon, nano-copper	Effective procedure, reduced fuel use, and reduced carbon emission	[75]
10.	Space	Durable spacecraft, sensors, propulsion system	aerogels, nanoceramics, quantum dots, graphene, nanosilver, carbon nanotubes	Durable pieces, less energy consumption hence low carbon emission	[76]

with CO_2, capturing, utilization and storage procedures can be achieved through various methods applied in combination such as separation, adsorption, absorption, membrane cryogenic distillation, chemical looping, biochemical and geothermal technologies. In this CO_2 absorption and adsorptions can be performed with the help of solvents modified with nanoparticles, mini, and microchannels, and nanosized zeolites [77]. Issues related to the environment like the elevation in temperature along with the energy crisis can be solved through the conversion of CO_2 to sustainable and renewable fuels (like green hydrogen and biofuels), and nanomaterials can be used in this process like carbon nanotubes, spherical nanoparticles, nanocubes, nanowires and nanosheets [77]. Nanoparticles of various shapes display numerous efficiencies for the catalytic process in the conversion of CO_2 to sustainable and renewable fuels.

The improvement of gas-liquid mass transfer in the process of CO_2 absorption with the help of TiO_2 nanoparticles showed the enhanced rate of CO_2 absorption and the rate of CO_2 absorption rate was high with the decrease in diameter of nanoparticles. TiO_2 nanoparticles improve CO_2 absorption capacity by modifying the mass transfer rate [78]. Nanotechnology can be defined as the cluster of emerging technologies that work on the nanometre scale in the range of 1–100 nm for the production of materials, systems, and devices with basic new properties and functions through controlling the shape and size of matter. Nanoparticles play a crucial role in improving the efficiency of low-carbon technology with their vast specific surface area, nanoparticles serve as excellent sorbents for capturing natural and anthropogenic CO_2 emissions. Nanocatalyst enables the conversion of CO_2 gas into valuable products (methanol and carbonates), nano-additives improve the combustion properties of low-carbon fuels, and nanoparticles boost the harvesting of renewable energy and energy storage technologies, in addition, there is a demand for green synthesis methods of nanoparticles to minimize carbon footprint (Table 13.2).

13.7 CONCLUSION

The environmental and health safety of nanomaterial is a serious concern at the current time because chances of unintentional environmental exposures of nanomaterial can pose a serious problem. Nanomaterials are used in various areas of human life that include agriculture, space technology, pharmaceutical, and environmental remediation. As accelerating use of nanomaterials are also increasing concern of the environmental and health. Nanomaterials hold promise for enabling low carbon emission technologies, their safe and responsible integration into the energy landscape requires careful consideration of environmental and health safety concerns, supported by robust regulation, research, and collaboration.

REFERENCES

[1] Dolez, P.I. 2015. Nanomaterials definitions, classifications, and applications. pp. 3–40. *In*: P.I. Dolez and A. Methner (eds). Nanoengineering: Global Approaches to Health and Safety. Amsterdam, Elsevier.

[2] Khan, I., Saeed, K. and Khan, I. 2019. Nanoparticles: Properties, applications and toxicities. Arabian Journal of Chemistry. 12(7): 908–931.

[3] Sajid, M. 2022. Nanomaterials: Types, properties, recent advances, and toxicity concerns. Current Opinion in Environmental Science & Health. 25: 100319.

[4] Adewuyi, A. and Lau, W.J. 2021. Nanomaterial development and its applications for emerging pollutant removal in water. pp. 67–97. *In*: W.J. Lau, K. Faungnawakij, K. Piyachomkwan and U.R. Ruktanonchai (eds). Handbook of Nanotechnology Applications. Amsterdam, Elsevier.

[5] Alhalili, Z. 2023. Metal oxides nanoparticles: General structural description, chemical, physical, and biological synthesis methods, role in pesticides and heavy metal removal through wastewater treatment. Molecules. 28(7): 3086.

[6] Rizwan, M., Shoukat, A., Ayub, A., Razzaq, B. and Tahir, M.B. 2021. Types and classification of nanomaterials. pp. 31–54 *In*: M.B. Tahir, Md. Sagir and A.M. Asiri (eds). Nanomaterials: Synthesis, Characterization, Hazards and Safety. Elsevier..

[7] Bhardwaj, H. and Jangde, R.K. 2023. Current updated review on preparation of polymeric nanoparticles for drug delivery and biomedical applications. Next Nanotechnology. 2: 100013.

[8] Hooshyar, P., Rezapour Kiani, D., Fakhraeelotfabadi, S. and Fardmanesh, M. 2023. Applications of graphene-based nanomaterials as contrast agents in biomedical imaging technologies: A review. Biomedical Materials & Devices. 2: 721–742.

[9] Chavda, V.P., Acharya, D., Hala, V., Vora, L.K. and Dawre, S. 2023. Sunscreens: A comprehensive review with the application of nanotechnology. Journal of Drug Delivery Science and Technology. 86: 104720.

[10] Gupta, S.K., Kumar, P. and Singh, D.P. 2022. Quantum dots for modern display devices. pp. 899–932. *In*: Y. Al-Douri (ed.). Graphene, Nanotubes and Quantum Dots-Based Nanotechnology: Fundamentals and Applications. Amsterdam: Woodhead Publishing (Elsevier).

[11] Mahmud, M.Z.A. 2023. A concise review of nanoparticles utilized energy storage and conservation. Journal of Nanomaterials. 2023(1): 5432099.

[12] Poorshakoor, E. and Darab, M. 2024. Advancements in the development of nanomaterials for lithium-ion batteries: A scientometric review. Journal of Energy Storage. 75: 109638.

[13] Shah, M.A., Pirzada, B.M., Price, G., Shibiru, A.L. and Qurashi, A. 2022. Applications of nanotechnology in smart textile industry: A critical review. Journal of Advanced Research. 38: 55–75.

[14] Saleem, H., Zaidi, S.J., Ismail, A.F. and Goh, P.S. 2022. Advances of nanomaterials for air pollution remediation and their impacts on the environment. Chemosphere. 287: 132083.

[15] Ibrahim, R.K., Hayyan, M., AlSaadi, M.A., Hayyan, A. and Ibrahim, S. 2016. Environmental application of nanotechnology: air, soil, and water. Environmental Science and Pollution Research. 23: 13754–13788.

[16] Mohajerani, A., Burnett, L., Smith, J.V., Kurmus, H., Milas, J., Arulrajah, A., et al. 2019. Nanoparticles in construction materials and other applications, and implications of nanoparticle use. Materials. 12(19): 3052.

[17] Zhang, W. and Xu, J. 2022. Advanced lightweight materials for Automobiles: A review. Materials & Design. 221: 110994.

[18] Al-Mamun, M.R., Hasan, M.R., Ahommed, M.S., Bacchu, M.S., Ali, M.R. and Khan, M.Z.H. 2021. Nanofertilizers towards sustainable agriculture and environment. Environmental Technology & Innovation. 23: 101658..

[19] Scalia, T. and Bonventre, L., 2020. Nanomaterials in space: Technology innovation and economic trends. Advances in Astronautics Science and Technology. 3: 145–155.

[20] Shalan, A.I., Barhoum, A., Elseman, A.M., Rashad, M.M. and Lira-Cantú, M., 2018. Nanofibers as promising materials for new generations of solar cells. pp. 1–33. *In*: Barhoum, A., Bechelany, M. and A. Makhlouf (eds). Handbook of Nanofibers. Springer, Cham.

[21] Jeevanandam, J., Barhoum, A., Chan, Y.S., Dufresne, A. and Danquah, M.K. 2018. Review on nanoparticles and nanostructured materials: history, sources, toxicity and regulations. Beilstein Journal of Nanotechnology. 9(1): 1050–1074.

[22] Hochella Jr, M.F., Mogk, D.W., Ranville, J., Allen, I.C., Luther, G.W., Marr, L.C., et al. 2019. Natural, incidental, and engineered nanomaterials and their impacts on the Earth system. Science. 363(6434): eaau8299.

[23] Lee, G.H., Choi, T.M., Kim, B., Han, S.H., Lee, J.M. and Kim, S.H. 2017. Chameleon-inspired mechanochromic photonic films composed of non-close-packed colloidal arrays. ACS Nano. 11(11): 11350–11357.

[24] Samyn, P. and Barhoum, A., 2018. Engineered nanomaterials for papermaking industry. pp. 245–277. *In*: A. Barhoum (ed.). Fundamentals of Nanoparticles. Elsevier.

[25] Missaoui, W.N., Arnold, R.D. and Cummings, B.S. 2018. Toxicological status of nanoparticles: What we know and what we don't know. Chemico-Biological Interactions. 295: 1–12.

[26] Tschiche, H.R., Bierkandt, F.S., Creutzenberg, O., Fessard, V., Franz, R., Greiner, R., et al. 2022. Analytical and toxicological aspects of nanomaterials in different product groups: Challenges and opportunities. NanoImpact, 28: 100416.

[27] Zielinska, E., Tukaj, C., Radomski, M.W. and Inkielewicz-Stepniak, I. 2016. Molecular mechanism of silver nanoparticles-induced human osteoblast cell death: Protective effect of inducible nitric oxide synthase inhibitor. PLoS One. 11(10): e0164137.

[28] Dhasmana, A., Firdaus, S., Singh, K.P., Raza, S., Jamal, Q.M.S., Kesari, K.K., Rahman, Q. and Lohani, M. 2017. Nanoparticles: Applications, toxicology and safety aspects. pp. 47–70. *In*: K. Kesari (ed.). Perspectives in Environmental Toxicology. Environmental Science and Engineering. Springer, Cham.

[29] Ramachandran, G. (ed.). 2016. Assessing Nanoparticle Risks to Human Health. 2nd Ed. William Andrew. Elsevier.

[30] Li, B., Bell, D.R., Gu, Z., Li, W. and Zhou, R. 2019. Protein WW domain denaturation on defective graphene reveals the significance of nanomaterial defects in nanotoxicity. Carbon. 146,: 257–264.

[31] Osman, E.M., 2019. Environmental and health safety considerations of nano-technology, nano safety. Biomedical Journal of Scientific & Technical Research. 19(4): 14501–14515.

[32] Reis, A.T., Costa, C. and Fraga, S. 2023. Editorial of special issue: The toxicity of nanomaterials and legacy contaminants: Risks to the environment and human health. International Journal of Molecular Sciences. 24(14): 11723.

[33] Iqbal, A., Raza, M.H. and Jadoon, H. 2022. Applications of nanomaterials for health and environment protection. MOJ Ecology & Environmental Sciences. 7(3): 84–87.

[34] Ganguly, P., Breen, A. and Pillai, S.C. 2018. Toxicity of nanomaterials: Exposure, pathways, assessment, and recent advances. ACS Biomaterials Science & Engineering. 4(7): 2237–2275.

[35] Asmatulu, E., Andalib, M.N., Subeshan, B. and Abedin, F. 2022. Impact of nanomaterials on human health: A review. Environmental Chemistry Letters, 20(4): 2509–2529.

[36] Turner, R.J., 2022. Toxicity of nanomaterials. pp. 221–247. *In*: J. Gailer and R.J. Turner (eds). Environmental and Biochemical Toxicology: Concepts, Case Studies and Challenges. Berlin, Boston: De Gruyter.

[37] Arfin, T. and Tarannum, A. 2018. Engineered nanomaterials for industrial application: An overview. pp. 127–134. *In*: Chaudhery M. Hussain (ed.). Handbook of Nanomaterials for Industrial Applications. Amsterdam: Elsevier.

[38] Liyanage, P.Y., Hettiarachchi, S.D., Zhou, Y., Ouhtit, A., Seven, E.S., Oztan, C.Y., et al. 2019. Nanoparticle-mediated targeted drug delivery for breast cancer treatment. Biochimica et Biophysica Acta (BBA)-Reviews on Cancer. 1871(2): 419–433.

[39] Raina, N., Sharma, P., Slathia, P.S., Bhagat, D. and Pathak, A.K., 2020. Efficiency enhancement of renewable energy systems using nanotechnology. pp. 271–297. *In*: I. Bhushan, V. Singh and D. Tripathi (eds). Nanomaterials and Environmental Biotechnology. Nanotechnology in the Life Sciences. Springer, Cham.

[40] Garcia-Segura, S., Qu, X., Alvarez, P.J., Chaplin, B.P., Chen, W., Crittenden, J.C., et al. 2020. Opportunities for nanotechnology to enhance electrochemical treatment of pollutants in potable water and industrial wastewater–a perspective. Environmental Science: Nano. 7(8): 2178–2194.

[41] Lespes, G., Faucher, S. and Slaveykova, V.I. 2020. Natural nanoparticles, anthropogenic nanoparticles, where is the frontier?. Frontiers in Environmental Science. 8: 71.

[42] Kaur, H., Kalia, A., Sandhu, J.S., Dheri, G.S., Kaur, G. and Pathania, S. 2022. Interaction of TiO_2 nanoparticles with soil: Effect on microbiological and chemical traits. Chemosphere. 301: 134629.

[43] Zia-ur-Rehman, M., Anayatullah, S., Irfan, E., Hussain, S.M., Rizwan, M., Sohail, et al. 2023. Nanoparticles assisted regulation of oxidative stress and antioxidant enzyme system in plants under salt stress: A review. Chemosphere. 314: 137649.

[44] Rodriguez-Garraus, A., Azqueta, A., Vettorazzi, A. and Lopez de Cerain, A. 2020. Genotoxicity of silver nanoparticles. Nanomaterials. 10(2): 251.

[45] Thomas, S.P., Al-Mutairi, E.M. and De, S.K. 2013. Impact of nanomaterials on health and environment. Arabian Journal for Science and Engineering. 38: 457–477.

[46] Gugulothu, D., Barhoum, A., Afzal, S.M., Venkateshwarlu, B. and Uludag, H. 2018. Structural multifunctional nanofibers and their emerging applications. pp. 693–732. *In*: A. Barhoum, M. Bechelany and A. Makhlouf (eds). Handbook of Nanofibers. Springer, Cham.

[47] Kundu, A., Sharma, S. and Pathania, A.R. 2023. Engineered nanomaterials and human exposure: An overview. Materials Today: Proceedings, 72: 1216–1223.

[48] Prasad, S., Kumar, V., Kirubanandam, S. and Barhoum, A., 2018. Engineered nanomaterials: Nanofabrication and surface functionalization. pp. 305–340. *In*: A. Barhoum and A.S.H. Makhlouf (eds). Emerging applications of Nanoparticles and Architecture Nanostructures: Current Prospects and Future Trends. Amsterdam: Elsevier.

[49] Lespes, G., Faucher, S. and Slaveykova, V.I. 2020. Natural nanoparticles, anthropogenic nanoparticles, where is the frontier?. Frontiers in Environmental Science. 8: 71.

[50] Bayda, S., Adeel, M., Tuccinardi, T., Cordani, M. and Rizzolio, F. 2019. The history of nanoscience and nanotechnology: From chemical–physical applications to nanomedicine. Molecules. 25(1): 112.

[51] Sahu, S.C. and Hayes, A.W. 2017. Toxicity of nanomaterials found in human environment: A literature review. Toxicology Research and Application. 1: 2397847317726352.

[52] Yetisgin, A.A., Cetinel, S., Zuvin, M., Kosar, A. and Kutlu, O. 2020. Therapeutic nanoparticles and their targeted delivery applications. Molecules. 25(9): 2193.

[53] Smolkova, B., El Yamani, N., Collins, A.R., Gutleb, A.C. and Dusinska, M. 2015. Nanoparticles in food. Epigenetic changes induced by nanomaterials and possible impact on health. Food and Chemical Toxicology. 77: 64–73.

[54] Abedin, F., Asmatulu, E. and Andalib, M.N. 2021. Nanomaterials and human health: An overview. pp. 165–180. *In*: V. Kumar, P. Guleria, S. Ranjan, N. Dasgupta and E. Lichtfouse (eds). Nanotoxicology and Nanoecotoxicology Vol. 2 . Environmental Chemistry for a Sustainable World, vol 67. Springer, Cham.

[55] Pelin, M., Fusco, L., Martín, C., Sosa, S., Frontiñán-Rubio, J., González-Domínguez, et al. 2018. Graphene and graphene oxide induce ROS production in human HaCaT skin keratinocytes: The role of xanthine oxidase and NADH dehydrogenase. Nanoscale. 10(25): 11820–11830.

[56] Beni, A.A. and Jabbari, H. 2022. Nanomaterials for environmental applications. Results in Engineering. 15: 100467.

[57] Zhao, L., Yu, C., Lv, J., Cui, Y., Wang, Y., Hou, C., et al. 2021. Fluoride exposure, dopamine relative gene polymorphism and intelligence: A cross-sectional study in China. Ecotoxicology and Environmental Safety. 209: 111826.

[58] Zerboni, A., Bengalli, R., Baeri, G., Fiandra, L., Catelani, T. and Mantecca, P. 2019. Mixture effects of diesel exhaust and metal oxide nanoparticles in human lung A549 cells. Nanomaterials. 9(9): 1302.

[59] Tolliver, L.M., Holl, N.J., Hou, F.Y.S., Lee, H.-J., Cambre, M.H. and Huang, Y.-W. 2020. Differential cytotoxicity induced by transition metal oxide nanoparticles is a function of cell killing and suppression of cell proliferation. International Journal of Molecular Sciences. 21(5): 1731.

[60] Fernández-Bertólez, N., Costa, C., Brandão, F., Kiliç, G., Duarte, J.A., Teixeira, J.P., et al. 2018. Toxicological assessment of silica-coated iron oxide nanoparticles in human astrocytes. Food and Chemical Toxicology. 118: 13–23.

[61] Jiménez-Chávez, A., Solorio-Rodríguez, A., Escamilla-Rivera, V., Leseman, D., Morales-Rubio, R., Uribe-Ramírez, M., et al. 2021. Inflammatory response in human alveolar epithelial cells after TiO_2 NPs or ZnO NPs exposure: Inhibition of surfactant protein a expression as an indicator for loss of lung function. Environmental Toxicology and Pharmacology. 86: 103654.

[62] Verdon, R., Gillies, S.L., Brown, D.M., Henry, T., Tran, L., Tyler, C.R., et al. 2021. Neutrophil activation by nanomaterials *in vitro*: Comparing strengths and

limitations of primary human cells with those of an immortalized (HL-60) cell line. Nanotoxicology. 15(1): 1–20.

[63] Mukherjee, S.P., Gupta, G., Klöditz, K., Wang, J., Rodrigues, A.F., Kostarelos, K., et al. 2020. Next-generation sequencing reveals differential responses to acute versus long-term exposures to graphene oxide in human lung cells. Small. 16(21): 1907686.

[64] Pérez, R.F., Soto Fernández, A.Y., Bousquets Muñoz, P., Sierra, M.I., Tejedor, J.R., Morales-Sánchez, P., et al. 2020. No genome-wide DNA methylation changes found associated with medium-term reduced graphene oxide exposure in human lung epithelial cells. Epigenetics. 15(3): 283–293.

[65] De Berardis, B., Marchetti, M., Risuglia, A., Ietto, F., Fanizza, C. and Superti, F., 2020. Exposure to airborne gold nanoparticles: a review of current toxicological data on the respiratory tract. Journal of Nanoparticle Research. 22: 1–41.

[66] Chausali, N., Saxena, J. and Prasad, R. 2023. Nanotechnology as a sustainable approach for combating the environmental effects of climate change. Journal of Agriculture and Food Research. 12: 100541.

[67] Lourenço, M.A., Fontana, M., Jagdale, P., Pirri, C.F. and Bocchini, S. 2021. Improved CO_2 adsorption properties through amine functionalization of multi-walled carbon nanotubes. Chemical Engineering Journal. 414: 128763.

[68] Dziejarski, B., Serafin, J., Andersson, K. and Krzyżyńska, R. 2023. CO_2 capture materials: A review of current trends and future challenges. Materials Today Sustainability. 24: 100483.

[69] Saleh, T.A. 2022. Nanomaterials and hybrid nanocomposites for CO_2 capture and utilization: Environmental and energy sustainability. RSC Advances. 12(37): 23869–23888.

[70] Ziental, D., Czarczynska-Goslinska, B., Mlynarczyk, D.T., Glowacka-Sobotta, A., Stanisz, B., Goslinski, T., et al. 2020. Titanium dioxide nanoparticles: prospects and applications in medicine. Nanomaterials. 10(2): 387.

[71] Haleem, A., Javaid, M., Singh, R.P., Rab, S. and Suman, R. 2023. Applications of nanotechnology in medical field: A brief review. Global Health Journal. 7(2): 70–77.

[72] Fytianos, G., Rahdar, A. and Kyzas, G.Z. 2020. Nanomaterials in cosmetics: Recent updates. Nanomaterials. 10(5): 979.

[73] Hossain, N., Mobarak, M.H., Mimona, M.A., Islam, M.A., Hossain, A., Zohur, F.T., et al. 2023. Advances and significances of nanoparticles in semiconductor applications–A review. Results in Engineering. 19: 101347.

[74] Belkhode, P., Giripunje, M., Dhande, M., Gajbhiye, T., Waghmare, S., Tupkar, R., et al. 2024. Nanomaterials applications in solar energy: Exploring future prospects and challenges. Materials Today: Proceedings.

[75] Shilar, F.A., Ganachari, S.V. and Patil, V.B. 2022. Advancement of nano-based construction materials-A review. Construction and Building Materials. 359: 129535.

[76] Mohite, D.D., Chaturvedi, V., De, S. and Jadhav, V.S. 2021. Nanomaterials in automotive applications: A review and its technical aspects. International Journal of Contemporary Architecture "The New ARCH". 8(2): 1450–1460.

[77] Usman, M., Farooq, M., Wakeel, A., Nawaz, A., Cheema, S.A., ur Rehman, H., et al. 2020. Nanotechnology in agriculture: Current status, challenges and future opportunities. Science of the Total Environment. 721: 137778.

[78] Scalia, T., Bonventre, L. and Terranova, M.L. 2023. From protosolar space to space exploration: The role of graphene in space technology and economy. Nanomaterials. 13(4): 680.

Index

P

R

S

T

W

Z

Brief Profiles of Editors

Dr. Tanuj Kumar is an Assistant Professor at the Department of Nanoscience and Materials, Central University of Jammu, Jammu & Kashmir, with 8 years of teaching experience. He completed his Ph.D. at Jawaharlal Nehru University (JNU), New Delhi, focusing on ion beam induced nanopatterning and their applications. Presently, his focus area is the importance of ion beam metal nanoparticles in SERS based chemical detection in water. Dr. Kumar has an impressive publication record, with 48 papers in international journals and 6 book chapters. Additionally, he has secured two sponsored projects from IUAC and DST-SERB and has guided 26 master students' dissertations along with three ongoing Ph.D. students. His research excellence has been recognized through CSIR-JRF, CSIR-SRF, and Fast-track young scientist fellowships.

Dr. Richa Kothari (M.Phil., Ph.D.) is currently a Professor at the Department of Environmental Sciences, Central University of Jammu, Jammu & Kashmir, India. She is a WARI (Water Advanced Research and Innovation) Fellow, Indo-U.S. Science and Technology Forum at the University of Nebraska-Lincoln, USA, with over 18 years of teaching and research experience. Recognized among Stanford University's World's top 2% scientists (2020–2024), her research interests include algal-based biofuels, bio-hydrogen, and renewable energy for wastewater treatment. She has been conferred with the Women Scientist's Award, 2025 by the Central University of Rajasthan, Ajmer, Rajasthan, India, in March 2025, and the ASSOCHAM J&K Women's Achiever Award on 10th March 2023. She has authored more than 140 journal articles, 5 patents, gained approximately 8907 citations, and holds an h-index of 45 and an i10-index of 106. She has also worked as a guest editor for special issues in the Journal of Biofuels, Journal of Biomass Conversion and Biorefinery, Environmental Sustainability, Renewable and Sustainable Energy Reviews. She has guided 8 Ph.D./M.Phil., 65 M.Sc. and approximately 40 M.Tech. students. She has also delivered a number of invited lectures at national and international level.

Dr. Anita Singh is currently working as Associate Professor, Department of Environmental Studies, Central University of Haryana, Haryana. She is having more than 12 years of teaching and research experience. Dr. Singh did her Ph.D. on bioethanol production from lignocellulosic biomass from Guru Jambheshwar University of Science & Technology, Hisar. She did her Post-doctoral research from University of Georgia, USA, on selective removal of inhibitors from lignocellulosic biomass. Dr. Singh, published 40 papers in international journals and 6 book chapters in various edited books. She has published three edited books with reputed publishers. Dr. Singh completed two sponsored projects from UGC and DST-SERB (ECR). She guided 36 master students' dissertation and supervised 2 Ph.D. students. Dr. Singh, awarded CSIR- SRF and CSIR-RA fellowships for research during Ph.D. and after Ph.D. Dr. Singh was awarded UGC Raman Post doctoral fellowship for doing research in USA.

Dr. Deepak Pathania is Professor and Dean Research Studies at Central University of Jammu, Jammu & Kashmir. He is also serving as President of Him Science Congress Association, Himachal Pradesh, a scientific society working for promotion of sciences in state. He also served as Dean Research Studies, Dean Students' welfare and Dean Life Sciences at central University of Jammu from 2017 to 2020. Dr. Pathania has served for 2 years on deputation as Dean Academic Affairs, Dean Physical Sciences and Dean Planning and Development at Sardar Patel University, Mandi, Himachal Pradesh. He has served as Chairman of several academic committees. He is member of University Court, Executive council, Academic Council and Board of Studies of different Universities. He has been awarded with D.Sc. Degree from Jaunpur University in 2021. He has about 23 years of teaching and research experience and has figured among 2% Scientists of the world continuously for 6th Year (2020-2025) as per survey Stanford University, USA. He served as court member (Nominated by Visitors) of University of Hyderabad from 2018-2021 and was the Member of Technical committee of Ministry of Environment, Govt. of India. He has received distinguished alumni award from National Institute of Technology, Jalandhar, Punjab for dedications, contribution and achievements in the field of teaching, innovations and research on 4th October, 2021. He has completed many research projects funded by AICTE and UGC. He is member of different professional scientific societies. Dr. Pathania has guided 17 Ph.D., 14 M.Phil and 52 M.Sc research projects for their respective degrees. He has about 202 publications in reputed Journals and 148 publications in conferences to his credit and has been awarded 3 international and 3 national patents. His research h-index is 58 and i10-index is 135. He has authorized 15 books in different areas of interest from believed publishers. Dr. Pathania is editorial member and reviewer of different national and international journals and has published 32 book chapters in reputed national and international books. He is also life member of different scientific societies.

For Product Safety Concerns and Information please contact our
EU representative GPSR@taylorandfrancis.com Taylor & Francis
Verlag GmbH, Kaufingerstraße 24, 80331 München, Germany